Lasers in the Preservation of Cultural Heritage
Principles and Applications

Series in Optics and Optoelectronics

Series Editors: **R G W Brown**, University of Nottingham, UK
E R Pike, Kings College, London, UK

Other titles in the series

Modeling Fluctuations in Scattered Waves
E Jakeman, K D Ridley

Fast Light, Slow Light and Left-Handed Light
P W Milonni

Diode Lasers
D Sands

Diffractional Optics of Millimetre Waves
I V Minin

Handbook of Electroluminescent Materials
D R Vij

Handbook of Moire Measurement
C A Walker

Next Generation Photovoltaics
A Martí

Stimulated Brillouin Scattering
M J Damzen

Laser Induced Damage of Optical Materials
Roger M Wood

Optical Applications of Liquid Crystals
L Vicari

Optical Fibre Devices
J P Goure

Applications of Silicon-Gremanium Heterostructure Devices
C K Maiti

Optical Transfer Function of Imaging Systems
Tom L Williams

Super-radiance
M G Benedict

Susceptibility Tensors for Nonlinear Optics
S V Popov

Series in Optics and Optoelectronics

Lasers in the Preservation of Cultural Heritage

Principles and Applications

C Fotakis
Institute of Electronic Structure and Laser, Foundation for Research and Technology-Hellas (IESL-FORTH) Heraklion, Crete, Greece

D Anglos
Institute of Electronic Structure and Laser, Foundation for Research and Technology-Hellas (IESL-FORTH) Heraklion, Crete, Greece

V Zafiropulos
Department of Human Nutrition and Dietetics, Superior Technological Educational Institute of Crete, Sitia, Crete, Greece
and
Institute of Electronic Structure and Laser, Foundation for Research and Technology-Hellas (IESL-FORTH) Heraklion, Crete, Greece

S Georgiou
Institute of Electronic Structure and Laser, Foundation for Research and Technology-Hellas (IESL-FORTH) Heraklion, Crete, Greece

V Tornari
Institute of Electronic Structure and Laser, Foundation for Research and Technology-Hellas (IESL-FORTH) Heraklion, Crete, Greece

CRC Press
Taylor & Francis Group
Boca Raton London New York

CRC Press is an imprint of the
Taylor & Francis Group, an **informa** business
A TAYLOR & FRANCIS BOOK

On the cover: Statue of god Hermes (2nd century BC) from the excavation of Ancient Mesene, Greece (courtesy of Yianna Doganis and Amerimni Galanos, conservation consultants).

CRC Press
Taylor & Francis Group
6000 Broken Sound Parkway NW, Suite 300
Boca Raton, FL 33487-2742

First issued in paperback 2019

© 2007 by Taylor & Francis Group, LLC
CRC Press is an imprint of Taylor & Francis Group, an Informa business

No claim to original U.S. Government works

ISBN-13: 978-0-7503-0873-1 (hbk)
ISBN-13: 978-0-367-39005-1 (pbk)

Library of Congress Cataloging-in-Publication Data

Lasers in the preservation of cultural heritage : principles and applications / C. Fotakis ... [et al.].
    p. cm. -- (Series in optics and optoelectronics ; v. 2)
    ISBN 0-7503-0873-7 (alk. paper)
    1. Lasers. 2. Cultural property--Protection. I. Fotakis, C. (Costas) II. Series.

TA1675.L392 2006
621.36'6--dc22   2006044567

Visit the Taylor & Francis Web site at
http://www.taylorandfrancis.com

and the CRC Press Web site at
http://www.crcpress.com

After a certain high level of technical skill is achieved, science and art tend to coalesce in aesthetics, plasticity, and form.

A. Einstein

Preface

Laser technology may well serve the future of our past, the preservation of our cultural heritage. The field has come to an era of maturity both in the diagnostic and restoration applications that have been developed and in the level of understanding of the fundamental aspects involved. The aim of this book is to provide an account of these recent developments in a way that is useful to conservation scientists, archaeologists, researchers in the field, and advanced science-oriented students. To achieve this goal, emphasis has been placed on addressing the basic principles on which laser applications for preserving the cultural heritage rely and on presenting case studies of analytical, structural diagnostic, and laser cleaning applications. Sections of the book that deal with celebrated cases of real-life applications may be particularly useful to art historians as well as to conservators and practitioners.

The authors are affiliated with the Foundation for Research and Technology-Hellas (FORTH), which has been active in this type of application for over 15 years. The book therefore covers in detail some of the investigations carried out at FORTH. There are obviously many other studies that could have been reported, had our treatment been all-inclusive. Instead, emphasis has been placed on basic principles and practical examples for which we had access to full documentation.

The chapters of the book are self-contained. This allows easy access according to the requirements and interests of individual readers. To this effect, there is a general introduction for each chapter, which briefly outlines the background knowledge of the methods and techniques discussed. Also, there is a reference list at the end of each chapter. The principal author for each chapter is as follows: Introduction, Costas Fotakis; Chapters 2 and 6, Savas Georgiou; Chapters 3 and 4, Demetrios Anglos; Chapter 5, Vivi Tornari; and Chapters 7, 8, and 9, Vassilis Zafiropulos.

A work of this kind benefits from the input of many people. We feel greatly indebted to Ed Teppo, founder and former president of Big Sky Laser Technologies, Inc., who critically proofread the manuscript and made many useful suggestions for its improvement.

Thanks are also due to Capucine Korenberg from the British Museum for providing the conservation scientist's perspective in several sections of the book. Marta Castillejo of CSIC, Madrid, and Sophia Sotiropoulou of the Ormylia Art Diagnosis Center are acknowledged for their constructive comments.

Michael Doulgeridis, director of conservation in the National Gallery in Athens and a close collaborator for many years, and Stergios Stassinopoulos from the Benaki Museum have proposed several of the test cases presented in this book. The input from Evi Papakonstantinou, head of the Surface Conservation Project of the Acropolis Monuments, and the late Theodore Skoulikidis of the Committee for the Conservation of the Acropolis Monuments is also acknowledged.

The critical remarks of Austin Nevin and Iacopo Osticioli, both Marie Curie doctorate students at FORTH, and of Olga Kokkinaki and Yiannis Bounos, who are doctorate students at the University of Crete, are also acknowledged. Also, the input from our colleagues at FORTH and, in particular, of Paraskevi Pouli and Kristalia Melessanaki has been most valuable. The assistance of Georgia Papadaki in preparing the manuscript, especially at the final stage, is greatly appreciated.

Finally, the authors would like to thank colleagues active in the field and in particular within the LACONA community (Lasers in the Conservation of Artworks Conference Series), whose work has promoted the use of lasers for the preservation of cultural heritage and has been the source of inspiration for writing this book.

Costas Fotakis

Authors

Costas Fotakis is director of the Institute of Electronic Structure and Laser (IESL) at FORTH (Foundation for Research and Technology-Hellas) and professor of physics at the University of Crete. He is also director of the European Ultraviolet Laser Facility operating at FORTH, which is currently part of the European Union project, LASERLAB–EUROPE, linking 17 major European laser infrastructures. His research interests include laser spectroscopy, molecular photophysics, laser interactions with materials and related applications for material processing and analysis. He has been chair or co-chair of several major international conferences on these topics. He has over 180 publications and is a member of the editorial boards of several scientific journals. He is the 2004 recipient of the Leadership Award / New Focus Prize of the Optical Society of America (OSA) "for decade-long leadership of, and personal research contribution to, the field of laser applications to art conservation and leadership in establishing and guiding the scientific excellence of the laser science programs at IESL/FORTH". He is also a life member and fellow of OSA and a member of the Fellows Committee of the European Optical Society.

Demetrios Anglos is a principal researcher at IESL-FORTH. He holds a Ph.D. in physical chemistry (1994) from Cornell University, Ithaca, New York, USA. His research activities include the study of laser-induced photochemical and photophysical processes (laser ablation, excitation energy and electron transfer, random laser action in scattering media) in the condensed phase, particularly polymeric environments, and the applications of laser spectroscopic techniques (LIF, LIBS, Raman spectroscopy) in the analysis of materials with emphasis on the development of novel methodology and instrumentation for the characterization of works of art and archaeological objects. The output of his research is presented in over 45 publications in refereed journals and in numerous talks (10 invited) at major international conferences. He has been involved in several European and national-funded research projects and currently coordinates a multi-site Marie Curie Early Stage Training Project offering advanced research training opportunities on laser sciences. Since 2001, he has been serving as the technical manager of the Ultraviolet Laser Facility operating at IESL-FORTH.

Vassilis Zafiropulos is associate professor of applied physics at the Department of Human Nutrition & Dietetics of the Superior Technological Educational Institute of Crete. In parallel, he has been associated with IESL-FORTH since 1989 and has taught at the Department of Chemistry of the University of Crete. He has contributed as well as coordinated many research projects on a national or European level.

He received his doctorate from the Department of Chemistry of the University of Iowa, Iowa City, USA, in 1988. His thesis was on "Laser spectroscopy and photodissociation dynamics of small molecules." He has worked on research areas of diverse or cross-disciplinary interest, ranging from laser-induced phenomena in gases to studies on laser–matter interactions and development of laser-based analytical and diagnostic techniques. He is author and co-author of more than 90 articles. For a number of years he was involved in research on the use of lasers in artworks conservation, and he was the president of COST Action G7 "Artwork Conservation by Laser" for three years.

Savas Georgiou received his B.Sc. in chemistry and mathematics from Knox College, Galesburg, Illinois, USA, in 1983 and his Ph.D. in physical chemistry from the University of Utah, Salt Lake City, USA, in 1988. He subsequently performed postdoctoral work at Princeton University, New Jersey, USA. In 1993, he joined the Institute of Electronic Structure and Laser of Foundation for Research and Technology-Hellas, IESL-FORTH, where he is now a senior researcher. He has also held position as a visiting assistant professor at the Chemistry Department of the University of Crete and as an assistant professor at the Physics Department of the University of Ioannina, Greece. He has received various awards and has participated in several European Union research projects. He has co-authored more than 70 articles in international scientific journals on photophysics/chemistry and in particular on laser–materials interactions. He was guest editor for the special issue of *Chemical Reviews* on "Laser Ablation of Molecular Substrates" (February 2003).

Vivi Tornari is researcher in charge of the Applied Optical and Digital Holography Laboratory operating within the Laser Applications Division of the Institute of Electronic Structure and Laser (IESL) at the Foundation for Research and Technology-Hellas (FORTH). Research interests include optical and wave physics, and holographic and holographic interferometry applications for structural diagnosis of artworks. She is author of over 20 relevant articles in journals, compiled volumes, and conference proceedings and belongs to the scientific boards of several related international conferences. She has participated in and coordinates several European Union, international, national and private projects. She has more than 10 years of research experience in modern laser technologies for artwork diagnostics and has contributed to trans-European research cooperations.

Table of Contents

1 Introduction

1.1 ART AND TECHNOLOGY

There is a widespread perception that art and technology are notions opposite to each other. However, they were closely linked in the ancient Greek world. In ancient (and in modern) Greek the word for art is τέχνη, and τεχνολογία, i.e. technology, meant the rationalization and understanding of art.

Modern science and technology are crucial for a better insight into art and the revelation of cultural heritage. Examples that easily come to mind are the use of radioactive carbon dating in archaeometry, and x-ray radiographs for understanding what lies beneath the surface layers of an artwork. Despite this, the applications of modern science and technology in art conservation and archaeology are rather limited in comparison to the applications in other fields, such as medicine. Empirical approaches, which can be effective, are still commonplace in everyday practice, and the process of adopting new technologies is often slow for two principal reasons. First is the unique character of the artifacts and archaeological objects, which leaves no room for errors. New technologies and in particular those based on invasive techniques require a high level of maturity, which in turn relies on deep knowledge of the processes involved and many times the existence of on-line monitoring and control to safeguard against potential damage. The second reason for the limited adoption of new technologies is related to the need for establishing common communication codes between professionals who have different educational backgrounds and perspectives on the preservation of our cultural heritage. The highly complex and diverse nature of the objects involved creates the need for an interdisciplinary approach, combining the expertise of chemists, physicists, and engineers with the knowledge and experience of art conservators, art historians, archaeologists, and conservation scientists. To this effect, we need interdisciplinary research for a better understanding of the fundamental aspects involved, and a higher level of scientific training and cultural heritage awareness, to overcome these obstacles and expand the scientific and technological tools available for the preservation of cultural artifacts.

1.2 LASER TECHNOLOGY AND APPLICATIONS

The second half of the twentieth century has witnessed a massive introduction of new scientific techniques and equipment in a variety of disciplines. For example, a great number of medical and industrial practices were revolutionized and optimized through the emergence and proliferation of novel technological tools. This has facilitated the establishment and evolution of scientific methods to replace previous reliance on speculation and deduction. A similar trend has appeared, although with some delay, in the fields of art conservation and validation as well as in archaeology. Nowadays, objects of art and antiquity may be diagnosed and treated using a variety of modern technologies including x-ray transmission, fluorescence and backscattering spectroscopies, magnetic resonance imaging (MRI), photoacoustic and optical spectroscopies, and, recently, remote sensing and monitoring techniques. Lasers have also contributed to the preservation of cultural heritage in a variety of ways, which will be discussed in this book.

Lasers are devices in which light amplification takes place. This amplified light is characterized by high intensity, monochromaticity (that is, emission in a very narrow wavelength band), and directionality. All these properties are interrelated and are largely due to the high degree of phase coherence of laser light. The emission of a laser may be continuous (continuous wave or cw operation) or pulsed. In the latter case, short pulses of nano (10^{-9}), pico (10^{-12}) or femto (10^{-15})-second duration may be obtained. This is another interesting feature of lasers, since picosecond and femtosecond pulses cannot be produced by conventional light sources (see insert explaining prefixes).

It is not within the scope of this book to discuss the fundamental physics and the technical aspects underlying laser operation. There are many dedicated textbooks addressing these issues in great detail [1–6]. Background knowledge required for understanding light amplification in a laser device is also included in these books. However, we give here a brief simplified description of some important topics related to light amplification in a laser device and introduce the basic relevant terminology.

The three main topics involved in laser action are

- Emission and absorption processes that occur when radiation interacts with matter and the conditions necessary for light amplification rather than absorption. The energy levels of atoms or molecules involved and therefore the quantum structure of matter are important in this respect.
- The preparation of a medium, the so-called active or amplifying medium, to obtain gain. To achieve this, it is necessary to excite the atoms and molecules of the active medium and create a state of population inversion, as it is called. A state of population inversion is characterized by a greater number of atoms or molecules in an excited energy level than that in a lower energy level. The corresponding mechanism is referred to as "pumping" and may occur through the excitation of the active medium by electrical, optical, or chemical power. There are many kinds of active media, corresponding to different types of lasers, for example the CO_2, Nd:YAG, and KrF lasers.
- The propagation of radiation in space and in optical resonators (optical feedback cavities), which are prerequisites for the practical realization of most laser systems, is necessary.

A schematic representation of the key components of a laser is shown in Figure 1.1. The active or amplifying medium, the pumping mechanism for achieving population

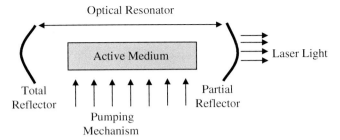

FIGURE 1.1 Key components of a laser: for each successive pass of light through an active medium in which population inversion exists, light amplification takes place.

inversion, and the optical resonator composed of two reflecting mirrors (reflectors) are illustrated in this figure. Of course, many other variables are involved in a laser system than those given in this simplified description. For example, matters related to the temporal and spatial profiles of the laser output are of prime importance for realistic applications and must be considered [1–3].

There is a great number of laser systems currently available, involving different types of active media in gas, liquid, or solid form. Different types of lasers are capable of emitting radiation at different wavelengths beyond the boundaries of the ultraviolet, visible, and infrared parts of the electromagnetic spectrum. Using nonlinear optics techniques, the emission of some lasers may be extended to shorter or longer wavelengths. Average powers of several kilowatts per square centimeter may be obtained by cw laser systems, while peak powers of petawatts per square centimeter can be achieved in high power pulsed lasers (see inset explaining prefixes). There are lasers that are tunable in wavelengths and lasers that are highly monochromatic, characterized by a high degree of coherence, that is, by a long coherence length. *The success of any application relies on the choice of the appropriate type of laser.* Details about the operation of various laser systems may be found in standard textbooks [1–6].

Prefixes Used for Small and Large Numbers	
Small numbers	
Factor	**Prefix**
10^{-3}	milli-
10^{-6}	micro-
10^{-9}	nano-
10^{-12}	pico-
10^{-15}	femto-
10^{-18}	atto-
Large numbers	
10^{3}	kilo-
10^{6}	mega-
10^{9}	giga-
10^{12}	tera-
10^{15}	peta-
10^{18}	hexta-

The inherent unique properties of laser light, such as *intensity, monochromaticity, directionality, and coherence,* have made lasers effective tools in a variety of applications in the industrial and biomedical fields. For example, there is a whole range of laser spectroscopic techniques that have been applied for sensitive and fast monitoring of industrial processes and materials, in many cases nondestructively and remotely [7]. Furthermore, the use of lasers for welding, drilling, cutting, and engraving of materials is now the basis of several established industrial applications, while laser cleaning applications in the microelectronics and aerospace industries have also been successfully applied [8–11]. In the latter case, promising laser techniques include dust removal from semiconductor devices and paint stripping from aircraft, especially from sensitive components consisting of composite materials [11]. Recently, a range of applications for the micro- and nanoengineering of materials has been realized. In these applications, materials processing by lasers may take place at micro- or nanometer scale, in two (surface) or three (volume) dimensions.

Along similar lines, there are several biomedical applications of lasers, which are now mature and have become part of standard practices. Tissue diagnostics by optical techniques using lasers have proven successful for the mapping of pathological lesions, such as atheromas and various types of tumors [12]. Selective labeling of pathological tissue at the molecular level has become particularly promising for noninvasive, high-resolution imaging applications. Currently, the use of laser-induced fluorescence and nonlinear optical effects (such as multiphoton excitation and second harmonic generation), combined with microscopy, open up new avenues for the exploration of biological processes at a cellular level, nonintrusively and *in vivo*. Last but not least, laser therapeutic applications are nowadays widespread in many cases of medical surgery. Cornea sculpturing by laser for the rectification of myopia is a good example, closely related to some of the cleaning applications described in this book [8].

Over the last 30 years, laser-based techniques have been used to determine the composition and internal structure of objects of cultural heritage, as well as for their restoration. It should be stressed at this point that the progress achieved has strongly relied on the transfer of know-how initially developed for industrial and biomedical applications of lasers. This was adapted to the specific needs associated with the preservation of cultural heritage. In particular, since 1990 several laser materials characterization (chemical analysis and structural diagnostics) and processing techniques have inspired new approaches and investigations. The overall advances in laser technology have led to the emergence of a range of new sophisticated and powerful techniques, superseding in performance the existing methods.

1.3 LASER INTERACTIONS WITH MATERIALS

A good understanding of the phenomena that take place when laser radiation interacts with matter is a key element for the success and optimization of any laser-based application. In general, let us assume that radiation having fluence, that is, energy per unit area, F_o, interacts with a material. Part of the energy may be absorbed by the material (F_{ab}), part may be scattered (F_{sc}), and part may be transmitted (F_{tr}). A pictorial representation of these processes is shown in Figure 1.2.

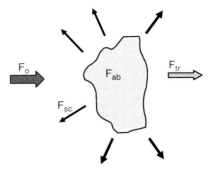

FIGURE 1.2 A simple scheme of processes that may take place when incident radiation of fluence, F_o interacts with a material. F_{ab}, F_{sc}, and F_{tr} are the absorbed, scattered, and transmitted parts of the radiation, respectively.

The overall energy balance is given by Equation (1.1):

$$F_0 = F_{ab} + F_{sc} + F_{tr} \tag{1.1}$$

In turn, the absorbed fluence F_{ab} may eventually cause thermal effects (F_{th}) or photochemical modifications (F_{ph}) in the material, and part of it may be reemitted as fluorescence or phosphorescence (F_{fl}). These processes are described by Equation (1.2):

$$F_{ab} = F_{th} + F_{ph} + F_{fl} \tag{1.2}$$

The values of the terms F_{sc}, F_{tr}, and F_{fl} may be used for obtaining diagnostic information for the identity and quantity of the material components. The interplay between the terms F_{sc} and F_{tr}, for example, is the basis for the imaging technique called reflectography, which depending on the wavelength of the incident radiation is either infrared (IR) or ultraviolet (UV) reflectography. Also, qualitative and quantitative analysis of the composition of a material may be obtained from the terms F_{tr} and F_{fl}. In contrast, F_{th} and F_{ph} may be exploited for material modifications and processing. For example, the cleaning or more generally the removal of unwanted materials from the surface of a target material may come about as the final consequence of such effects. The relative importance of the various terms involved in the above simplified scheme, which describes how radiation interacts with matter, may be controlled to a large extent when a laser is used as the radiation source. Effectively there are three important considerations, whose interplay determines the final outcome in this case.

The laser parameters: wavelength, fluence (energy per unit area), intensity, pulse duration, pulse repetition rate, mode of operation (continuous or pulsed), beam quality, and coherence length
The material parameters: absorption coefficient, heat capacity, thermal conductivity, and other physical properties

The ambient environment: air, inert atmosphere, or vacuum. This is important for the presence of secondary effects following the interaction, for example, oxidation processes, which may influence the final outcome

The appropriate combination of these parameters is critical for the success of any laser-based application. In the laser cleaning of artifacts, for example, depending on the nature of the materials to be removed (e.g., black encrustations, organic materials such as polymerized or oxidized varnish) and the underlying original surface (e.g., stonework, painted layer), it is crucial to define the optimal laser parameters so that the dirt or pollutants will be removed (ablated) while any thermal or photochemical effects due to the absorbed laser radiation [Equation (1.2)] are confined as much as possible to the outermost layer occupied by these unwanted materials, without affecting the original underlying surface. As will be described later in this book, a viable option for the cleaning of polymerized or oxidized varnish layers from painted artworks is the use of pulsed lasers, which emit pulses of nanosecond duration at ultraviolet wavelengths ($\lambda < 250$ nm), which are strongly absorbed by these layers.

There are many processes and effects that may take place as a result of the interaction between laser light and a material. These are determined by the properties of the material and the laser parameters, as will be discussed in Chapter 2. Considering that to date a large number of laser sources is available and that each of them causes different effects depending on the parameters of its operation, the optimization of any specific application relies largely on the selection of the appropriate laser. Different types of laser sources that are employed in cultural heritage applications and the wavelengths that they emit are included in Table 1.1. A primer for the parameters characterizing optical radiation is given in the insert at the end of this section.

Finally, it is not only the laser source but also the design and engineering of the overall workstation that will determine the degree of success of a laser application.

TABLE 1.1
Characteristics of Laser Systems Commonly Used in Art Conservation

Laser type	Wavelength (nm)	Photon energy (e_V)	Pulse duration
ArF excimer	193	6.4	10–20 nsec
KrF excimer	248	5.0	10–20 nsec
Nd:YAG (4ω)	266	4.6	5–20 nsec
XeCl excimer	308	4.0	20–300 nsec
Nd:YAG (3ω)	355	3.5	5–20 nsec
Ar$^+$	488/514	2.5/2.4	cw
Nd:YAG (2ω)	532	2.3	5–20 nsec
HeNe	633	1.9	cw
Nd:YAG (fundamental)	1064	1.2	5–25 nsec *or* 20–150 μsec
Er:YAG	2940	0.42	100–150 μsec
CO_2	10600	0.12	100–150 nsec

For example, the availability of the appropriate laser beam delivery system, either in the form of stable adaptive optical components, usually incorporated in articulated arms, or through optical fibers, is a key element for the practical implementation of several applications. In addition, the integration of several different optical techniques may be useful for monitoring and controlling the quality of the laser interaction, especially in cases where this is intrusive. For example, laser spectroscopic analytical techniques may be incorporated in laser processing apparatus for monitoring the quality of processing in real time. This may form a smart workstation enabling optimal use. Altogether, the maturity and proliferation of possible applications depend strongly on the advances achieved in laser sources, beam delivery systems, and monitoring techniques and their potential combination into robust and user-friendly systems.

A Primer on Optical Radiation

Optical radiation, electromagnetic waves (radiation), can be divided into different ranges depending on its wavelength or frequency as shown in Table 1.2. The *wavelength* (λ) is defined as the length of the cycle or the distance between successive maxima or minima. The usual employed units of wavelength are 1 $\mu m = 10^{-6}$ m, 1 nm $= 10^{-9}$ m, 1 Å $= 10^{-10}$ m (Å = Angstrom), or the *frequency* (ν) which represents the number of cycles per unit time (in cycles per second or Hertz): $\nu = \frac{1}{t}$, where t is time it takes for one complete wave to pass. Wavelength and frequency are related by: $\lambda = \frac{u}{\nu} = \frac{c}{\nu}$, where u is the velocity of propagation. All electromagnetic radiation (light) travels through vacuum with the same velocity (c), which has the value $\sim 3 \times 10^8$ m s^{-1}. The frequency (ν) is the only true characteristic of a particular radiation; whereas, the velocity (u) and the wavelength (λ) depend on the nature of the medium through which the electromagnetic wave travels. In spectroscopic studies, it is common to use instead as a unit the *wavenumber* ($\bar{\nu}$) which is the number of waves per centimeter: $\bar{\nu} = \frac{1}{\lambda} = \frac{\nu}{c}$ i.e., $\overline{\nu(cm^{-1})} = \lambda(\frac{1}{\mu m}) \times 10^4$). Typical values for the optical spectrum such as wavelength (λ), frequency (ν) and wavenumber ($\bar{\nu}$) are given in Table I.1.

TABLE I.1
Typical Values for the Optical Spectrum

	X-rays	Vacuum ultraviolet	Near-ultraviolet	Visible	Near-infrared		Far-infrared
λ, Å	1	2000	4000	7000	10,000		
nm	0.1	200	400	700	1000		
μm			0.4	0.7	1	50	500
ν, Hz		1.5×10^{15}	7.5×10^{14}		3×10^{14}	6×10^{12}	6×10^{11}
$\bar{\nu}$, cm^{-1}		50,000	25,000	14,300	10,000	200	20

Alternatively the characteristics of the radiation are described in terms of photons, which represent the quantized form of electromagnetic wave. The energy of a photon is given by:

$$E = h\nu$$

where: h = Planck's constant (6.626×10^{-34} Js), ν = photon frequency (s^{-1}). The energy of a photon can be calculated in various energy units using the following equations in which λ is expressed in Å:

$$E = \frac{1.24x10^{4}}{\lambda}(eV) = \frac{1.986x10^{-15}}{\lambda}(J)$$

1.4   TYPES OF LASER APPLICATIONS IN CULTURAL HERITAGE

Lasers contribute in a variety of ways to the preservation of cultural heritage: They may reveal the composition of materials of art and antiquity, facilitating their understanding and historic provenance. They may be further used for assessing the condition of an object with regard to previous interventions, environmental pollution, and internal structural defects and weaknesses, which is of great interest to conservators, art historians, and scientists. Lasers may also be used for the restoration of the original artifacts including cleaning them of pollutants or other materials that may reduce their lifetime and aesthetic impact. Altogether they may provide valuable information for the identity, technology, and historic evolution of an artifact as well as contribute to its protection and revelation.

In categorizing the use of lasers for the preservation of cultural heritage, one may distinguish three major types of applications.

Analytical, for the determination of elemental or molecular compositions of materials. These rely on laser spectroscopic techniques such as laser-induced breakdown spectroscopy (LIBS), laser-induced fluorescence (LIF), laser Raman spectroscopy, and laser mass spectroscopy (LMS).

Structural diagnostics, for the detection and mapping of defects (e.g., cracks, detachments, stress concentrations) in the bulk of an artifact, which may not be visible or may induce invisible effects. These rely on laser interferometric techniques based on optical and digital holographic interferometry, including the techniques of speckle pattern interferometry and, more recently, laser shearography. Laser Doppler vibrometry also serves this purpose.

Restoration, including primarily laser cleaning. Lasers are effective tools for the removal of unwanted surface materials, such as black encrustations, pollutants, chemically altered surface layers (for example, oxidized and polymerized varnishes), unwanted overpaints, and so forth.

The first two types of applications are mostly noninvasive, since very little or no sampling is required. In this context, LIF, Raman spectroscopy, and interferometric techniques may be considered noninvasive. In contrast, LIBS is an invasive technique based on high precision laser microsampling. It may be applied, however, locally (*in situ*) and directly without any special sample preparation. This does not apply for the case of laser mass spectroscopy, for which sampling and sample handling is required for the analysis. This disadvantage is counterbalanced by the high sensitivity of this technique, which may be important for the understanding of an object in detail. An additional interesting aspect of the optical analytical techniques such as LIBS and LIF is that they may be used for on-line monitoring and controlling the effects of restoration interventions, laser based or not, thus optimizing the outcome and safeguarding from potential damages. Laser interferometric techniques are also nonintrusive, and some of them are characterized by a high degree of sensitivity, like the technique of holographic interferometry for nondestructive testing (HINDT).

In contrast, several important general considerations must be made when considering the cleaning of an object, an intervention that is completely irreversible. These include

- Maintenance of the chemical and physical state of the original object
- The possibility for extreme control of the cleaning process, as it is often desirable to achieve only partial cleaning
- Safe use for the conservator

In order to satisfy these considerations, it is crucial to have a good understanding of the fundamental processes associated with laser cleaning and assess the consequences of any effects that may come about in the short or long term. In some cases, it is also useful to exploit process monitoring techniques integrated in feedback control systems. In this way, the conservator has the opportunity to assess the outcome in real time and determine the desirable level of cleaning. The specialized advanced training of the conservation professionals involved and the availability of user-friendly systems are also major concerns. To date, there are commercial systems that fulfill a combination of these requirements for a range of demanding applications. Figure 1.3 shows a commercial Nd:YAG laser system appropriate for the cleaning of stonework, incorporating a hand-held optical fiber for beam manipulation. A successful example of the laser cleaning of an ancient sculpture is shown in Figure 1.4.

Lasers are not only analytical, diagnostic, or cleaning tools. They can also be used for other applications, such as microwelding applications of artifacts [13]. There are also laser-based techniques, as for example the three-dimensional (3D) prototyping of artifacts, which are well established and will not be discussed in this book.

In conclusion, the differences in the degree of maturity of the various applications of lasers in cultural heritage work depend strongly on the advances of laser sources, optics, and monitoring systems. Several new possibilities appear on the horizon in

FIGURE 1.3 A commercial Nd:YAG laser cleaning system.

this respect, which are based on the use of pulsed lasers, capable of emitting ultrashort pulses of several pico- or femtosecond duration, for the surface treatment of artifacts. The examination of these possibilities is current.

   Some general remarks regarding laser applications for the preservation of cultural heritage are worthy at this point. At first it should be emphasized that lasers are by no means a panacea for meeting all possible conservation challenges. Lasers should be considered as useful tools in the hands of a conservator or an archaeologist, offering several distinct advantages in comparison to other standard techniques but also having their own drawbacks. For example, as already mentioned, among the advantages are the monochromaticity of laser light, which allows the *in situ* noninvasive and dynamic chemical analysis of pollutants embedded in the original materials of an artwork, without the need for sampling by mechanical means or transferring the sample to the analytical instrument, as happens in many standard analytical techniques. The directionality of laser light may be used for localized spot or remote analysis, while the high coherence of laser beams may be exploited in highly sensitive interferometric techniques, providing a picture of the internal state of an artifact with better spatial resolution and sensitivity than acoustic techniques. Laser spectroscopic techniques may also be used for stratigraphic analytical

FIGURE 1.4 Nd:YAG laser cleaning (left, before; right, after) of an ancient statue of the god Hermes (2nd century B.C.) from the excavation of ancient Messene. (Courtesy Yianna Doganis and Amerimni Galanos, Conservation Consultants).

applications. Material removal by laser light, either in laser cleaning or in overpaint removal applications, is a noncontact, controllable, and environmentally friendly technique as opposed to the standard mechanical and chemical means, for which there is limited control, a high probability for human error, and the risk of chemical hazards.

However, despite these advantages, the use of laser technology in conservation applications also has limitations and should be applied with care. Determining the composition of complex art materials by lasers relies on the existence of relevant data banks and calibration data, which are currently missing in many cases. In restoration applications, the misuse of lasers may cause undesirable and irreversible effects either immediately obvious or in the long term. Good training of the laser users and, most important, a thorough preliminary investigation of the materials involved for selecting the appropriate laser parameters are crucial in this respect. Table 1.3 summarizes some important advantages and drawbacks in using current laser technology for the preservation of cultural heritage.

As will become evident in this book, it is the combination of laser with traditional conservation techniques that leads to optimal results in the majority of cases. Furthermore, the role of ongoing research is to define the potential and limitations of laser technology for the preservation of cultural heritage.

TABLE 1.3
Laser Technology for the Preservation of Cultural Heritage

Analytical applications	Advantages	Disadvantages
LIF (analysis of molecular materials, e.g., pigments, media, binders, fungi)	noninvasive *in situ* remote	spectral complexity limited compound sensitivity no databanks
Raman Spectroscopy (analysis of organic and inorganic molecular materials, e.g., pigments, stone composition)	noninvasive *in situ* microscopic	interference from fluorescence limited applicability
LIBS (direct elemental analysis and indirect inorganic compound determination, e.g., inorganic pigments, pollutants, overpaints)	no sample preparation spot analysis on-line control stratigraphic analysis	invasive (microsampling) no compound information
LMS (molecular and elemental analysis, e.g., pigments, media, binders, isotopic distributions)	high sensitivity	intrusive sample handling limited applicability
Interferometric applications		
Doppler velocimetry	easy to apply transportable dynamic characterization	spread acoustic excitation massive or in contact excitation source primarily surface layer characterization
Holographic interferometry nondestructive testing (HINDT)	high resolution surface and in bulk information transportable *in situ* application	interpretation of results need for vibration isolation *in situ* applications depend on photosensitive medium

Cleaning applications
(e.g., stonework, paintings, wood, paper, ivory, leather)

noncontact	intrusive
possible on-line monitoring	possible irreversible short-and/or long-term
controllable	effects
high spatial resolution	more expensive than traditional technique
	a preliminary study is necessary

1.5 BRIEF HISTORY OF LASER USE IN PRESERVING CULTURAL HERITAGE

Laser cleaning applications are usually associated with the historic origin of lasers in preserving cultural heritage. A comprehensive overview is given in [14]. John Asmus of the University of California San Diego is recognized as the pioneer in this field. Asmus and his collaborators in Italy, in the early 1970s, during tests of the applicability of holography to the conservation of Venetian marble sculpture, discovered by chance that high-power pulsed lasers can be used for the cleaning of black crust from stonework [15,16]. Later, he also investigated the use of pulsed lasers for the cleaning of stained glass, frescoes, leather, and other materials, establishing the laser as a viable solution for a range of conservation applications [17]. Other research groups, including that of Orial and Verges-Belmin and their colleagues at the Laboratoire de Recherche des Monuments Historiques in France and the teams of Larson, Cooper, and Emmony at the Conservation Centre, National Museums Liverpool, and the University of Loughborough in the U.K., made vital contributions to the further establishment of the field through their detailed studies and case applications using the Nd:YAG laser emitting at 1064 nm in the near infrared (IR) spectrum for the conservation of marble sculpture, limestone, terracotta, wood, metal, and other materials [18–20]. Emmony and Pouli have also investigated laser discoloration effects on polychromes [21]. The use of laser cleaning for preserving the patina on marble sculpture was a particularly important contribution in this respect. Based on the works of these groups and others, commercial laser cleaning systems were produced by the companies Laserblast, a subsidiary of Quantel in France, Quanta System in Italy, and Lynton Lasers in the U.K. Also in the U.K., Watkins at the University of Liverpool addressed fundamental aspects of the interactions involved in laser cleaning [22]. The Italian researchers Salimbeni, Siano, and Pini at CNR undertook case studies using the Nd:YAG laser and showed the importance of optimizing the laser pulse duration for cleaning applications [13]. Based on their work, the Italian company El.En. produced a commercial laser cleaning system capable of emitting longer laser pulses than those usually applied. These researchers also examined the applicability of lasers for micro-welding and other applications [13]. In parallel, Calcagno, also in Italy, developed practical methods for the laser cleaning of stonework monuments employing a laser built by Big Sky Laser Technologies in the U.S. for the Italian company Altech. Further extensive applications were carried out on important monuments in Austria by the team of Nimmrichter in Vienna [23]. Following these studies, the group of Fotakis and Zafiropulos at FORTH in Greece investigated details of the mechanisms operating in the laser cleaning of stonework, defining the laser parameters for eliminating undesirable consequences that may appear in certain cases, such as the so-called yellowing effect. This is a discoloration that has been reported to appear occasionally in some types of stone following cleaning by the Nd:YAG laser at 1064 nm [24,25]. Depending on the type of encrustation, this effect may be eliminated by using a dual wavelength laser system, emitting simultaneously at 1064 and 355 nm [26]. They also showed how LIBS may be used for controlling the cleaning process [27]. In general, the Nd:YAG laser has been used extensively in a variety of applications, for the cleaning of not only stonework but also other materials. For example, the groups of Abraham at LACMA in the U.S., Dignard at the Canadian Conservation Institute, and Koss and Strzelec in Poland have demonstrated several applications along these lines [28–30]. Significant contributions for the understanding of the interactions involved were made by Watkins at the University of Liverpool and in the theoretical work of Luk'yanchuk at DSI in Singapore [11,22,31].

The cleaning of parchments and paper by using pulsed lasers emitting in the visible, for example using a frequency-doubled Nd:YAG laser emitting at 532 nm, was established in detailed studies by Kautek and his group at the Federal Institute for Materials Research and Testing in Germany. Significant contributions were also made by Kolar and Strlic at the National and University Library of Slovenia [32,33]. Later, Cefalas in Athens used an F_2 laser emitting at 157 nm also for paper cleaning [34]. The group of Sliwinski in the Polish Academy of Sciences studied the post-processing effects due to laser cleaning of paper [35]. In the mid-1990s, two German teams led by Dickmann at the Fachhochschule Münster and Leissner, Fuchs, and Römich at the Fraunhofer Institute in Würzburg studied the performance of excimer lasers for the removal of organic and inorganic corrosion from stained glass [36–38]. Wiedemann at the Fraunhofer Institute in Dresden examined the laser cleaning of various materials such as metal and wood [39,40].

A particularly demanding application has been the laser cleaning of painted artworks. Initial studies were undertaken at the Canadian Conservation Institute by Leslie Carlyle, and the results were published in an internal report. Significant contributions to this application were made by the group of FORTH in Crete, Greece, in the early 1990s in a collaboration with the Conservation Department of the National Gallery of Athens led by Michael Doulgeridis [41,42]. Based on the work at FORTH, the first commercial system for the cleaning of paintings by KrF lasers emitting at 248 nm (in the UV) was produced in The Netherlands by Art Innovation b.v. Later, the American team of de Cruz and Wolbarsht reported the use of an Er:YAG laser emitting at 2.94 μm for the cleaning of paintings, work that also led to the establishment of an easy-to-use commercial product in the US, which has been tested in Italy by Matteini and his colleagues [43]. Finally, the group of Castillejo and Martin at CSIC in Spain undertook detailed investigations for understanding the laser removal of polymeric and other materials from various substrates [44].

Table 1.4 includes materials that have been cleaned by lasers successfully according to data presented in LACONA conferences only.

Historically, the first reports of laser-based analytical applications in cultural heritage appear in the early 1980s. LIF studies of oil colors and for the identification of pigments in oil paintings were first reported in Japan by Miyoshi et al. using N_2 laser excitation [45]. Later the FORTH group in Greece developed LIF further [46]. Cubeddu at the Polytechnic of Milan in Italy exploited the time evolution of the LIF spectra, expanding the potential of the technique [47]. Remote applications of LIF were first demonstrated by Svanberg et al. at the University of Lund in Sweden and by Pantani and his team at CNR in Italy [48].

TABLE 1.4
Summary of Successful Laser Cleaning Applications

Material/Medium	Nd:YAG 532 nm psec	CO$_2$ TEA	Er:YAG 2940 nm	Nd:YAG 1064 nm	Nd:YAG 1320 nm	Nd:YAG SFR	Nd:YAG 532 nm	Ti:S 780 nm	Nd:YAG 355 nm	Nd:YAG 1064+355	Excimer 308 nm	Nd:YAG 266 nm	Excimer 248 nm
Adhesives (removal)													x
Biological (removal)			x										
Brick				x				x					
Ceramics						x							
Consolidants (removal)									x				
Daguerreotype	x						x						
Dyes/inks (removal)							x						
Feathers							x						
Frescoes				x			x						
Grafitti (removal)				x			x						
Ivory							x						
Leather				x								x	
Metals (Various)													
Archeological iron		x		x		x							
Old ironworks		x		x									
Silver/threads				x		x	x		x				
Copper/bronze					x	x		x					
Old coins				x		x							
Lead				x		x							
Gold leaf/gilding			x	x		x							
Old paint (removal)									x				
Overpaints (removal)			x										x
Paleofossils				x			x						

Paper	x	x			x	
Parchment	x	x			x	x
Picture postcards					x	x
Plaster/stucco		x			x	x
Polychrome	x	x				
Silk				x		
Stone	x	x	x		x	
Stained glass						x
Terracotta		x				
Textiles	x	x				
Varnish (removal)	x	x				x
Wood		x				

*Based on LACONA Conferences reports (Compiled by Ed Teppo).

LIBS for the elemental analysis of works of art was used by Asmus in his early work and was established as a promising analytical technique by Anglos and his collaborators at FORTH [49]. Fantoni et al. at ENEA and Palleschi et al. at CNR in Italy contributed further to the maturity of this technique [50–52]. The use of LIBS for stratigraphic analytical applications was the outcome of the collaborative work of Dickmann, Klein, Zafiropulos, Fotakis, and their colleagues, who also demonstrated that it can be used for the monitoring of laser cleaning in real time [53,54]. Figure 1.5 shows an example of artifacts analyzed successfully by LIBS. The LIBS spectra indicate that the black pigment in the Minoan period was iron based and that in the Byzantine period manganese based.

The first report of Raman spectroscopy for the identification of pigments appeared in an article by Guineau in 1984 [55]. This technique was combined with microscopy by Moreno in Spain [56] and primarily through the work of Clarc and his collaborators at University College in the UK [57,58]. Recent studies by Castellucci at LENS in Italy and the collaborative work of Vandenabeele in Belgium and Grant in the US have improved the prospects of laser Raman spectroscopy and expanded the basis for its applicability [59–61]. Furthermore, the work of the last

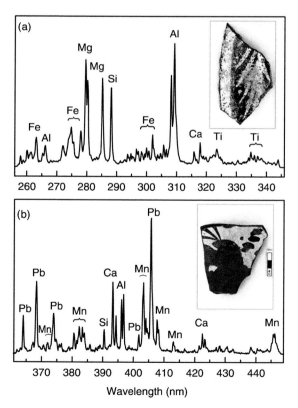

FIGURE 1.5 Black pigment analysis in pottery by LIBS. (a) Minoan period (Fe-based); (b) Byzantine period (Mn-based).

two researchers led to the establishment of a versatile mobile laser Raman spectrometer [61]. Laser mass spectrometric applications in cultural heritage work have been scarce. The work of Miller in the US for the analysis of daguerreotypes is a good example [62].

A variety of noninvasive optical techniques has been used for monitoring the shape and surface roughness relief of artworks. Several examples of laser scanning profilometry used in this respect have revealed painting detachments and deformation of materials [63]. Laser holographic techniques are extremely sensitive in this respect. The first holographic recording of an artwork was made by Asmus et al. in Venice [15], and very soon thereafter the technique of HINDT for diagnosing the structural state of artifacts was established at universities in Aquila and Firenze in Italy [64–66]. An important advance in this field has been made by the use of digital image acquisition for the recording of holographic and speckle interferograms [67]. Today there is a wide range of variations of these techniques providing surface stress information or in-volume structural mapping of an artifact. For example, techniques such as digital speckle photography (DSP) or electronic speckle pattern interferometry (ESPI) have been applied by Hinsch and his associates at the University of Oldersburg in Germany for the study of the mechanical response of artworks subjected to external loads [68,69]. Among others, the group of von Bally at the University of Münster has demonstrated the combination of ESPI with stereoscopic 3D coordinate measurements for determining deformation in depth at submicrometer levels [70]. The technique of scanning laser Doppler vibrometry (SLDV) has been mastered by the team of Tomassini, Paone, and Esposito et al. at the University of Ancona in Italy. This technique has been also applied for detecting and characterizing deficiencies in the plaster support of frescoes and paintings [71,72]. Double exposure or comparative holography is an extremely sensitive interferometric technique, appropriate for high-resolution mapping of internal defects in the bulk of artifacts. Tornari and her group at FORTH in Greece have developed this technique to a high degree of sophistication and established methods for a wide range of applications [42,73,74]. The capabilities of this technique have been demonstrated by a number of examples. Recently, Targowski and his students in Poland demonstrated the use of optical coherence tomography (OCT) for structural diagnosis of artworks [75].

An important milestone for the promotion of all the above activities was the establishment of the conference series "Lasers in the Conservation of Artworks" (LACONA). The first conference was organized by FORTH in Crete, Greece, in 1995, followed by LACONA conferences in Liverpool (1997), Florence (1999), Paris (2001), Osnabrück (2003), and Vienna (2005). The next LACONA conference is scheduled for Madrid in 2007. A vibrant interdisciplinary community has been formed around the LACONA conferences.

The impact lasers have made for the preservation of cultural heritage has been greatly facilitated by the support received by various national and international programs. In particular, European Union (EU) Research and Training Programs have provided the means for bringing scientists, engineers, conservators, archaeologists, and art historians together for an effective use of resources. Supporting measures, such as the COST Action G7, also contribute to harmonizing the efforts in this field.

1.6 STRUCTURE OF THIS BOOK

Following this first introductory chapter, Chapter 2 provides an overview of the fundamental aspects of laser interactions with matter, with emphasis placed on issues important in laser diagnostics and laser processing applications. Chapters 3 and 4 are devoted to laser-based spectroscopic techniques for the analysis of the composition of materials of art and archaeology. In Chapter 3, emphasis is placed on the LIBS technique, which besides the elemental surface composition may also be used for stratigraphic analytical studies and the control of cleaning applications in real time. Chapter 4 is focused on Raman, the applications of Raman spectroscopy in cultural heritage.

Chapter 5 describes nondestructive diagnostic techniques for the assessment of the structural state of an artifact by exploiting the high coherence of laser light.

An overview of the basic principles determining laser processing applications is given in Chapter 6, and laser applications for the cleaning of painted artworks are presented in Chapter 7. In particular, laser parameters for the safe removal of polymerized or oxidized surface layers of organic materials and overpaints are discussed in detail.

In Chapter 8, fundamental aspects of laser cleaning of stonework are discussed together with methods for the removal of encrustations, placing emphasis in the presentation of case studies.

Case studies for the restoration of other materials such as parchment, paper, metal, ivory, and wood, including approaches for the restoration of modern paintings, are summarized in Chapter 9.

REFERENCES

1. Wilson, J. and Hawkes, J.F.B., *Lasers, Principles and Applications,* Prentice Hall International, 1987.
2. Silfvast, W.T., *Laser Fundamentals,* Cambridge University Press, Cambridge, 1996.
3. Svelto, O., *Principles of Lasers,* Plenum Press, New York, 1998.
4. Siegman, A.E., *Lasers,* Oxford University Press, Oxford, 1986.
5. Verdeyen, J.T., *Laser Electronics,* Prentice Hall, London, 1995.
6. Yariv, A., *Quantum Electronics,* John Wiley, New York, 1989.
7. Demtroeder, W., *Laser Spectroscopy: Basic Concepts and Instrumentation,* Springer-Verlag, Berlin, 1996.
8. Finlayson, D.M. and Sinclair, B.D., *Advances in Lasers and Applications,* Proceedings of the Fifty-Second Scottish Universities Summer School in Physics, SUSSP and Institute of Physics, London, 1999.
9. Schuoecker, D., *Handbook of the Eurolaser Academy,* Chapman and Hall, London, 1998.
10. Steen, W.M., *Laser Material Processing,* Springer-Verlag, London, c1991.
11. Luk'yanchuk, B.S., *Laser Cleaning,* World Scientific, Singapore, 2002.
12. Svanberg, S., *Atomic and Molecular Spectroscopy. Basic Aspects and Practical Applications,* Springer-Verlag, Berlin, c1991.
13. Salimbeni, R., Pini, R., and Siano, S., A variable pulse width Nd:YAG laser for conservation, LACONA IV, *Journal of Cultural Heritage,* 4, 72, 2003, and Innocenti, C., Pieri, G., Yanagishita, M., Pini, R., Siano, S., and Zanini, A., Application of laser

welding to the restoration of the ostensory of the martyr St. Ignatius from Palermo, LACONA IV Proceedings, *Journal of Cultural Heritage,* **4**, s362, 2003.

14. Cooper, M., *Laser Cleaning in Conservation: An Introduction,* Butterworth Heinemann, Oxford, 1998.

15. Asmus, J.F., Guattari, J., and Lazzarini, L., Holography in the conservation of statuary, *Stud. Conserv.,* **18**, 49, 1973.

16. Asmus, J.F., Light cleaning: laser technology for surface preparation in the arts, *Technology and Conservation,* **3**, 14, 1978.

17. Asmus, J.F., Murphy, C.G., and Munk, W.H., Studies on the interaction of lasers with art artifacts, *SPIE,* **41**, 19, 1973.

18. Verges-Belmin, V., Comparison of three cleaning methods — microsandlasting, chemical pads and Q-switched YAG laser — on a portal of the cathedral Notre-Dame in Paris, France, LACONA I Proceedings, *Restauratoreblatter,* **1**, 17, 1997.

19. Verges-Belmin, V., Pichot, C., and Orial, G., Elimination de croutes noires sur marbre et craie: a quel niveau arreter le nettoyage? International Congress on the Conservation of Stone and Other Materials, 534, 1993.

20. Cooper M.I., D.C. Emmony, and J. Larson, Characterisation of laser cleaning of limestone, *Optics and Laser Technology,* **27**(1), 69, 1995, and Cooper, M.I. and Larson, J.H., The use of laser cleaning to preserve patina on marble sculpture, *Conservator,* **20**, 28, 1996.

21. Pouli, P., Emmony, D.C., Madden, C.E., and Sutherland, I., Analysis of the laser induced reduction mechanisms of medieval pigments, *Applied Surface Science,* **173**, 252, 2001.

22. Waltkins, K.G., A review of materials interaction during laser cleaning in art restoration, *LACONA I Proceedings,* **1**, 7, 1997.

23. Calcagno, G., Koller, M., Nimmrichter, H., Laser based cleaning on stonework at St. Stephen's Cathedral Vienna, *LACONA I Proceedings,* 39, 1997.

24. Skoulikidis, T., Vassiliou, P., Papakonstantinou, E., Moraitou, A., Zafiropulos, V., Kalaitzaki, M., Spetsidou, I., Perdikatsis, V., and Maravelaki, P., Some remarks on Nd:YAG and excimer UV lasers for cleaning soiled sulfated monument surfaces, LACONA I Conference, *Book of Abstracts,* 1995.

25. Verges-Belmin, V. and Dignard, C., Laser yellowing: myth or reality? *LACONA IV Proceedings,* **4**, 238, 2003.

26. Pouli, P., Frantzikinaki, K., Papakonstantinou, P., Zafiropulos, V., and Fotakis, C., Pollution encrustation removal by means of combined ultraviolet and infrared laser radiation: the application of this innovative methodology on the surface of the Parthenon West Frieze, LACONA V, *Springer Proceedings in Physics,* Dickmann, K., Fotakis, C., and Asmus, J.F., Eds., **100**, 333, 2004.

27. Govenardo-Mitre, I., Prieto, A.C., Zafiropulos, V., Spetsidou, Y., and Fotakis, C., On-line monitoring of laser cleaning of limestone by laser induced breakdown spectroscopy, *Appl. Spectroscopy,* **57**(1125), 1997.

28. Dignard, C., Lai, W.F., Binnie, N., Young, G, Abraham, M., and Scheerer, S., Cleaning of soiled white feathers using the Nd:YAG laser and traditional methods, LACONA I Conference, Lasers in the Conservation of Artworks, *Book of Abstracts,* Heraklion, Greece, 1995, LACONA V Proceedings, *Springer Proceedings in Physics*, Dickmann, K., Fotakis, C., and Asmus, J.F., Eds., **100**, 227, 2004.

29. Abraham, M., Madden, O., and Scheerer, S., The use added matrix elements such as chemical assists, colorants and controlled plasma formation as methods to enhance laser conservation of works of art, *LACONA IV,* **4**, 92, 2003.

30. Strzelec, M., Marczak, J., Ostrowski, R., Koss, A., and Szambelan, R., Results of Nd:YAG laser restoration of decorative ivory jug, LACONA V Proceedings, *Springer Proceedings in Physics,* Dickmann, K., Fotakis, C., and Asmus, J.F., Eds., **100**, 163, 2004.

31. Watkins, K.G., Curran, C., and Lee, J.M., Two new mechanisms for laser cleaning using Nd:YAG sources, *LACONA IV Proceedings,* **4**, 59, 2003.

32. Kolar, J., Strlic, M., Pentzien, S., and Kautek, W., Near-UV, visible and IR pulsed laser light interaction with cellulose, *Appl. Phys. A,* **71**, 87, 2000.

33. Puchinger, L., Pentzien, S., Koter, R., and Kautek, W., Chemistry of parchment — laser interaction, LACONA V, *Springer Proceedings, Physics,* Dickmann, K., Fotakis, C., and Asmus, J.F., Eds., **100**, 51, 2004.

34. Kollia, Z., Sarantopoulou, E., Cefalas, A.C., Kobe, S., and Samardzija, Z., Nanometric size control and treatment of historic paper manuscript and prints with laser light at 157 nm, *Appl. Phys. A.,* **A79**, 379, 2004.

35. Kaminska, A., Sawczak, M., Cieplnski, M., and Sliwinski, G., The post-processing effects due to pulsed laser ablation of paper, LACONA V Proceedings, *Springer Proceedings in Physics,* Dickmann, K., Fotakis, C., and Asmus, J.F., Eds., **100**, 2004.

36. Leissner, J. and Fuchs, D.R., Examination of excimer laser treatments as a cleaning method for historical glass windows, in 3rd European Society of Glass Science and Technology Conference, Fundamentals of Glass Science and Technology, 1995.

37. Fekrsanati, F., Hildenhagen, J., Dickmann, K., Trolf, C., and Olaineck, C., UV-laser radiation: basic research of their potential for cleaning stained glass, LACONA III Conference, *Book of Abstracts,* 34, 1999.

38. Romich, H., Dickmann, K., Mottner, P., Hildenhagen, J., and Muller, E., Laser cleaning of stained glass windows — final results of a research project, *LACONA IV Proceedings,* **4**, 112, 2004.

39. Wiedemann, G., Pueschner, K., Wust, H., and Kempe, A., The capability of the laser application for selective cleaning and the removal of different layers on wooden artworks, LACONA V, *Springer Proceedings in Physics,* Dickmann, K., Fotakis, C., and Asmus, J.F., Eds., **100**, 179, 2004.

40. Mottner, P., Wiedemann, G., Haber, G., Conrad, W., and Gervais, A., Laser cleaning of metal surface — laboratory investigations, *Springer Proceedings in Physics,* Dickmann, K., Fotakis, C., and Asmus, J.F., Eds., **100**, 79, 2004.

41. Fotakis, C., Anglos, D., Balas, C., Georgiou, S., Vainos, N.A., Zergioti, I., and Zafiropulos, V., Laser technology in art conservation, Optical Society of America, *OSA TOPS on Lasers and Optics for Manufacturing,* Tam, A.C., Ed., **9**, 99, 1997.

42. Georgiou, S., Zafiropulos, V., Anglos, D., Balas, C., Tornari, V., and Fotakis, C., Excimer laser restoration of painted artworks: procedures, mechanisms and effects, *Appl. Surface Science,* **127–129**, 738, 1998.

43. Bracco, P., Lanterna, G., Matteini, M., Nakahara, K., Sartiani, O., de Cruz, A., Wolbarsht, M.L., Adamkiewicz, E., and Colombini, M.P., Er:YAG laser: an innovative tool for controlled cleaning of old paintings: testing and evaluation, LACONA IV Proceedings, *Journal of Cultural Heritage,* **4**, 202s, 2003.

44. Castillejo, M., Martin, M., Oujja, M., Silva, D., Torres, R., Manousaki, A., Zafiropulos, V., van den Brink, O., Heeren, R.M.A., Teule, R., Silva, A., and Gouveia, H., Analytical study of the chemical and physical changes induced by KrF laser cleaning of tempera paints, *Anal. Chem.,* **74**, 4662, 2002.

45. Miyoshi, T., Fluorescence from varnishes for oil paintings under N_2 laser excitation, *Jpn. J. Appl. Phys.,* **26**, 780, 1987.

46. Anglos, D., Solomidou, M., Zergioti, I., Zafiropulos, V., Papazoglou, T.G., and Fotakis, C., Laser induced fluorescence in artworks diagnostics: an application in pigment analysis, *Appl. Spectroscopy,* **50**, 1331, 1996.

47. Cornel, D., d'Andea, C., Valentini, G., Cubeddu, R., Colombo, C., and Toniolo, L., Fluorescence lifetime imaging and spectroscopy as tools for nondestructive analysis of works of art, *Appl. Optics,* **43**, 2175, 2004.

48. Weibring, P., Johansson, T., Edner, H., Svanberg, S., Sudner, B., Raimondi, V., Cecchi, G., and Pantani, L., Fluorescence lidar imaging of historical monuments, *Appl. Optics,* **40**, 6111, 2001.

49. Anglos, D., Couris, S., and Fotakis, C., Laser diagnostics of painted artworks: laser induced breakdown spectroscopy in pigment identification, *Appl. Spectroscopy,* **51**, 1025, 1997, and Melessanaki, K., Mateo, M.P., Ferrence, S.C., Betancourt, P.P., and Anglos, D., The application of LIBS for the analysis of archaeological ceramic and metal artifacts, *Appl. Surf. Sci.,* 197–198, 156, 2002.

50. Ciussi, A., Pallesci, V., Rastelli, S., Salvetti, A., Tognoni, E., Fantoni, R., and Borgia, I., Fast and precise determination of painted artwork composition by laser induced plasma spectroscopy, *Springer Series of the International Society of Optics within Life Sciences,* **5**, 138, 2000.

51. Colao, F., Fantoni, R., Lazic, V., Morona, A., Santagata, A., and Giardini, A., LIBS used as a diagnostic tool during the laser cleaning of ancient marble from the Mediterranean areas, *Appl. Phys. A.,* **79**, 213, 2004.

52. Colao, F., Fantoni, R., Lazic, V., and Spizzichino V., Laser induced breakdown spectroscopy for semi-quantitative and quantitative analyses of artworks — application on multi-layered ceramics and copper based alloys, *Spectrochimica Acta Part B,* **57**, 1219, 2002.

53. Klein, S., Stratoudaki, T., Zafiropulos, V., Hildenhagen, G., Dickmann, K., and Lehmkuhl, Th., Laser induced beakdown spectroscopy for on-line control of laser cleaning of sandstone and stained glass, *Appl. Phys. A.,* **69**, 441, 1999.

54. Maravelaki, P.V., Zafiropulos, V., Kylikoglou, V., Kalaitzaki, M.P., and Fotakis, C., Laser induced breakdown spectroscopy as a diagnostic technique for the laser cleaning of marble, *Spectrochimica Acta Part B,* **52**, 41, 1997.

55. Guineau, B., Analyse non destructive des pigments par microseconde Raman laser: exemples de l' azurite et de la malechite, *Stud. Conserv.,* **29**, 35, 1984.

56. Ruiz-Moreno, S., Perez-Pueyo, R., Gabaldon, A., Soneira, M.J., and Sandalinas, C., Raman laser fibre optic strategy for non-destructive pigment analysis. Identification of a new yellow pigment (Pb, Sn, Sb) from the Italian XII century painting, *LACONA IV Proceedings,* **4**, 309, 2004.

57. Clark, R.J.H., Raman microscopy: application to the identification of pigments of medieval manuscripts, *Chem. Soc. Rev.,* **24**, 187, 1995.

58. Burgio, L.H., Clark, R.J.H., Stratoudaki, Th., Doulgeridis, M., and Anglos, D., Pigment identification in painted artworks: a dual analytical approach employing LIBS and Raman microscopy, *Appl. Spectroscopy,* **54**, 463, 2000.

59. Bussott, L., Carboncini, M.P., Castellucci, E., Giuntini, L., and Mando, P.A., Identification of pigments in a fourteenth century miniature by combined micro-Raman and PIXE spectroscopic techniques, *Stud. Conserv.,* **42(2)**, 83, 1997.

60. Weis, T.L., Jiang, Y., and Grant, E.R., Toward the comprehensive spectrochemical imaging of painted works of art: a new instrumental approach, *J. Raman Spectroscopy,* **35**, 813, 2004.

61. Vandenabeele, P., Weis, T.L., Grant, E.R., and Moens, L.J., A new instrument adapted to in situ Raman analysis of objects of art, *Anal. Bioanal. Chem.,* **379**, 137, 2004.

62. Hogan, D.L., Goloviev, V.V., Gresalfi, J., Chaney, J.A., Feigerle, C.S., Miller, J.C., Romer, G., and Messier, P., Laser ablation mass spectroscopy in nineteenth century daguerreotypes, *Appl. Spectroscopy,* **53**, 1161, 1999.

63. Hinsch, K.D. and Gulker, G., Lasers in art conservation, *Physics World,* **14**, 37, 2001.

64. Ambrosini, D. and Paoletti, D., Holographic and speckle methods for the analysis of panel paintings. Developments since the early 1970's, *Reviews in Conservation,* **5**, 38, 2004.

65. Bertani, D., Cetica, M., and Molesini, G., Holographic tests on the Ghiberti panel: the life of Joseph, *Stud. Conserv.,* **27**, 61, 1982.

66. Boone, P.M., Use of close range objective speckles for displacement measurement, *Opt. Engineering,* **21(3)**, 407, 1982.

67. Boone, P.M. and Markow, V.B., Examination of museum objects by means of video holography, *Stud. Conserv.,* **40**, 103, 1995.

68. Gulker, G., Hinsch, K.D., Holscher, C., Kramer, A., and Neunaber, H. 1. *In situ* application of electronic speckle pattern interferometry (ESPI) in the investigation of stone decay. 2. Laser interferometry, *Quantitative Analysis in SPIE,* 1990.

69. Hinsch, K.D., Gulker, G., Hinrichs, H., and Joost, H., Artwork monitoring by digital image correlation, LACONA V, *Springer Proceedings in Physics,* Dickmann, K., Fotakis, C., and Asmus, J.F., Eds., **100**, 459, 2004.

70. Dirksen, D., Kemper, B., Guttzeit, A., Bischoff, G., and von Bally, G., Parallel acquisition of 3-D surface coordinates and deformations by combining electronic speckle pattern interferometry and optical topometry, LACONA V, *Springer Proceedings in Physics,* Dickmann, K., Fotakis, C., and Asmus, J.F., Eds., **100**, 493, 2004.

71. Castellini, P., Esposito, E., Marchetti, B., Paone, N., and Tomasini, E.P., New applications of SLDV to non-destructive diagnostics of artworks: mosaics, ceramics, inlaid wood and easel paintings, LACONA IV, *Journal of Cultural Heritage,* **4**, 321, 2003.

72. Esposito, E., Castellini, P., Paone, N., and Tomerrini, E.P., Laser signal dependence on artworks surface characteristics: a study of frescoes and icons samples, LACONA V, *Springer Proceedings in Physics,* Dickmann, K., Fotakis, C., and Asmus, J.F., Eds., **100**, 327, 2004.

73. Tornari, V., Zafiropulos, V., Bonarou, A., Vainos, N.A., and Fotakis, C., Modern technology in artwork conservation: a laser based approach for process control and evaluation, *Optics and Lasers in Engineering,* **34**, 309, 2000.

74. Tornari, V., Bonarou, A., Zafiropulos, V., Fotakis, C., and Doulgeridis, M., Holographic applications in evaluation of defect and cleaning procedures, LACONA III Proceedings, *Journal of Cultural Heritage,* **4**, S325, 2000.

75. Targowski, P., Rouba, B., Wojtkowski, M., Kowalczyk, A., The application of optical coherence tomography to non-destructive examination of museum objects, *Studies in Conservation,* **49**(2), 107, 2004.

2 Fundamentals of the Laser Diagnostics and Interactions with Matter

2.1 INTRODUCTION

Besides technical characteristics of different lasers, the most important factors for the success of laser diagnostic and processing techniques relate to the physical and chemical processes that underlie the interaction of laser radiation with materials. An understanding of the fundamental processes of laser–material interaction is necessary for the full exploitation of laser techniques in the challenges that conservators may address. For this reason, the emphasis in this chapter is placed on basic photophysical and chemical concepts. Although some of these concepts are well established [1], their importance in the optimization of laser diagnostics and restoration is often overlooked.

The importance of this point cannot be understated. It is best illustrated in medical applications where some initially much-heralded capabilities of lasers have never materialized, whereas other powerful laser schemes were only developed after careful assessment of laser–tissue interaction. In particular, the complexity of the problems in conservation and restoration of artworks requires that a number of issues be carefully addressed and that laser parameters be appropriately optimized. Besides the question of efficiency and effectiveness of the laser as a diagnostic or processing tool, the most important one is the nature and extent of any deleterious effects that may be done on the substrate [2]. This issue is crucial for artworks, since minor chemical or structural modifications may result in their accelerated aging and deterioration over time.

2.2 BASIC PHOTOPHYSICAL CONCEPTS

2.2.1 ABSORPTION OF RADIATION

In most diagnostic techniques (e.g., laser-induced fluorescence), as well as in laser processing schemes, the radiation must be absorbed by the substrate. Thus the probed depth and the extent of various effects (thermal, mechanical, chemical) done to the substrate upon laser irradiation are primarily determined by the spatial distribution of the absorbed laser energy in the substrate. This distribution is specified by the optical absorption and scattering properties of the substrate (Figure 2.1). In nonturbid media, the optical transmission follows the Beer–Lambert law, according to which: $F_{transmitted} / F_{incident} = (1 - R_s) = 10^{-\varepsilon cl} = e^{-\alpha l}$, where R_s is the sample reflectivity,

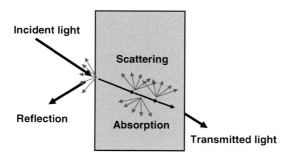

FIGURE 2.1 Schematic of the various effects on the incident optical radiation on a substrate.

$F_{\text{transmitted}}$ is the transmitted fluence through a path length (or depth) l in the sample, ε (in $M^{-1}cm^{-1}$) is the extinction coefficient, c (in mol/liter -M-) is the concentration; and α (in cm^{-1}) is the absorption coefficient. Typically, the absorption co-efficient α is used to express the absorption of a sample, while ε is used in reference to specific chromophores.

The absorption process can be written as an elementary reaction, $M + h\nu \rightarrow M^*$, where M^* indicates molecules in an excited state. The rate of absorption per unit volume is given by

$$-\frac{d[M]}{dt} = \frac{d[M^*]}{dt} = \sigma I [M] \tag{2.1}$$

The absorption cross section is $\sigma = 3.82 \times 10^{-21} \varepsilon \, cm^2/molecule$.

2.2.2 Electronic Properties of Materials

A good understanding of laser–matter interaction requires a description of the processes on an atomic or molecular level. Although this description is heavily mathematical, there are simple representations that adequately describe the interaction. Absorption and emission processes are commonly described either in terms of Jablonski energy level diagrams or in terms of electronic surfaces. Detailed information can be found in a number of references [3–11]. The major aspects can be summarized as follows.

The *Jablonski diagram* (Figure 2.2) emphasizes the energetic arrangement of the electronic states. In the case of an atom or molecule, the states are represented by simple lines. The *electronic ground state of the atom/molecule* is indicated by S_0, and the *excited electronic states, singlet (S)* or *triplet (T)*, by S_1 or T_1, S_2 or T_2. Higher electronic excited states differ successively, as shown, by progressively smaller increments of energy. Each excited state has: definite energy, lifetime, and structure. In addition, the excited states are different chemical entities from the singlet electronic ground state and behave (react) differently.

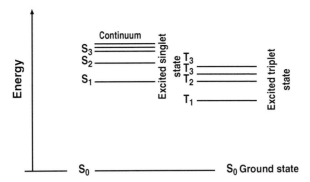

FIGURE 2.2 Schematic of Jablonski diagram for a molecule.

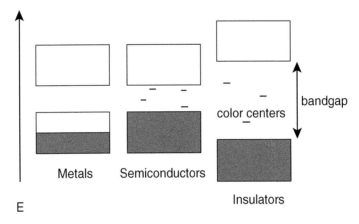

FIGURE 2.3 Schematic of the energy bands in the case of metals, semiconductors, and insulators. For the insulators and the semiconductors, the lower band is called the valence band and the higher band is called the conduction band. The lower energy band in metals is patially filled with electrons.

In the case of periodic arrays of atoms, e.g., semiconductors and metals, the interaction between the atomic orbitals of the very large number of atoms results in the formation of very closely spaced states. Thus in this case, the electronic states are best described by bands. The bands filled with electrons constitute the *valence bands*, whereas unfilled bands constitute *conduction bands*. Electrons in these bands conduct electricity within the crystal. (Figure 2.3).

Excitation from the valence band to the conduction band results in the formation of an electron–hole pair (electron in the upper band, hole in the lower). The conduction band in molecular crystals is very high in energy, since excitation of an electron into this band entails essentially molecular ionization.

In actual materials, however, the periodic arrangement of the crystal is not perfect, and defects such as vacancies occur. For instance, a common defect in ionic insulating crystals (e.g., NaCl) is the so-called *F-center*, in which an electron occupies an anion

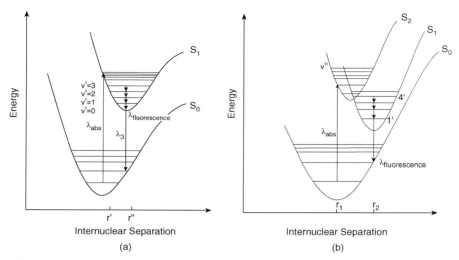

FIGURE 2.4 (a) Illustration of the dynamic surfaces for a molecule. (b) A typical situation in an actual system.

vacancy. Such defects give rise to energy states within the valence-conduction band gap (Figure 2.3). Many other defect centers are found in ionic crystals, which are collectively called color centers, because they give rise to characteristic absorption bands.

For molecules, the energy state depends on the nuclear arrangement and the internuclear distances. Thus a more accurate representation is provided by plots of the potential energies of the electronic states as a function of the nuclei separation (for polyatomic molecules, the plot is, more accurately, along the normal coordinates)—Figure 2.4. The curves for the upper electronic states are usually displaced to larger internuclear separations, because they have a higher antibonding character than the ground state. In such figures, the lines drawn within each potential surface represent the quantized vibrational levels of the corresponding electronic state. At each vibrational level, the energy content is constant and the internuclear separation oscillates between two extreme values. Overlapping of vibrational energy levels of different excited electronic states is a common feature. The upper limit of the potential surfaces of the molecule is determined by the energy required for ionization or for rupture (dissociation) of chemical bonds between nuclei.

2.2.3 EXCITATION PROCESSES/FLUORESCENCE

Because light is an electromagnetic wave, it can exert force on the electrons in the substrate, thereby altering their motion. The strength of interaction depends on the frequency (energy) of the light compared to the binding energy of the electrons. If these energies match (resonance condition), then the light is absorbed and the electron is excited to a higher state. For the different types of materials, the absorption properties (Table 2.1) can be summarized as follows.

TABLE 2.1
Optical Properties of Selected Materials

Material	Wavelength (μm)	Absorption coefficient (cm⁻¹)	Reflectivity R
Ag	0.25	5 E5	0.30
	0.532	8.1 E5	0.91
	1.06	8.33 E5	0.99
	10.6	8.33 E5	0.99
Al	0.248	1 E6	0.93
	0.532	1.5 E6	0.92
	1.06	1 E6	0.94–0.62
	10.6	8.33 E5	0.98
Cu	0.266	7.8 E5	0.23
	0.5	7.14 E5	0.43
	0.8	7.7 E5	0.86
	1.06	7.7 E5	0.98–0.71
	10.6	7.7 E5	0.99
PMMA	0.193	2 E3	
	0.248	400	0.06
	0.308	<20	
	0.351	<10	
PS	0.193	8 E5	
	0.248	6.5 E3	
	0.308	80	
	0.351	≈10	

Metals

The interaction of electromagnetic radiation with the free electrons in metals is so strong that the penetration depth of radiation is limited to a few wavelengths (referred to as skin depth). Typically, the absorption coefficient of metals from the near UV through visible (VIS) and near IR spectral range is between 10^5 and 10^7 cm⁻¹. As shown theoretically [3–4,6], a high absorption also implies a high reflectivity. Thus the reflectivity of metals over UV through the VIS spectral region ranges between 0.25 and 0.95. In the IR, typical values are between 0.9 and 0.99. In contrast, reflectivity decreases considerably at wavelengths below 300 nm, because the electrons cannot respond to the high frequency of the UV light. Certain metals such as gold and copper present, in addition, selective absorption (related to excitation of electrons in d-orbital) and, therefore, selective reflection (which is in fact responsible for the characteristic color of these metals).

Semiconductors

Here the band gap is small enough that, at room temperature, an appreciable number of electrons is thermally excited to the conduction band. Thus there is appreciable free carrier absorption at room temperature. Band gap excitation for semiconductors takes place generally in the near IR and visible spectral region.

Insulators

For insulators, because of the very wide band gap, nearly no carriers are thermally excited to the conduction band at room temperature. Thus there is no free carrier absorption. Generally the band gap is so wide that interband transitions become important only with excitation at UV or VUV wavelengths. In ionically bonded materials, e.g., alkali-halides such as KCl, the valence electrons are quite strongly localized at the negative ion (for KCl, this would be the Cl atom), and hence the optical spectrum contains some atomic-like features (resonances).

In actual materials, insulators and semiconductors are not ideal crystalline; they exhibit various defects. These defects give rise to states within the band gap. Absorption to these states is generally in the visible. As a result, although these crystals should be transparent in the visible, the absorption by defect centers causes them to appear colored. Upon heating of the crystal to sufficiently high temperatures, these defects anneal and the absorption bands disappear.

Molecules

For molecules, irradiation in the VIS or UV entails excitation of the molecule/chromophore from the electronic ground state to some excited state. Because of the much higher mass of the nuclei than that of electrons, the nuclei do not have time to change their position upon electron excitation. Thus the initial molecular arrangement and intermolecular distances do not change upon excitation. Therefore electronic excitation (transition) occurs only vertically in the diagram of potential surfaces [3–9]. However, in quantum mechanics, the position of particles, for example, nuclei, is not fixed (uncertainty principle) but instead is described by a probability distribution. As a result, there is a probability of transitions to various vibrational states of the excited electronic state (given by the square of the net overlap of the vibrational wavefunctions of the initial and final states (Franck–Condon overlap, $<\Psi_i/\Psi_f>$). The Frank-Condon factor governs the relative intensities of vibrational bands in electronic absorption and emission spectra. The classical analogue of this is to view the molecules as oscillating in the vibrational states of ground electronic state between the extreme positions specified by the potential surface, thus interaction with the photon excites the molecules to different vibrational states of the excited electronic state according to the internuclear separation at the "moment of oscillation."

The textbook example is illustrated in Figure 2.4a. Photon absorption raises the molecule at the most probable internuclear distance to some vibrational levels of the excited electronic state. The excited molecules from the initial configuration subsequently relax to the lowest vibrational state of S_1 (internal conversion). From there, a proportion of molecules may return to the ground state via fluorescence, with the remaining decaying to the ground or other states via radiationless transitions. Typically, the fluorescent decay times (radiative lifetimes) are 10^{-9} to 10^{-7} sec [7–9]. The probability of emission decay per excited molecule is expressed by the fluorescent quantum yield.

Most often, the dynamics of excited-state transitions of molecules is much more complex than the ideal case depicted in Figure 2a, as two or more excited electronic states may interact with each other (Figure 2.4b). In that case, the molecule may very quickly decay to an electronic state different from the initially excited one. For molecules in condensed phases, because of the interaction of the excited molecules with nearby molecules, the deactivation from higher electronic or vibrational states to the lower vibrational state of S_1 occurs within picoseconds, i.e., well before fluorescence occurs. Thus in condensed phases, fluorescence emission (as well as photochemical reactions) occurs exclusively from the lowest vibrational level of S_1 (Kasha's rule). Although not widely recognized, this is an important feature, because it results in significant simplification of the fluorescence spectra, thereby enabling their direct use for characterization and providing the basis for the laser-induced fluorescence methods discussed in Chapter 3. For molecules with heavy atoms (e.g., iodine), a large percentage may also decay into the triplet state-intersystem crossing. Dissociation will also occur (even for a stable excited electronic state) if excitation occurs to sufficient energies.

2.2.4 LIGHT SCATTERING PROCESSES

2.2.4.1 Optical Scattering

Besides absorption, the spatial distribution of laser energy in the sample is determined by the scattering of the light as it propagates within the substrate. Scattering is due to variations of the refractive index within the sample. Thus the intensity of scattering depends on the magnitude of the refractive index variation and on the size distribution and concentration of the scattering sites. For particles of size much smaller than the incident wavelength, scattering follows the well-known Rayleigh form where scattering intensity is proportional to $1/\lambda^4$. However, as the size of the scattering sites increases, the mathematical description of scattering becomes increasingly complex. In practice, approximately, the scattering coefficient μ scales as $\sim\lambda^{-b}$, where $b \approx 0.5$ to 2.

Scattering can severely affect the quantitative information derived from laser diagnostic techniques, since it modifies the optical penetration depth of the probing beam (evidently, the effective optical penetration depth is in this case $(\alpha + \mu)^{-1}$) and thus the depth over which chromophores are detected. Various methods of different levels of complexity have been developed, especially in relationship with medical diagnostic problems [12], for the accurate quantification of laser spectroscopic signals recorded from turbid media, but they have been employed very little in the case of artwork diagnostics has been rather limited.

2.2.4.2 Raman Scattering

In the previous case, the scattered light has the same frequency as the incident light. It was found already in 1932 by Raman that in addition, a small percentage of the light is scattered at different frequencies. Their differences from the incident light frequency correspond to vibrations of molecules.

This phenomenon is due to the polarizability of atoms/molecules. Thus even if they do not absorb at the wavelength of the incident light, they become temporarily polarized by the light (i.e., a dipole moment is induced in the atom/molecule). The polarized molecule acts as an antenna, reemitting the light. Such a process can be represented in a Jablonski diagram as indicated in Figure 2.5 where the temporarily polarized condition of the atom/molecule is represented by the dotted line and is denoted as "virtual" state. The sensitivity of Raman scattering to molecular vibrations makes it a powerful

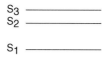

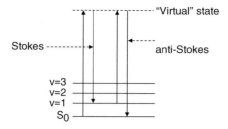

FIGURE 2.5 The Jablonski diagram for a Raman process.

tool for the chemical and structural characterization of pigments and materials in artworks. This advantage crucially relies on the availability of intense lasers, because Raman signals are very weak and their excitation by conventional light sources is impracticable. A detailed description of Raman spectroscopy is presented in Chapter 3.

2.2.5 AN EXAMPLE: REFLECTOGRAPHY

A concrete example of the above considerations to artwork diagnostics is illustrated by broadband reflectography [1], employed for the structural diagnostics of paintings. This technique exploits the fact that the optical penetration depth varies widely in the various spectral regions (e.g., varnish absorbs strongly in the UV, $\alpha \approx 10^5$ cm^{-1}, while it is almost transparent in the 760 to 2500 nm range). Thus the specularly reflected and backscattered light in the corresponding spectral regions provides information about different layers of the substrate. Thus multispectral imaging techniques can be used for mapping the composition and coloration of a painted artwork. To this end, a specialized single unit multifunctional detection system for spectral imaging in the region of 0.35 to 1.6 µm has been developed [13]. Furthermore, the UV absorptivity of varnish increases strongly upon degradation/oxidation [14], and it is characterized by a much lower reflectivity than nondegraded, "fresh" varnish. Thus reflectography provides a powerful tool for optimizing any conservation methods, by detecting the underlying layers that are exposed as dirt and debris are removed.

Although this technique does not rely on lasers, it represents an example of likely areas of advance by exploiting laser technology. For instance, the development of easily wavelength-tunable, robust optical parametric oscillators may significantly enhance the capabilities of reflectography.

2.2.6 DYNAMIC OPTICAL PROCESSES: MULTIPHOTON PROCESSES

Typically, the absorption and scattering coefficients are measured at low light intensities on a spectrophotometer. However, upon pulsed laser irradiation, the dynamic optical properties may deviate widely from the small-signal values, resulting in a significant change of the laser propagation depth.

Upon laser pulsed irradiation, strong thermal and chemical transients may be generated that can alter the optical properties of the substrate. For semiconductors, the temperature elevation upon pulsed laser irradiation may be high enough to result in an increase of the number of electrons in the conduction band, thereby resulting in a strong enhancement of the absorptivity [15]. Similarly, pronounced changes of absorptivity are often noted in the laser irradiation of molecular substrates [16]. These may be due to the formation of intermediates that absorb differently from the precursor, or even to simple structural changes due to the laser-induced temperature increase. Illustrative examples of the latter possibility have been encountered in medical applications. For instance, the depth of thermal damage in Q-switched Er:YAG laser irradiation ($\lambda = 2.94$ µm) of tissues is nearly 2 to 3 times larger than expected on the basis of the room-temperature absorption coefficient [16]. It turns out that the peak absorption of H_2O (the main tissue component) in the IR shifts away from 2.94 µm as its temperature increases. On the other hand, in the UV (e.g., at 193 nm), the effective absorption increases much with increasing laser fluence. In these cases, the absorption changes are due to the strong influence of the laser-induced temperature increase on the hydrogen-bond network of H_2O, with a consequent change of the energy of the electronic states of water.

In most cases however, the change in the absorptivity can be related to changes in the excitation step. To this end, time-resolved absorption and luminescence spectroscopies have been used to probe the dynamics of electronic excitation and deexcitation processes. Depending on the specific properties of the electronically excited states, saturation or multiphoton processes may dominate [17]. In the former case, because of the high laser irradiances used, a high percentage of the molecules may be excited from the electronic ground state. Because of the high reduction of the number of molecules in the ground state, absorption "saturates." Consequently, a larger portion of the incident light penetrates deeper into the material. As a result, thermal and chemical modifications are induced much deeper in the bulk of the substrate.

In other cases, multiphoton processes may dominate (i.e., absorbance increases during the excitation pulse), limiting light propagation into the substrate. In the simplest case, multiphoton processes entail the absorption of successive photons exciting the molecules or chromophores to higher electronic states. The rate for such processes scales as $d[M^*]/dt = \sigma(\lambda)[M]I^\gamma$, where I is the laser intensity, γ is the order of the multiphoton absorption, $\sigma(\lambda)$ is the wavelength-dependent excitation cross section for the process, and $[M]$ is the concentration of molecules. The probability of a multiphoton process is much enhanced by the presence of intermediate states (e.g., intermediate defect states in the case of wide-band insulators) at near resonance with the first or subsequent photon excitation steps.

For irradiation with intense nanosecond pulses, different variants of multiphoton processes have been indicated, for instance, the so-called cyclic multiphotonic absorption processes [18]. According to this, species in high excited states are produced upon multiphoton excitation (or via annihilation of S_1 excited molecules); however, in condensed phases, deactivation of these higher electronic states occurs extremely fast (~psec), resulting in the rapid conversion of absorbed electronic energy into heat. Following deactivation, the recovered S_1 state can participate in subsequent excitation and deexcitation cycles (Scheme 1). Such processes turn out to be particularly pronounced in the pulsed UV irradiation of doped polymers at high laser fluences.

Scheme 1 Cyclic multiphoton scheme

$$A + h\nu \rightarrow A^*(S_1)$$

$$A^*(S_1) + h\nu \rightarrow A^{**}(S_n) \rightarrow A^*(S_1) + heat$$

$$A^*(S_1) + A^*(S_1) \rightarrow A^{**}(S_n) \rightarrow A^*(S_1) + heat$$

Concerning applications, the important implication is that the optical penetration depth, and thus the extent of potential substrate damage, may differ substantially from that expected on the basis of the small-signal absorption coefficient. Unfortunately, there is not a sufficiently general model to predict the relative importance of such processes for different substrates. In fact, even for the same system, dynamic changes of the absorption coefficient may differ significantly according to the wavelength used. Thus the extent of the contribution must be evaluated experimentally in each case individually.

At fluences where material is ejected (ablation), additional factors become important in affecting light propagation in the substrate (see Chapter 6).

2.3   DESCRIPTION OF LASER-INDUCED PROCESSES

Following light absorption and excitation, the absorbed energy can result in a number of different processes. It is customary to delineate the induced processes and effects into three types, namely thermal, photochemical, and photomechanical. At low irradiances, this delineation of processes is generally justified. At high irradiances, however, there can be significant coupling between the different processes; and the delineation, though useful, becomes somewhat questionable (Chapter 6).

In a number of cases, these processes can be used to advantage. For instance, the laser-induced mechanical waves (ultrasound) can be used for diagnostic purposes (structural characterization) of substrates, including artworks. At high fluences, these processes are exploited for effecting material removal, which constitutes the basis of laser cleaning and conservation methods described in Chapter 6. On the other hand, these same processes determine the extent and nature of any side effects of laser techniques. For instance, in diagnostic uses, ideally the laser irradiation is so weak that it has no influence on the substrate. In practice, however, signal-to-noise ratio considerations may necessitate the use of higher laser irradiances, averaging over several pulses, or of long dwell times (if cw laser is used). In this case, photochemical and/or thermal effects may limit the use of the diagnostic technique and compromise the quality of the recorded spectra. Similarly, in laser restoration, the same processes determine the nature and extent of side effects caused to the substrate and thus specify the appropriateness of laser processing in specific situations.

2.3.1 THERMAL PROCESSES AND EFFECTS

2.3.1.1 Basic Concepts

Deexcitation in condensed phases (liquids/solids) is very fast (picoseconds for the vibrational states and usually nanoseconds for electronic states). Thus following light absorption, a part of the absorbed energy — at least, for irradiation with nanosecond and longer laser pulses — decays into heat. Consequently, thermal side effects are of major concern in the optimization of laser applications, since pigments, paper, and parchment tend to be highly sensitive to heating.

In the absence of any mechanisms that may "consume" laser energy (i.e., photodissociation of bonds), the temperature attained at depth z within the substrate at the end of the laser pulse can be estimated from

$$\Delta T(z) = \frac{\alpha F_{\text{LASER}}}{\rho c_P} e^{-\alpha z} \tag{2.2}$$

where the numerator represents the energy absorbed per unit volume at depth z. In the case of metals, absorption can be assumed to be uniform within the skin depth δ_s, so:

$$\Delta T = \frac{\alpha F_{\text{LASER}}}{\rho c_P \delta_s} \tag{2.3}$$

However, with time, the temperature distribution changes due to heat diffusion. Assuming a planar heat source, the time evolution of the temperature at a distance z into the substrate is described by

$$T(z,t) = T(z,0) e^{\frac{z^2}{-2D_{\text{th}}t}} \tag{2.4}$$

where D_{th} is the thermal diffusivity and t is the time after the pulse. [This equation results from the fact that heat flow q between two points is proportional to the difference (gradient) of their temperatures; $q = -\xi (\Delta T/\Delta x)$ — where Δx is the distance between the points and $\xi = \rho c_P D_{\text{th}}$ is the thermal conductivity. In turn, this implies that the temperature at a specific point changes as a result of heat conduction, $\partial T / \partial t = \xi (\partial^2 T / \partial x^2)$.] Consequently, during time t, the thermal wave advances into the bulk at a distance $l_{\text{th}} = 2(D_{\text{th}}t)^{1/2}$. Since the laser-heated depth is α^{-1}, the time necessary to reach thermal equilibrium within this depth is

$$t_{\text{th}} = \frac{1}{D_{\text{th}} \cdot \alpha^2} \tag{2.5}$$

This t_{th} is the characteristic *thermal diffusion time*.

TABLE 2.2
Thermophysical Properties of Selected Materials

Material	c_p* (J/g-K)	D_{th}* (cm²/sec)
Ag	0.23	1.72
Al	0.90	1.03
Al$_2$O$_3$ (ceramic)	0.9	0.09
C (graphite)	0.71	12.58
Cu	0.39	1.14
NaCl	0.83	
PI (Kapton)	1.09	10×10^{-4}
PMMA	1.41	1.1×10^{-3}

(*) The values for these parameters depend sensitively on temperature. The ones reported here refer to room temperature.

Typical α and D_{th} values for representative materials are given in Table 2.1 and Table 2.2. For organic substrates, $D_{th} \approx 10^{-3}$ to 10^{-4} cm²/sec, and $\alpha \approx 10^2$ cm^{-1} (weak absorption, e.g., acrylates at 308 nm) up to $\alpha \approx 10^5$ cm^{-1} (very strong absorption). Therefore t_{th} ranges from a few tens of microseconds (strongly absorbing organics) to milliseconds (weakly absorbing organics). Therefore, for shorter times, heat remains confined within the irradiated volume, which can thus reach quite high temperatures even for moderate laser fluences. To illustrate the importance of heat diffusion, Figure 2.6 shows the temperature profiles as a function of depth within a polymer at different times after the laser pulse. On the other hand, for metals, D_{th} is so high that significant heat transport out of the irradiated area occurs for irradiation even with nanosecond laser pulses (Figure 2.7). For 10 to 100 nsec pulses, $l_{th} \sim 1$ µm, i.e., the heat is actually distributed over this depth. Thus for metals, the temperature attained at the surface with nsec pulses is much lower than estimated from Equation (2.3) and irradiation can be performed at much higher fluences without causing damage. Detailed discussion and modeling of heat effects can be found in [15,20].

When multiple pulses are applied, it is important to ensure that the repetition rate is low enough to avoid progressive heat accumulation that will result in a larger thermal damage zone. Thus the time between pulses must be sufficient for complete heat diffusion. If the available laser system does not permit this condition, an alternative strategy involves scanning the laser beam over the substrate in order to increase the time interval between subsequent exposures at each location.

2.3.1.2 Temperature Measurements

Given the importance of temperature evolution upon pulsed laser irradiation, it is worth considering the proper experimental techniques for its determination. Experimental assessment of temperature evolution upon pulsed laser irradiation is far from trivial. Unfortunately, the commercial availability of temperature-measuring systems may lure

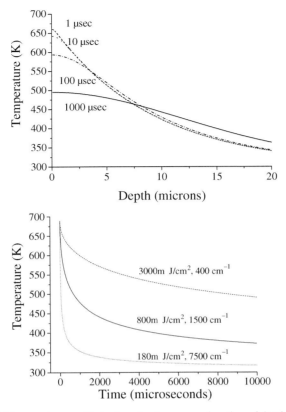

FIGURE 2.6 (a) Temperature profile in doped polymer as a function of depth at different times after the laser pulse. The absorption coefficient is assumed to be $\alpha = 1100$ cm^{-1}; $F_{LASER} = 0.5$ J/cm^2. (b) Temperature profile at the surface of doped *PMMA* for different combinations of fluence and absorption coefficients. The details of the simulation are described in Bounos et al., *J. Phys. Chem. B*, **108**, 7050, 2004. With permission.

us into quick determinations. Such measurements can be unreliable because the temperature changes upon pulsed laser irradiation are very fast, whereas the conventional instruments have much slower response times. Further errors can derive from the use of inappropriate calibration methods. On the other hand, accurate measurement of temperature can be attained via thermocouples attached to the substrate, but even so it is crucial to ensure close proximity of the thermocouple attachment to the irradiated area, and a fast enough temporal response of the sensor [23–25]. Highly accurate measurements can also be obtained from the ratio of Stokes/anti-Stokes lines in Raman measurements [24].

2.3.1.3 Laser Melting: Implications

As indicated by Figure 2.6 and Figure 2.7, the temperatures upon attained pulsed laser irradiation can be high enough to cause transient substrate melting. Laser-induced melting has been extensively studied in metals and semiconductors. Experimentally, these studies exploit the fact that optical properties such as reflectivity of these substrates may change

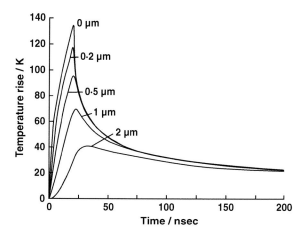

FIGURE 2.7 Temperature evolution in aluminum at different depths as a function of time. (square laser pulse 20 nsec long, absorbed energy = 4 mJ, $D_{th} = 10^{-4}$ $m^2 sec^{-1}$). (From Scruby and Drain, *Laser Ultrasonics:Techniques and Applications*, Adam Hilger, Bristol, 1990. With permission.)

substantially upon melting. In contrast, optical techniques are not successful in the case of organics because optical changes are much smaller and scattering much higher. Nevertheless, melting of polymers has recently been studied by an interferometric technique [26]. Another sensitive method is based on the formation of recombination products of radicals of dopants/probe compounds incorporated in the substrate (Section 2.3.3.3).

The melting induced upon laser irradiation can have deleterious effects on the appearance and integrity of the irradiated substrate. For multicomponent systems with constituents of different binding energies to the matrix, the enhanced diffusion of species within the molten zone can result in significant segregation effects. Furthermore, this can result in the preferential desorption of species weakly bound to the matrix, with a consequent change in the chemical composition of the substrate [27]. These effects can be especially pronounced for multipulse irradiation protocols.

Melting may also result in changes to the morphology and texture of the surface upon resolidification. These changes may be particularly pronounced at high laser fluences at which material ejection occurs (see Chapter 6). Thus they constitute an important limiting factor in laser processing methods. Furthermore, such morphological features are used extensively as a diagnostic tool of the mechanism of the laser interaction. This issue is discussed in detail in Chapter 6.

2.3.2 PHOTOMECHANICAL PROCESSES AND EFFECTS

2.3.2.1 Basic Concepts

As shown before, laser irradiation can result in rapid heating of the substrate. Normally, these high temperatures suggest a high thermal expansion of the substrate. But this may not be feasible at the very high heating rates involved with nanosecond or shorter laser pulses. Therefore heating may occur under nearly constant volume (isochoric) condition. This situation results in a pressure rise given by [25,28]

$$\Delta P = \frac{\beta \alpha F_{\text{LASER}}}{\rho \kappa_{\text{T}} c_{\text{V}}} \left(\frac{1 - e^{-\theta}}{\theta} \right) \qquad (2.6)$$

where β is the thermal expansion coefficient, c_{V} is the heat capacity at constant volume, κ_{T} is the isothermal compressibility, and $\theta = \tau_{\text{pulse}} / \tau_{\text{ac}}$, where $\tau_{\text{ac}} = 1 / C_{\text{S}} \alpha$ (C_{S} is the speed of sound) is the time required for an acoustic wave to traverse the irradiated thickness. The factor in parenthesis corrects for the reduction in the stress amplitude due to wave propagating out of the irradiated volume during the laser pulse (assumed to have a rectangular time profile).

This pressure rise results in three waves (appropriately, termed thermoelastic) propagating through the material: a radially propagating cylindrical wave, which can usually be neglected for beam diameters (mm to cm) substantially wider than the light penetration depth (typically a few micrometers), and two plane waves counterpropagating along the beam axis (one toward the surface and the other into the sample). The wave that travels toward the free surface (substrate/air interface) suffers a change of amplitude sign upon reflection from it, due to the higher acoustic impedance of the irradiated medium, ρC_{S}, than that of air. Physically, the thermal expansion directed into the medium generates compression stress, whereas the outward expansion generates tensile stress (rarefaction wave) (Figure 2.8a). Thus the axial (i.e., along the laser beam axis) wave produced by this mechanism is bipolar as evidenced in measurements by piezoelectric transducers attached to the substrate (Figure 2.8b). The faster the heating, the higher the magnitude of the generated thermoelastic stress in the medium. Thus the ultimate efficiency, with the highest value, is attained for a heating duration much shorter than the time required for stress to propagate through the irradiated volume (this time being simply $t = \alpha^{-1}/C_{\text{S}} = 1/\alpha C_{\text{S}} \ll \tau_{\text{pulse}}$, where C_{S} is the acoustic velocity). This condition yields the so-called stress confinement regime.

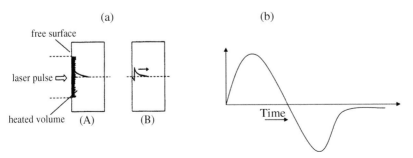

FIGURE 2.8 (a) Schematic illustrating the laser-induced thermoelastic stress-generation mechanism. (A) Thermal expansion of the volume heated by a laser pulse causes mechanical stress; (B) upon propagation from the free surface a tensile component develops with gradually increasing amplitude. (b) Schematic illustrating the stress wave signal developed by thermoelastic mechanism. (From Paltauf and Dyer, *Chem. Rev.*, **103**, 487, 2003. With permission.)

Especially in the case of organics, another important source of stress waves derives from the expansion of gases that may be produced by thermal or photochemical decomposition within the substrate [25]. This factor, for instance, has been invoked to account for the transient stresses of about 0.1 MPa detected in the UV irradiation of polyimide below the ablation threshold [29].

Several studies, largely relying on the use of piezoelectric crystals attached to the substrates, have demonstrated the development of stress waves with amplitudes of several hundred bars in the excimer laser irradiation of tissue and polymers.

2.3.2.2 Diagnostics and Implications

Ultrasound waves (i.e., acoustic wave frequencies higher than 20 MHz) can provide an important diagnostic tool of the structural properties of substrates since they can propagate over relatively large distances in solids and liquids. Thus they can be used to image structure and defects hidden within opaque substrates. The best known example of this capability is provided from medical applications, where ultrasound is employed to image internal organs [31].

Because of the ease of implementation, laser-induced ultrasound methods are used extensively in the industry for inspection purposes. Bulk and surface defects are detected in a wide range of materials, including metals, concrete, ceramics, and composites. Detailed description of the theoretical basis and of experimental aspects of laser-induced ultrasonics is provided in [31]. Thus laser-induced ultrasound may be a powerful technique in the structural diagnostics of artworks, although this potential has not been exploited much. Cooper [32] used a piezoelectric crystal to detect the waves developed upon laser irradiation of a thin marble sample at fluences below any visible damage.

However, at high laser fluences, the produced waves may induce structural modifications to the substrate, even outside the irradiation area. This can be an important consideration in the laser diagnostics and restoration of mechanically fragile substrates. This aspect is described in detail in Chapter 6. Furthermore, if the pressure wave amplitude exceeds the tensile strength of the substrate (the minimum tension required to cause fracture), then it can result in material ejection essentially via fracture. This is one of the mechanisms (the so-called photomechanical mechanism) exploited in effecting material ejection in laser conservation procedures.

2.3.3 CHEMICAL PROCESSES AND EFFECTS

A most crucial question in laser irradiation concerns the nature and extent of chemical modifications that may be made to the substrate. Especially organic substrates include a wide variety of chromophores, which upon UV excitation may dissociate into highly reactive fragments [33]. Additional species may be formed by the thermal or stress-induced breakage of weak bonds. These species may form in the short-or long-term by-products (e.g., oxidation by-products) with detrimental effects for the substrate's integrity. Thus minimization of chemical modifications induced to the substrate is crucial for the optimization of laser diagnostics and restoration applications.

2.3.3.1 Thermally Activated Reactions

The rate R_{rxn} of thermal reactions is well described by $R_{rxn} = k(T)[A]^x[B]^y$, where [ ] represents the concentration of the reactant (usually expressed in mol/l) and the exponents x, y are characteristic of the specific reaction mechanism. The simplest cases include a unimolecular process in which case $-d[A]/dt = k(T)[A]$ or a bimolecular reaction $-d[A]/dt = k(T)[A][B]$. The dependence of the reaction rate on the temperature is included in the reaction rate constant $k(T)$ and is often well described by the Arrhenius law, according to which $k = A\,e^{-E_{act}/R_G T}$, where the pre-exponential factor A is a measure of the probability of reaction, R_G is the universal gas constant and E_{act} the activation energy (which represents essentially the minimum energy per mole required for the reaction to take place). Maximum A values are 10^{12} to 10^{15} sec^{-1} for simple unimolecular reactions; whereas much smaller values in the 10^6 to 10^8 sec^{-1} range are usually found for bimolecular reactions. E_{act} ranges from a few kJ/mol to values approaching bond dissociation energies.

2.3.3.2 Photochemical Reactions

Everybody is familiar with the discoloration and photoaging effects induced upon prolonged irradiation of an organic substrate. These effects relate to photochemical processes, such as bond photodissociation, with the formation of reactive radicals, or bond rearrangement, induced on the absorbing units (chromophores) of the substrate. Although the final result may be a rather complex chemical modification, the initial step of the process is characterized by specificity (e.g., dissociation or other reactions initiated by the excited chromophores). A detailed discussion of the principles of "conventional" photochemistry (i.e., underlying the interaction of light with organic materials at low irradiances) is given in [34].

Photochemical reactions are initiated by the activation of a molecule to an excited electronic state by the absorption of a photon. It is customary to differentiate the subsequent processes into the primary ones, which involve excited electronic states, and the secondary or thermal reactions (described in 2.3.3.1) of the chemical species produced by the primary processes.

Upon laser irradiation, if the absorbing chromophores in the substrate undergo photofragmentation or other photoexcited reactions with a quantum efficiency q_d, the number of photomodified molecules scales with depth z in the substrate as

$$N_D(z) = q_d \frac{\sigma N F_{LASER} e^{-az}}{h\nu} \tag{2.7}$$

where σ is the absorption cross section, N is the number density of the chromophores, and $h\nu$ is the photon energy. The photodissociation of a simple sigma bond entails either a homolytic cleavage (resulting in two radicals) or a heterolytic one (resulting in an ion pair or zwitterion). [There are also certain chromophores (e.g., diazo compounds, azides) that dissociate into a molecule (e.g., N_2 or CO) and a diradical or zwitterions.] Depending on the nature of the radicals and of the matrix, at low temperature (at which the substrate structure is not disrupted), the radicals formed by the photodissociation may be prevented from diffusing away, so that they recombine finally to reform the precursor (geminate recombination) [35]. Thus the photodissociation efficiency in rigid media (e.g., polymers) is often much lower than that in solution or gas phase.

The weakest (single) bonds in organic molecules have a binding energy of 145 kJ/mol (e.g., O–O single bond), while strong (single) bonds such as C–H are of 420 kJ/mol. Note that these bond energies are comparable to or even smaller than the photon energy of UV (excimer) lasers commonly used in art conservation (Table 2.1). However, even if the photon energy exceeds the bond energy, for high enough photon energies the bonds will not be unselectively broken, because of the localization of the excitation on specific chromophores/groups of the molecule. Thus the nature of photochemical activation differentiates it from thermal activation. In contrast, thermal activation of the same bond can only be achieved by an increase in the overall molecular energy of the environment.

2.3.3.3 Chemical Processes upon Pulsed Laser Irradiation

The nature of the by-products formed, of course, depends crucially on the specific chemical characteristics of the irradiated material. Detailed information on organic and polymer photochemistry can be found in many reviews and books [9,11,36]. Here, we focus in particular on the likely effects of the laser-induced temperature increase on the evolution of chemical processes. Many chromophores encountered in artworks dissociate into reactive radicals [33]. Since the reactivity of such radicals depends sensitively on temperature, it may be much affected by the laser-induced temperature increase (Section 2.3.1). Reactions of radicals within polymers usually follow a pseudo-unimolecular Arrhenius equation whereby $[\text{Product}] = [R](1 - e^{-kt})$, where $[R]$ represents the concentration of radicals produced by the photolysis and k is the reaction rate constant. At low fluences (i.e., low attained temperatures), only a small percentage of the photoproduced radicals may react to form by-products; because of the low attained temperatures reached, reactivity is quenched relatively quickly (i.e., the exponential in the Arrhenius equation $k = Ae^{-E_{act}/R_G T(t)}$ becomes too small relatively fast). With increasing laser fluence, however, the reaction efficiency may increase sharply. In parallel, heat diffusion may result in a higher percentage of the radicals reacting in the sublayers of the substrate. Accordingly, thermally-activated reactions should be limited by the heat relaxation time $t_{th} \approx 1/(a_{eff}^2 D_{th})$. Thus for a low D_{th}, temperatures remain high, thereby enabling such activated processes (e.g., abstraction reactions by radicals) to proceed longer.

In addition, if the substrate melts, diffusion in the irradiated volume increases greatly, so that formation of by-products via recombination reactions can occur. Radical diffusion within melts can be approximately described as Fickian-like, in which case diffusion length $\sim 6(D_{sp}t)^{1/2}$ with D_{sp} given by the Einstein–Stokes equation, $D_{sp} = \kappa_B T / (6\pi\eta R_{radical})$, where κ_B is Boltzmann constant, η is the medium viscosity, and $R_{radical}$ is the radical radius. Qualitatively, the temperature dependence of the viscosity η of melts, at least at high enough temperatures, follows an exponential: $\eta = \eta_0(T_{ref})e^{E_V/R_G T}$, where $\eta_0(T_{ref})$ is the viscosity at a reference temperature (T_{ref}) and E_V is the activation energy for the viscosity–temperature dependence [37]. Thus radical diffusivity is much enhanced with increasing temperature and so is the formation of recombination by-products.

The previous effects have been illustrated in the examination of the products formed in the irradiation of polymers doped with the photolabile iodonaphthalene or iodo-phenanthrene [19]. (Doped systems constitute a good, even if idealized, model of the paint layer in artworks, which essentially consists of chromophores dissolved or dispersed within an organic medium.) Upon excitation, these compounds dissociate to aryl radical and iodine. Thermal decomposition for these compounds is insignificant because of the high C–I bond energy (250 kJ/mol). The aryl radicals may abstract a hydrogen atom from nearby polymer units to form an ArH by-product (Scheme 2) (detectable in laser-induced fluorescence spectra by emission bands at $\approx$330 nm for NapH and $\approx$375 nm for PhenH). The experiment is of the pump-probe type, in which an excimer laser pump pulse irradiates the film, and laser pulses of very low fluence are used to induce by-product fluorescence [19].

Scheme 2

Pathways of aryl product formation in the irradiation of ArI doped polymers

$$\text{ArX (Ar=Nap, Phen)} \xrightarrow{\;\;h\nu\;\;} \text{Ar} + \text{X}$$

$$\text{Ar} + \text{polymer} \longrightarrow \text{ArH}$$

$$\text{Ar} + \text{Ar} \longrightarrow \text{Ar}_2$$

Figure 2.9 depicts the dependence of the ArH by-product emission intensity following a single pump pulse on virgin polymer as a function of laser fluence. This intensity is essentially proportional to the amount of the by-product that is formed upon irradiation. At low fluences, the amount of by-product scales linearly with F_{LASER}, consistent with 1-photon photolysis of the ArI compound. However, at higher fluences, the dependence of the product amount becomes supralinear, in other words, there is an enhancement in ArH by-product formation per unit volume. The increase in ArH formation at these fluences is due to the increased substrate temperatures attained with increasing F_{LASER}. However, at even higher fluences, the ArH by-product amount in the substrate is found to reach a limiting value due to material removal (ablation) (Chapter 6).

The examination of the doped system also illustrates the formation of by-products via recombination reactions that may occur via diffusion in molten zone. Thus in the irradiation (at 248 nm and 308 nm) of NapI/PMMA samples with a dopant concentration higher than 1% wt. at low fluences, NapH-product is the exclusive dopant-deriving by-product. However, for irradiation at higher fluences, formation of Nap_2 and perylene (perylene is a concatenated bi-aryl species) is demonstrated, respectively, by the broad emission at $\approx$375 nm and by the double peak structure around 450 nm (Figure 2.10). These by-products are formed through the reaction of diffusing Nap radicals (Nap + Nap $\rightarrow$ Nap_2 in Scheme 2). The formation of these by-products increases with increasing

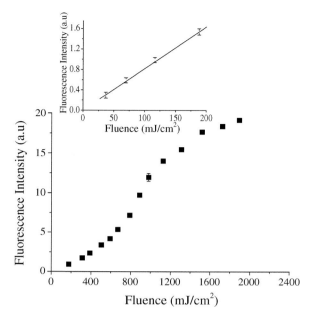

FIGURE 2.9 The figure illustrates the F_{LASER}-dependence of aryl (NapH) product fluorescence intensity in the UV irradiation of NapI-doped *PMMA* films (0.4% wt concentration in the dopant).

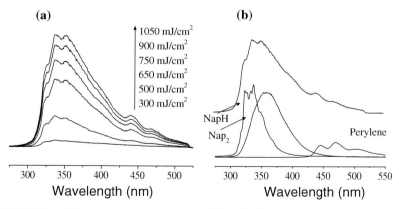

FIGURE 2.10 (a) Fluorescence spectra recorded from NapI-doped *PMMA* samples after their irradiation with a single UV laser pulse (at 248 nm) at the indicated laser fluences. (b) Deconvolution of the spectrum into the contributing fluorescence compounds. (From Bounos et al., *J. Phys. Chem.B*, **108**, 7050–7060, 2004. With permission.)

laser fluence, because of the sharp drop of the polymer viscosity with increasing temperature. Although the formation of such species is evidently deleterious and must be avoided in laser applications, it provides, on the other hand, a direct experimental probe of the degree of substrate disruption during laser irradiation.

The extent of these effects depends crucially on material and laser parameters and especially on the substrate absorptivity. Further discussion is postponed to Chapter 6, in relationship with the effects of laser conservation methods.

In most cases, a significant number of laser pulses is required either for improving the S/N ratio in laser diagnostic methods or for enhanced material removal in laser conservation methods. Thus the dependence of chemical effects on the number of pulses is a crucial factor in laser techniques.

At low and moderate laser intensities (i.e., at which material desorption/removal is minimal), the by-products accumulate in the substrate with successive pulses. For a simple reaction, the accumulation can be described by an exponential of the form $[1 - \exp(-\sigma q_d (F_{LASER} / h\nu)N_{pulse})]$, where σ is the absorption cross section, q_d: quantum yield of photo-reaction and N_{pulse} is the number of pulses. Thus product accumulates in the substrate until finally a plateau is reached [38]. In practice, in the irradiation of chemically complex substrates (such as polymers and varishes), the situation may be more complex, since the by-products formed in the initial pulses may initiate further (different) chemical reactions. Thus in the irradiation of various polymers, e.g., *PMMA* (doped or neat) or polystyrene, formation of conjugated polymeric species (via the elimination of side groups) occurs, as is demonstrated by a number of characterization techniques.

The accumulating chemical modifications may result in the stepwise change of the physical and chemical properties of the substrate. For example, in weakly absorbing systems (e.g., *PMMA* at 248 nm), irradiation results in highly conjugated species that absorb more strongly than the initial substrate. In the case of semicrystalline polymers, volume changes may be observed, and they have been ascribed to the amorphization of the crystalline domains and the accumulation of trapped gases. These physical and chemical changes result in refractive index changes, a fact that has exploited for the fabrication of waveguides and gratings. However, in laser diagnostic procedures, such changes evidently must be minimized by using the least possible number of laser pulses.

With increasing F_{LASER}, the effects become more pronounced. However, at higher fluences, there are new phenomena setting in, as described below, namely desorption and ablation. These result in deviations from the predictions of conventional photochemistry and provide the basis for the success of laser restoration methods.

2.4 DELINEATION OF PROCESSES WITH INCREASING LASER IRRADIANCE

In Chapter 1, a very broad delineation of laser-induced processes was given. With the detailed background developed in this chapter, a more accurate presentation of the basis of laser applications can be provided. It is clear that the most important parameter for process classification is laser intensity irradiance (or for a fixed laser pulse width, laser fluence, i.e., laser light energy per unit area). This is the determining parameter establishing the extent of perturbation effected to the substrate.

At low intensities, the processes relate to simple excitation of absorbing chromophores. Furthermore, at low fluence irradiances photochemical and thermal effects are minimal, so that the laser radiation may be considered to effect only minor perturbation to the substrate. Thus, it is largely this low irradiance/fluence range that is employed for diagnostic purposes. The main advantages of laser use in this respect are short pulse duration (thereby enabling time-resolved recording), high tunability, focusability, and monochromaticity (thereby permitting selective excitation of molecules, pigments etc.). In the case of scattering methods, e.g., Raman, the high intensity of the laser sources is also crucial to ensure that the spectroscopic signals are sufficiently intense for detection. (In this case, side effects to the substrate are avoided because the light is not absorbed but only scattered). Further diagnostic capabilities derive from the high degree of polarization of the laser light. In the case of holographic techniques, the coherence of the laser irradiation is crucial. The diagnostic techniques are discussed in detail in subsequent chapters.

On the other hand, at high fluences, material removal occurs, resulting in the formation of a well-defined beam imprint in the substrate. Depending on laser parameters and material properties, the removed depth may range from tens of nanometers up to a few micrometers (μm) [38–43]. This effect is observed only at the high irradiances that are attainable by short pulse durations. With increasing fluence, the amount of material removed generally increases. The phenomenon giving rise to this pronounced material ejection is called laser ablation (from the Latin word *ablatio* meaning material removal) a complex phenomenon described in detail in Chapter 6. It is exactly this efficient material removal process with intense laser pulses that provides the basis for laser conservation methods discussed fully in Chapter 6 to Chapter 8.

It should be clarified that the transition from the simple photochemistry at low fluences to the ablative regime is not abrupt. As will be shown in Chapter 6 to Chapter 8, different mechanisms can contribute to material ejection. In most cases, a fluence range below the ablation threshold can be delineated resulting in morphological changes and even mass loss. Generally, at these intermediate fluences, for nanosecond pulses, a thermal desorption evaporation process operates [15,43]. In other words, in this fluence range, the ejection of compounds dispersed or dissolved in the substrate correlates with their binding energy to the substrate with signal being proportional to $e^{-E_{\text{binding}}/\kappa_B T}$. For a number of different polymers, similarly at moderate fluences, some mass loss is observed [22]. The loss is due to the desorption of by-products weakly bound to the polymer matrix that are formed upon laser irradiation. Although this complicates the description of the mechanisms and processes involved, it contributes to the versatility of the laser restoration methods. Different mechanisms of material removal can be exploited in dealing with different conservation challenges.

The material ejection observed at high fluences constitutes the basis not only for laser processing and conservation methods but also for a number of powerful analytical and diagnostic techniques (In addition to the ones relying on the use of low laser intensities).

For instance, by enabling the ejection of material to the gas phase, it enables the implementation of powerful techniques that operate only on gas phase species for the characterization of the material. In particular, the combination of laser with mass spectrometry has found extensive use as an analytical technique in industry, environmental studies, geology, and most importantly matrix-assisted laser desorption in biology. In addition, mass spectroscopy provides a powerful method for characterizing laser-induced desorption and ablation processes. This has been used extensively in ablation studies [44]. The technique permits identification of ejected material, their kinetic energies, and by more exacting instrumentation their angular distribution. This information can be most useful in understanding the underlying processes. Its use, however, in the case of artworks is limited by the requirement of placing the substrate under study into a vacuum system.

Another technique that has turned out highly successful in diagnostics of artworks is based on the spectral analysis of the radiation emitted by the plasma formed during ablation. Specifically, because of the high light intensities, some of the ejected atoms or molecules become ionized, resulting in the formation of a plasma (consisting of neutrals/ions and free electrons). Upon cooling of the plasma, the recombination of the ions with free electrons results in excited species that decay by fluorescence emission. This emission is characteristic of the elements contained in the sample yielding direct analytical information [45]. Laser-induced breakdown spectroscopy (LIBS) [45] is discussed in detail in Chapter 3. LIBS features high sensitivity and selectivity and can be performed *in situ,* thereby eliminating the need for sample removal. Since only a minute amount of sample is consumed in the process of atomization (achieved by tightly focusing the laser beam onto the surface), the technique is characterized as microdestructive. The insensitivity of the LIBS process to surface charging effects (thus to substrate conductivity, thus permitting its use on almost any different material) in combination with its high sensitivity for pigment constituents makes it a powerful tool for monitoring laser conservation of painted artworks.

Symbols and Abbreviations

A	Measure of the probability reaction
α	Absorption coefficient (cm^{-1})
α_{eff}	Effective absorption coefficient
β	Thermal expansion coefficient at constant temperature
c	Concentration (mol/L)
c_P	Specific heat capacity at constant pressure
c_V	Heat capacity at constant volume
C_S	Speed of sound
δ	Ablation (etched) depth per pulse
δ_S	Skin depth
$E_{binding}$	Binding energy to the substrate
ε	Extinction coefficient ($M^{-1}\ cm^{-1}$)
D_{th}	Thermal diffusivity
D_{sp}	Species diffusion
E_{act}	Reaction activation energy
E_{cr}	Critical energy density for ablation
E_V	Activation energy for the temperature dependence of viscosity
F_{LASER}	Laser fluence
F_{thr}	Threshold fluence for material removal
$F_{transmitted}$	Transmitted fluence through a path length
ΔH_{vap}	Evaporation enthalpy
ΔH_{sub}	Sublimation enthalpy
η	Medium viscosity
η_0	Viscosity at a reference temperature
θ	Ratio $\tau_{pulse}\ /\ \tau_{ac}$
$h\nu$	Photon energy
I	Laser intensity
$k\left(T\right)$	Reaction rate constant at temperature T
κ_B	Boltzmann constant
κ_T	Isothermal compressibility
l	Length (depth) in the sample
l_{th}	Thermal diffusion length
λ	Wavelength

M	Molecular mass
M^*	Molecules in an excited state
$[M]$	Concentration of molecules
μ	Scattering coefficient
N	Number density of chromophores
N_D	Number of photomodified molecules
N_{pulse}	Number of laser pulses
n_R	Refractive index
γ	Order of multiphoton absorption
ν_{ac}	Acoustic wave frequency
k	Thermal conductivity
P	Pressure
PI	Polyimide (Kapton)
$PMMA$	Polymethyl-methacrylate
PS	Polystyrene
q	Heat flow
q_d	Quantum efficiency
R	Reflectivity
$[R]$	Concentration of radicals
R_G	Universal gas constant
$R_{radical}$	Radical radius
R_S	Sample reflectivity
ρ	Density
$\sigma(\lambda)$	Wavelength dependent excitation/absorption cross section
T	Temperature
t	Time after pulse
t_{th}	Thermal diffusion time
τ_{pulse}	Laser pulse duration
τ_{ac}	Time for an acoustic wave to traverse the irradiated optical volume penetration depth
x, y	Exponents characteristic of the specific reaction mechanism
z	Depth

REFERENCES

1. Taft, W.S. and Mayer, J.W., *The Science of Paintings*, Springer Verlag, New York, 2000.
2. Georgiou, S., Zafiropulos, V., Anglos. D., Balas, C., Tornari, V., and Fotakis, C., Excimer laser restoration of painted artworks: procedures, mechanisms and effects, *Appl. Surf. Sci.*, **129**, 738–748,1998.
3. Jenkins, F.A. and White, H.E., *Fundamentals of Optics*, McGraw-Hill, New York, 1976.
4. Young, H.D., *Fundamentals of Waves, Optics and Modern Physics*, McGraw-Hill, New York, 1976.
5. Francon, M., *Laser Speckle and Applications in Optics*, Academic Press, New York, 1979.
6. Hecht, E., *Optics*, 2nd ed. Addison-Wesley, Reading, MA, 1987.
7. Wehry, E.L., *Modern Fluorescence Spectroscopy*, Plenum Press, New York, 1976.
8. Lakowicz, J.R., *Principles of Fluorescence Spectroscopy*, Plenum Press, New York, 1983.
9. Turro, N.J., *Modern Molecular Photochemisty*, Benjamin/Cummings, Menlo Park, CA, 1978.
10. Rabek, J.F., *Mechanisms of Photophysical Processes and Photochemical Reactions in Polymers: Theory and Applications*, John Wiley, Chichester, 1987.
11. Barltrop, J.A. and Coyle, J.D., *Principles of Photochemistry*, John Wiley, New York, 1978.
12. Tuchin, V., *Handbook of Optical Biomedical Diagnostics*, International Society for Optical Engineering, Orlando, FL, 2002.
13. Balas, C., An imaging colourimeter for noncontact tissue color mapping, *IEEE Trans. Biomed. Eng.*, **44**, 468–474, 1997.
14. de la Rie, R.E., Old master paintings, *Anal. Chem.*, **61**, 1228 A, 1989.
15. Bauerle, D., *Laser Processing and Chemistry*, 3rd ed., Springer-Verlag, Berlin, 2000.
16. Vogel, A. and Venugopalan, V., Mechanisms of pulsed laser ablation of biological tissues, *Chem. Rev.*, **103**, 577–644, 2003.
17. Pettit, G.H., Ediger, M.N., Hahn, D.W.. Brinson, B.E., and Sauerbrey, R., Transmission of polyimide during pulsed ultraviolet-laser irradiation, *Appl. Phys. A*, **58**, 573–579, 1994.
18. Fujiwara, H., Fukumura, H., and Masuhara, H., Laser ablation of pyrene-doped poly-(methyl methacrylate) film: dynamics of pyrene transient species by spectroscopic measurements, *J. Phys. Chem.*, **99**, 11844–11853, 1995.
19. Bounos, G., Athanassiou, A., Anglos, D., Georgiou, S. and Fotakis, C., Product formation in the laser irradiation of doped poly(methyl) methacrylate at 248 nm: implications for chemical effects in UV ablation, *J. Phys. Chem. B*, **108**, 7050–7060, 2004.
20. Venugopalan, V., Nishioka, N., and Mikic, B.B., The thermodynamic response of soft biological tissues to pulsed ultraviolet laser irradiation, *Biophys. J.*, **69**, 1259–1271, 1995.
21. Bityurin, N., Arnold, N., Luk'yanchuk, B., and Bauerle, D., Bulk model of laser ablation of polymers, *Appl. Surf. Sci.*, **127–129**, 164–170, 1998.
22. Bityurin, N., Luk'yanchuk, B.S., Hong, M.H., and Chong, T.C., Models for laser ablation of polymers, *Chem. Rev.*, **103**, 519–552, 2003.
23. Kuper, S., Brannon, J., and Brannon, K., Threshold behavior in polyimide photoablation-single-shot rate measurements and surface-temperature modeling, *Appl. Phys. A*, **56**, 43–50, 1993.

24. Lee, I.-Y.S., Wen, X., Tolbert, W.A., Dlott, D.D., Doxtader, D., and Arnold, D.R., Direct measurement of polymer temperature during laser ablation using a molecular thermometer, *J. Appl. Phys.,* **72**, 2440–2448, 1992.

25. Paltauf, G. and Dyer, P.E., Photomechanical processes and effects in ablation, *Chem. Rev.,* **103**, 487–518, 2003.

26. Tokarev, V.N., Lazare, S., and Belin, C., Viscous flow and ablation pressure phenomena in nanosecond UV laser irradiation of polymers, *Appl. Phys. A Mater.,* **79**, 717–720, 2004.

27. Koubenakis, A., Labrakis, J., and Georgiou, S., Pulse dependence of ejection efficiencies in the UV ablation of bi-component van der Waals solids, *Chem. Phys. Lett.,* **346**, 54–60, 2001.

28. Gusev, V.E. and Karabutov, A.A., *Laser Optoacoustics,* American Institute of Physics, New York, 1993.

29. Zweig, A.D., Venugopalan, V., and Deutch, T.F., Stress generated in polyimide by excimer-laser irradiation, *J. Appl. Phys.,* **74**, 4181–4189, 1993.

30. Hare, D.E., Franken, J., Dlott, D.D., Doxtader, D., and Arnold, D.R., Coherent Raman measurements of polymer thin-film pressure and temperature during picosecond laser ablation, *J. Appl. Phys.,* **77**, 5950–5960,1992.

31. Scruby, C.B. and Drain, L.E., *Laser Ultrasonics: Techniques and Applications,* Adam Hilger, Bristol, 1990.

32. Cooper, M.I. and Loton, A., Light years ahead? *Natural Stone Specialist,* 24–30, 1994.

33. de la Rie, E.R., Photochemical and thermal degradation of films of dammar resin, *Stud. Conserv.,* **33**, 53–71, 1988.

34. Smith, F.G. and Thomson, J.H., *Optics,* John Wiley, London, 1971.

35. Noyes, R.M., Effects of diffusion rates on chemical kinetics, *Prog. React. Kinet. Mech.,* **1**, 129–160, 1961.

36. Scaiano, J.C., *CRC Handbook of Organic Photochemistry,* CRC Press, Boca Raton, FL,1988.

37. Harvard, R.N. and Young, R.J.E., *The Physics of Glassy Polymers,* 2d ed., Chapman and Hall, London, 1997.

38. Srinivasan, R. and Braren, B., Ultraviolet laser ablation of organic polymers, *Chem. Rev.,* **89**, 1303–1316, 1989.

39. Snirivasan, R., *Laser Ablation Principles and Applications,* Springer-Verlag, Berlin, 1994.

40. Dyer, P.E., Excimer laser polymer ablation: twenty years on, *Appl. Phys. A Mater.,* **77**, 167–173, 2003.

41. Schroeder, K. and Schuoecker, D., Ultrahigh-power lasers and their industrial applications, *Laser Phys.,* **8**, 38–46, 1998.

42. Laude, L.D., Ed., *Excimer Lasers,* Kluwer Academic Publishers, Dordrecht, 1994.

43. Zhigilei, L.V., Leveugle, E., Garrison, B.J., Yingling, Y.G., and Zeifman, M.I., Computer simulations of laser ablation of molecular substrates, *Chem. Rev.,* **103**, 321–348, 2003.

44. Lippert, T. and Dickinson, T.J., Chemical and spectroscopic aspects of polymer ablation: special features and novel directions, *Chem. Rev.,* **103**, 453–485, 2003.

45. Anglos, D., Laser-induced breakdown spectroscopy in art and archaeology, *Appl. Spectroscopy,* **55**, 186–205, 2001.

3 Laser-Induced Breakdown Spectroscopy (LIBS): Cultural Heritage Applications

3.1 INTRODUCTION: WHY LIBS IN CULTURAL HERITAGE?

The aim of this chapter is to provide the reader with a basic understanding of the main physical principles and analytical features of laser-induced breakdown spectroscopy (LIBS) and give an overview of its use in the field of cultural heritage. Laser-induced breakdown spectroscopy is an analytical technique that enables the determination of the elemental composition of materials on the basis of the characteristic atomic fluorescence emitted from a microplasma produced by focusing a high-power laser on or in a material [1–6].

In general a large number of elemental analysis techniques are well established in the field of material science and certainly quite a few of them have enjoyed recognition in archaeometry and artwork analysis, providing useful qualitative and/or quantitative information on the composition of materials [7–11]. To mention a few, x-ray fluorescence (XRF) spectrometry, based on the characteristic fluorescence originating from materials when they are excited by an x-ray source, has become almost a routine technique in the analysis of materials in archaeological, historical, and art objects or monuments [12–15]. The analysis of ancient jewelry and metal or glass objects as well as the identification of pigments in ancient frescoes or modern paintings are representative examples of the use of XRF [16–19]. Likewise, particle induced x-ray emission (PIXE) permits, in a nondestructive way, the determination of the concentration of the elements in a material by means of the emission of characteristic x-rays, following bombardment of the sample or object with an ion beam [20]. Both major and minor (down to the ppm level) components are analyzed with high spatial resolution as, for example, in the case of pigment analysis on ancient pottery or the determination of metal composition in ancient coins [21–25]. Another widely established technique is ICP-MS (inductively coupled plasma mass spectrometry) based on the mass spectrometric analysis of the elements produced by introducing a small sample of the material to be analyzed in a steady-state plasma. ICP-MS is destructive (it consumes the sample) but offers superb detection limits and accuracy and is ideal for the analysis of minor and trace elements [26,27]. On the basis of trace element concentration patterns, different types of obsidian have been discriminated [28] and the provenance of various objects has been established [29]. More examples from the use of various analytical techniques to determine the elemental composition of materials in ancient historical and art objects are listed in Table 3.1.

Given this wide range of powerful techniques, an obvious question arises: why LIBS in cultural heritage? The answer will, we hope, become obvious by the end of this chapter, especially through the test cases that are presented. Key points, indicating the potential of LIBS in applications related to cultural heritage, will be outlined here.

LIBS is, in principle, a straightforward and simple analytical technique that can be employed even by nonspecialist users. It is a quick and portable measurement technique providing results practically instantaneously after analysis. Also, it is applicable *in situ* — that is, on the object itself — and under certain conditions is nearly nondestructive. In the context of analysis related to objects of cultural heritage, the above features are considered very important. The simplicity of the technique and its speed permit the analysis of a relatively large number of objects in a short time. For example, archaeological excavations often produce large numbers of artifacts, and their timely characterization or simple screening is required. The possibility of using the technique *in situ* eliminates the need for sampling, a process that is time-consuming and sometimes damaging to the object. This is important because the sensitivity and value of most works of art and archaeological objects often preclude sampling, thus preventing the use of analytical techniques such as atomic absorption spectrometry (AAS) and ICP-MS that

TABLE 3.1
Elemental Analysis Techniques in Cultural Heritage

Analytical method	Applications	References
Atomic absorption/emission spectroscopy	Elemental analysis of pottery, metal, and glass	[30,31]
Inductively coupled plasma-optical emission spectrometry (ICP-OES)	Major and trace element analysis of metals and minerals	[32,33]
Inductively coupled plasma-mass spectrometry (ICP-MS)	Trace element and isotope analysis of metals and minerals	[26–29]
Secondary ion mass spectrometry (SIMS)	Elemental analysis of pigments, pottery, metals, alloys, and minerals	[34]
Scanning electron microscopy (SEM, energy dispersive x-ray (EDX) analysis)	Mapping and elemental analysis of pigments, pottery, metals, and minerals	[35,36]
X-ray fluorescence spectrometry (XRF)	Elemental analysis of pigments, metals, and minerals	[12–19,37]
Particle-induced x-ray emission (PIXE)	Major and trace element analysis of pigments, pottery, metals, and minerals	[20–25]
Neutron activation analysis (NAA)	Analysis of major and trace elements in pigments, pottery, and minerals. Provenance.	[38,39]
Isotope analysis	Dating and provenance	[40,41]

require a small quantity of sample that is consumed during the measurement. Obviously, nondestructive techniques are preferred over destructive ones, and even though LIBS is not strictly a nondestructive technique it is considered minimally invasive given the very small area of interaction of the laser pulse with the sample surface. On the basis of these features and research done to date, LIBS appears as a useful alternative (or complement) to other sophisticated techniques for obtaining information on the elemental composition of materials in cultural heritage objects. The potential of LIBS in this field is shown by several research papers that have appeared in the last few years, describing its use for the analysis of works of art and objects of archaeological importance [42–63]. Earlier reports can also be found on the use of laser microspectral analysis in the determination of the elemental content of metal, pottery, and paint samples from different objects [65,66]. The analytical capabilities of LIBS are dealt with in more detail in the following sections and in the examples presented.

3.2 PHYSICAL PRINCIPLES

As already mentioned, LIBS is the spectroscopic analysis of the plasma formed as a result of the interaction of a focused light pulse from a high-power laser with a material. It was proposed as an analytical technique to provide compositional information on the elemental content of materials and developed shortly after the discovery of the laser in 1960 [1–3]. It received much more attention in the 1980s with advances in reliable pulsed laser sources and multichannel detectors. LIBS has been used in a wide variety of analytical applications for the qualitative, semiquantitative, and quantitative analysis of materials and is currently in the focus of research efforts world wide, which aim at producing versatile LIBS systems for field analysis (outside the laboratory) and units for industrial process control [6, 67–71].

The analysis by laser-induced breakdown spectroscopy starts with the deposition of laser energy in a small volume of material (less than 0.1 mm^3) and within a short time period (5 to 20 nsec). This rapid energy deposition achieved by focusing a laser pulse on the surface of the object being analyzed leads to material breakdown and generation of a microplasma or plume [4,5,72] as shown in Figure 3.1. The processes involved are described in a detailed manner in Chapters 2 and 6.

The laser ablation plasma consists of electrons, neutral and ionized atoms, small molecules, and larger clusters and that move away from the object surface, with typical initial velocities in the order of 0.5 to 50 km/sec, colliding with the surrounding atmosphere. Immediately upon its formation the plasma is characterized by high temperature and electron density, and as a result it shows intense broadband spectral emission in the ultraviolet and visible arising from highly excited species. This broadband continuum carries no analytical information, but as the plasma expands in space, the emission evolves with time to a spectrum with sharply peaked features, corresponding to distinct electronic transitions of the different species in the plume. Recording of this emission on a spectrometer produces the LIBS spectrum, which, through a straightforward analysis, yields compositional information about the material examined. More specifically, the characteristic sharp atomic emission peaks in the spectrum lead to the identification of the elements contained in the minute amount of material ablated,

FIGURE 3.1 Laser ablation plasma (plume) formed by focusing a pulse (from a 10 nsec Nd:YAG laser at 1064 nm) on the surface of a brass sample. The focusing lens and the optical fiber collecting the emitted light are shown.

reflecting the local elemental composition of the sample (qualitative analysis). The peak intensity or the integrated emission can, in principle, be associated with the number density of each emitting species in the plume, and this in turn with the concentration of specific elements in the ablated material (quantitative analysis).

The integrated intensity of an emission spectral line from a single species present in the plasma, in a state of local thermodynamic equilibrium, is given in Equation (3.1) as a function that relates the emission line intensity to the number density of the species, relevant spectroscopic parameters, and plasma electron temperature:

$$I_\omega^{ik} = N_S \frac{h\omega}{8\pi^2} \frac{g_i A_{ik} \exp(-\dfrac{E_i}{k_B T})}{Z_S(T)} \tag{3.1}$$

where I_ω^{ik} is the intensity of emission line at frequency ω corresponding to a transition from the upper quantum state i to the lower k; N_S is the number density of species S in the point of observation within the plasma; h is Planck's constant; g_i is the statistical weight of state i; A_{ik} is the transition probability for spontaneous emission from energy level i to k; E_i is the energy of quantum state i with respect to the ground state of the emitting species; k_B is Boltzmann's constant; T is the plasma temperature; and $Z_S(T)$ is the partition function for the quantum states of the emitting species.

On the basis of Equation (3.1), the number density for each emitting species can be associated with the intensity of its spectral lines, thus providing a link between signal intensity and elemental concentration. A discussion on semiquantitative analysis by LIBS follows in Section 4.4.

3.3 INSTRUMENTATION

The main components of LIBS analysis include laser-induced plasma formation, optical sampling of the plasma emission and interpretation of the spectral data collected. The LIBS instrumentation and its basic components are shown diagrammatically in a typical experimental setup (Figure 3.2). The laser source most commonly used is Q-switched Nd:YAG. Excimer lasers have also been used. Table 3.2 lists common types of lasers used in LIBS measurements with their corresponding wavelength and pulse duration. The wavelength is a critical parameter that determines the coupling of irradiation to the surface.

Focusing of the laser beam is done by a convergent lens of appropriate focal length, typically 50 to 500 mm. The focusing distance can be much longer in cases of remote analysis [73,74] or very short in cases of analysis through a microscope objective [75,76]. Typical values of the laser pulse energy lie in the range of 1 to 30 mJ, which translate to energy density levels on the surface in the range of 1 to 50 J/cm^2.

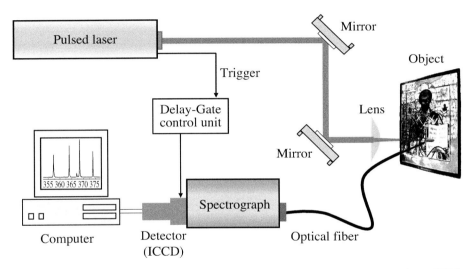

FIGURE 3.2 Diagrammatic representation of typical instrumentation used in a LIBS experiment.

TABLE 3.2
Laser Sources Used in LIBS Analysis

Analytical method	Pulse duration	Wavelength (nm)	
Neodymium:YAG[a] (Q-switched)	5–20 nsec	1064	(532, 355, 266)[b]
Neodymium:YAG (Q-switched, mode-locked)	20–100 psec	1064	(532, 355, 266)
Argon fluoride excimer (ArF)	10–35 nsec	193	
Krypton fluoride excimer (KrF)	10–35 nsec	248	
Xenon chloride excimer (XeCl)	10–35 nsec	308	
Titanium-sapphire	50–250 fsec	800	

[a] Neodymium:yttrium aluminum garnet (Nd:YAG).

[b] 2nd, 3rd, and 4th harmonics, respectively, by using nonlinear crystals.

The fluence level is adjusted either by varying the energy per pulse or by changing the working distance (i.e., lens-to-sample distance) and therefore the spot size on the sample. The quality of the laser beam is a critical parameter in achieving tight focusing, which is important both for minimizing the affected surface area and for obtaining good spatial resolution.

The collection of the emitted light is done by using either a proper lens or lens system or directly through an optical fiber placed near the plume. Lens systems lead to improved collection efficiency, but optical fibers offer simplicity in signal collection. The spectral analysis of the plume emission is done with an imaging spectrograph employing a diffraction grating as the dispersing element. The spectrum is projected onto the image plane of the spectrograph and is recorded on the detection system.

Detection nowadays is based solely on intensified photodiode arrays or intensified CCD (charge-coupled device) detectors. These detection devices offer high sensitivity with variable gain. They also permit adjustable time gating in order to achieve discrimination of the useful atomic emission signal from the broadband continuum background that is present immediately following sample irradiation. A pulse generator is commonly used to control the timing of the measurement, while in new generations of detectors software control of the timing parameters is possible. By employing small or medium-size standard spectrographs one has the option of recording a wide spectral range (typically 100 to 400 nm) with medium- to low-resolution gratings or a narrower range with the use of high-resolution gratings. Obviously the benefit in the former case is the ability to maximize spectral coverage and detect more elements or more lines for each element. However, even in relatively simple spectra, lines can be overlapping and spectral information rather uncertain and ambiguous. In the latter case, the high-resolution spectrum provides a clean recording of even very closely spaced emission lines, permitting, with few exceptions, unambiguous discrimination between different elements. However, at the same time, the limited spectral coverage can result in the loss of the emission of other elements. To obtain both high resolution and wide spectral coverage, multiple spectrographs can be used, which obviously increases the instrumentation cost substantially [77,78]. Recently, significant technological progress has been achieved employing spectrographs based on échelle gratings, which when coupled to two-dimensional CCD arrays provide wide spectral coverage simultaneously with excellent spectral resolution [55,56,79–82]. Figure 3.3 shows spectra obtained on a regular spectrograph at low and high resolution and on an échelle spectrograph illustrating these effects. Monochromators coupled to photomultiplier tube detectors have been used in some studies, but they are not practical mainly because of the single wavelength detection mode of operation, which requires scanning in order to record the spectrum. This implies exposure of the object to a large number of pulses, which in many cases is simply not wise, as it can lead to extensive surface disruption.

Compact or medium-sized LIBS systems, which lend themselves to portability or at least to easy transportation, have appeared in the market or as prototypes [81,83–85]. With these developments, more tests will be possible by archaeologists, historians, and conservators interested in employing LIBS analysis. Technological advances are expected to lower the cost of such units, which may still be prohibitive for small-and medium-scale laboratories.

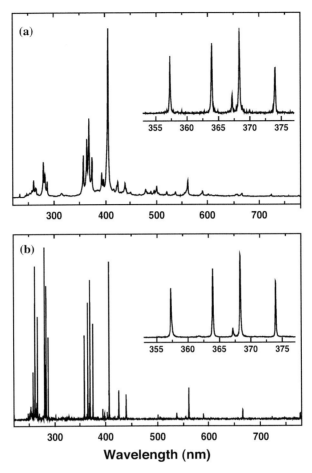

FIGURE 3.3 LIBS spectra of lead white pigment recorded (a) with a standard grating spectrograph at low spectral resolution (spectrograph: 0.19 m; grating: 300 grooves/mm; the spectrum shown is a composition of 4 spectra each covering a wavelength range about 200 nm) and high spectral resolution (Inset, spectrograph: 0.32 m; grating: 1800 grooves/mm) and (b) with an échelle spectrograph that captures the emission across a wide wavelength range, 250–800 nm, featuring high spectral resolution (inset: expanded view of the 355–375 nm range).

Finally, the success of the analysis relies on correct interpretation of spectra and reliable identification of elements present. While this is rather straightforward for an experienced spectroscopist, it can be a challenging and tedious task for a nonspecialist. In this case, the use of proper software, which provides spectral line data for each element or even reference spectra, can be helpful in allowing the users to compare their LIBS spectra against those of the library and identify qualitatively the elemental composition of their samples. Several commercially available LIBS instruments provide some spectra recognition features, while more sophisticated approaches relating to correlation methods have also been proposed in the context of specific applications [86].

3.4 ANALYTICAL PARAMETERS AND METHODOLOGY

Several important factors governing LIBS analysis relate to both material and instrument parameters and depend on the specific analytical issues being addressed. To ensure success of the analysis, an appropriate approach must be developed and optimal working parameters determined. This is even more important in the field of cultural heritage, as the analysis often deals with objects of great historical and artistic value; any modification of the surface analyzed has to be limited to the minimum possible, ensuring, in parallel, the capture of a clean and reliable spectrum.

3.4.1 MATERIAL PARAMETERS

From the historical and artistic point of view, the object's value is a critical factor, which often imposes limitations on the analysis. The approach for analyzing a valuable painting can be very different from that used for examining a pottery shard or a piece of metal scrap from an archaeological excavation. For example, in the case of a sensitive miniature painting, the LIBS measurement must be performed on carefully selected points, representative of the pigments used, and under exposure conditions that minimize adverse effects to the painted surface.

Apart from the obvious dependence on the value of the object, certain physical and chemical properties of the materials examined are critical for the analysis. First of all, the optical properties of the surface probed (absorptivity) determine the material–irradiation interaction, namely the ablation process [5,72]. Obviously, different materials have different ablation and plasma formation threshold levels for a given irradiation wavelength. For instance, irradiation parameters at the fundamental frequency of the Nd:YAG laser (1064 nm) may be appropriate for a metal object while inadequate for other materials, such as pottery or glass, which are often weakly absorbing at that wavelength. Differences in optical properties across the object's surface can also lead to similar testing challenges.

The macroscopic nature of the object's surface is also important to the extent that it can affect the result of the qualitative or quantitative analysis. Thus in the analysis of ancient metal objects, care has to be taken to avoid analyzing environmental deposits on the surface or corrosion products instead of the metal itself. In that case, the result can be significantly different and not representative of the bulk metal composition. Even when the metal itself is probed, compositional heterogeneity due to poor mixing of the alloy components during production, or the presence of different thermodynamic phases, may result in data far from representative of the average material composition. The same applies in pottery analysis, as clay is a highly heterogeneous mixture of materials. Furthermore, surface fragility may pose the danger of causing damage to a more extended surface area than that irradiated. This is because the pressure shock wave that accompanies ablation can lead to the detachment of poorly adhering parts of the surface at or around the spot analyzed. Therefore a thorough examination of the areas to be analyzed is highly recommended in order to avoid unwanted side-effects of the laser pulse on the object surface.

3.4.2 Irradiation–Detection Parameters

Important parameters associated with irradiation include laser wavelength, fluence, and pulse duration. As already discussed, the laser wavelength is a critical factor that, along with the material's optical properties, determines the efficiency of light absorption, and thus energy deposition, in a very thin layer on the surface. The fluence, in turn, must be sufficient to create plasma and determines the intensity of the emitted signal. A short pulse duration, typically in the range of a few nanoseconds, is required to ensure high peak power density as well. Therefore, given the wide variety and complexity of objects and surfaces, it is essential to know the threshold fluence level for ablation of different materials and wavelengths in order to establish optimal parameters for each case. In general, because of the sensitivity and value of the objects examined, work is carried out at the lowest possible fluence, in the range of 1 to 50 J/cm^2, corresponding to peak power density regimes of 0.1 to 5 GW/cm^2. However, when working at fluence levels close to the ablation threshold, care should be taken to ensure that the emission is truly representative of the sample composition, and one must avoid, if possible, the regime of nonstoichiometric ablation. The term stoichiometric ablation implies that the composition of the solid is maintained in the plasma plume, which is further assumed to be homogeneous with respect to the spatial distribution of various species in its volume.

Proper focusing contributes to a greater signal intensity as it optimizes the fluence for a given pulse energy while at the same time it limits the effect of the laser pulse on the surface. The minimization of the beam waist depends largely on laser beam quality (its divergence), proper alignment, and optical component quality. For a laser beam with a gaussian profile and aberration-free optics, a beam diameter as little as a few microns at the focal plane is predicted by diffraction theory of light and is given by

$$d = \frac{4 \cdot f \cdot \lambda}{\pi \cdot D} \qquad (3.2)$$

where d is the beam diameter at the focal plane, f the focal length of the lens, λ the laser wavelength, and D the laser beam diameter.

According to Equation (3.2), a laser beam at 1064 nm having a diameter of 5 mm can be focused down to a spot diameter of 13 μm by using a lens of 50 mm focal length. In practice, the beam diameter of a laser pulse, achieved at the focal plane of the lens, is usually somewhat larger because of imperfect beam quality and lens aberrations. Furthermore, the interaction of the beam with the surface gives rise to photothermal effects, which often result in a crater with a larger diameter than that of the focused beam. Figure 3.4a and Figure 3.4b show examples of craters formed using one and ten pulses at 1064 nm, during the LIBS analysis of a daguerreotype [59]. Crater diameters were in the range of 70 to 80 μm. The crater's surface profile shows a deeper central part indicative of the particular beam intensity distribution and a thermally affected zone at the periphery. Around the area irradiated with ten pulses, ejected metal droplets are observed, while the crater profile (measured with a profilometer) indicates the presence of a rim formed by material melting and redeposition. For a painting, craters produced in a depth-profiling study

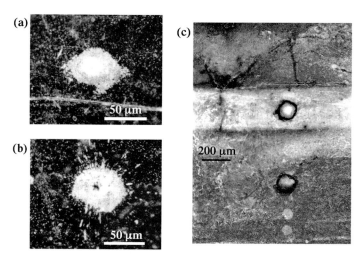

FIGURE 3.4 Craters formed in measurements carried out on 19th century daguerreotype upon irradiation with (a) one laser pulse and (b) ten laser pulses at 1064 nm; (c) craters formed on Byzantine icon upon irradiation with 50 (top) and 40 (bottom) pulses at 355 nm.

(Section 4.3) using multiple pulse irradiation (355 nm) are shown in Figure 3.4c for two locations on the painted surface. As a result of the large number of pulses per site, the craters are wider and much deeper than those observed with a single pulse. In both model and realistic samples, irradiation with one or two pulses results in average crater diameters less than 100 µm, the exact diameter depending upon laser beam profile and focusing, the specific ablation characteristics of the sample (pigment and medium), and the local inhomogeneity of the surface.

Optimization of detection parameters such as signal collection, spectral resolution, delay time, and gate pulse duration are very important. This is a consequence of the lower fluence levels used, particularly in valuable objects, yielding relatively weak plasma emission.

For this reason, relatively short delay times, even as short as 100 nsec, are employed in certain cases of weak emission in order to keep the signal intensity at sufficient levels to obtain satisfactory spectra. Typically, the time delay between laser pulse and gating pulse ranges from 500 to 1000 nsec. This range has been shown to be adequate to suppress the continuum background emission and minimize Stark broadening of spectral lines.

Also, the selection of the proper spectral range and resolution, so that detection of certain elements is optimum, is essential for recording clean spectra and optimizing the analytical information obtained from each spectrum. The advantage of the échelle spectrograph, which provides wide spectral coverage without sacrificing resolution, is obviously very important in this respect.

3.4.3 ANALYZING OBJECTS OF CULTURAL HERITAGE

As described briefly in Section 3, the LIBS spectrum provides immediate information on the identity of elements present on the surface. The analysis is carried out on the object itself and requires no sample removal or preparation. The qualitative and semiquantitative information provided in such an analysis can lead to valuable results such as the identification of pigments used on a painting or the determination of components in a metal alloy. In view of the value and sensitivity of art objects, the analysis should be carried out under conditions where a single laser pulse measurement is adequate for providing the information needed. The single-pulse experiment is important for minimizing material removal and also avoiding possible side effects such as thermally induced pigment discoloration [87–89]. While averaging of multiple pulse measurements improves signal quality, it should be avoided because it can result in excessive material removal. In the case of multiple layer materials such as paints, multiple pulses can even result in the sampling of successive material layers of varying composition, thus producing a more complex superposition of spectral information. LIBS can be characterized as microdestructive (nearly nondestructive), since tight focusing of the laser beam leads to very small affected areas that in most cases are invisible to the naked eye.

The tight focusing of the laser beam is a very important feature of LIBS not only because it results in minimization of the area affected but also because it gives rise to superb spatial resolution. Thus it is possible, by using a single focusing lens, to discriminate between neighboring features and analyze very fine paint lines (less than 100 μm) [90–92]. In fact, recent developments in micro-LIBS analysis show that much smaller dimensions can be probed, extending the range of spatial resolution to a few micrometers [75,76,86,92].

In the same context, LIBS has the ability to perform depth profile analysis. This can be very important if layered materials are probed [93–96]. Such cases are common in painted works of art, which are always composed of multiple paint layers. Also, probing of environmental deposits on stone or marble or corrosion products on metal surfaces is another situation where depth profiling can provide important details of the cross-sectional distribution of different species. Conventional stratigraphic analysis is done by examining, under an optical or electron microscope, the cross section of a properly prepared sample, removed from the object. By using LIBS, an alternative approach to depth profile analysis is achieved. It essentially avoids the removal of a sample from the object while performing, in effect, minute sampling *in situ*. Each laser pulse in the process of analysis ablates a thin layer of material, and as a result the next pulse probes a fresh surface. Therefore, after delivering a certain number of laser pulses on the same spot and recording the spectra produced by each pulse, the analyst can monitor the elemental content of successive layers (Figure 3.5).

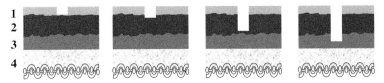

FIGURE 3.5 Schematic representation of a multilayer painting indicating the basic principle behind the LIBS depth profile analysis. Progressive material removal by successive laser pulses results in mapping of the layers of varnish (1), paint layers (2, 3), and ground (4).

In the case of a painting, typical thickness for paint layers ranges from 5 to 50 μm. The depth resolution depends on the characteristics of the ablation process, as already mentioned. Measurements on model samples and real paintings have shown that on average the ablation depth per pulse ranges from 0.5 to 2 μm. Assuming a known and constant etch depth per laser pulse for a certain type of material and irradiation source, one could effectively measure the thickness of each layer by counting the number of pulses required to reach the interface between two successive layers. In most cases however, variation in paint properties affects the ablation etch depth, and so depth resolution studies remain qualitative. Also, as mentioned earlier, local imperfections within the paint layers often cause the removal of larger amounts of material as a result of the shock wave induced by the ablation process. In such cases, craters as deep as 10 μm have been observed.

For an accurate depth profiling, the ideal laser beam would have a top-hat profile, and each succeeding laser pulse would impinge on a fresh surface at a new depth. However, the profile of a typical laser is at best gaussian and often may have irregularities and even hot spots. Under these circumstances, each successive laser pulse will sample the sides of the evolving crater in addition to the bottom. Some fresh surface will also be sampled, since the crater diameter typically increases with the number of pulses. So the observed intensities of the spectral components will be a complex mixture of signals from different depths, which is hard to calibrate. At higher laser fluence, some partial melting may also complicate the picture. In spite of these problems with the analysis of elemental composition in the interior of samples, if the thickness of a layer is determined by noting the interface, where a new element appears, for instance, the results can be quite accurate and reproducible. A good example of this type of depth analysis is given by a recent study of the thickness of the silver layers in old daguerreotypes [58]. Several examples of depth profile analysis are shown in Section 3.5.

3.4.4 QUANTITATIVE ANALYSIS

In many cases, quantitative information about the composition of the material analyzed is important, as for example in the determination of minor and trace elements in glass objects, coins, metallic objects, or minerals. Such data, apart from material characterization and classification, can provide provenance information.

An important issue with LIBS is the ability of the technique to produce reliable quantitative information [97]. In this respect, a lot of work has been published in which methodologies have been proposed for getting quantitative results that rely mainly on calibration curves for selected analytes, produced with reference samples of known composition [4–6,67–71,98,99]. Such an approach is in principle transferable to the analysis of art objects; but the following considerations have to be seriously taken into account. The vastly different types of materials encountered and their highly varying composition require a large number of calibration samples to cover all cases, which willl obviously complicate the analysis. Furthermore, in order to minimize the effects of material heterogeneity and other matrix effects [99] to the analytical result, averaging the signal of several pulses is required, which may not be possible in all cases, especially for valuable and fragile objects.

A promising approach to the problem appears to be the use of the calibration-free LIBS (CF-LIBS) method, which can in principle provide quantitative elemental analysis with no need of calibration curves or internal standards [48,49,56,62,100]. It relies on the assumption that stoichiometric ablation takes place and that an optically thin plasma is produced, which is at local thermodynamic equilibrium (LTE approximation). From the practical point of view, it requires recording of the emission intensity at high resolution and across a wide spectral range for identifying all or most of the elements present in the sample and for improving the accuracy in the calculations. According to the CF-LIBS method, the intensity of the spectral lines for each emitting species (different states of ionization for each element are treated as separate species) is represented as a Boltzmann plot on the basis of Equation (3.3), which is derived by taking the logarithm of Equation (3.1). This procedure produces a set of parallel straight lines, one for each species. The slope of these lines yields the plasma temperature, while the intercept relates to the logarithm of species concentration within a factor F [Equation (3.3)], which incorporates experimental parameters related to optical collection and plasma density and volume.

$$\ln \frac{I_\omega^{ik}}{g_i A_{ik}} = \ln \frac{FN_S}{Z_S(T)} - \frac{E_i}{k_B T} \qquad (3.3)$$

Elimination of the unknown instrumental factor in a final normalization step, based on the requirement that the sum over all species concentrations equals unity, leads to the calculation of the concentration of each species within the sample. Furthermore, the Saha–Boltzmann equation [Equation (3.4)] is used, if needed, to calculate species concentrations for a given element in adjacent states of ionization.

$$N_S(II) = 6.0 \times 10^{21} \frac{N_S(I)}{N_e} \frac{g^{II} T^{3/2}}{g^I} \exp(-\frac{E_S}{T}) \qquad (3.4)$$

where $N_S(I)$ and $N_S(II)$ are the ground state concentrations of the neutral and singly ionized atomic species S, N_e is the electron density, g^I and g^{II} are the ground state degeneracy values, E_S is the ground state ionization potential, and T is the plasma temperature.

Despite the obvious advantages of the CF-LIBS approach, data interpretation must be done with caution as in all cases of chemical analysis. Regarding the method itself, it is of concern that several emission lines are affected owing to self-absorption (unexcited species in the plume absorb the radiation emitted from their excited counterparts) appearing at an apparently lower intensity. This requires special corrections to be applied to the data in order to avoid unreliable measurements of plasma temperature and errors in the calculated species concentration values. Regarding the materials analyzed, in cases of highly heterogeneous substrates (e.g., pigment mixtures), analytical results are representative of only the local microscopic composition of the sample and not of the average composition, which is of most interest.

3.5 EXAMPLES OF LIBS ANALYSIS IN ART AND ARCHAEOLOGY

Specific examples involving analysis of real objects are presented to illustrate the application of LIBS to various types of analytical challenges encountered in art history, conservation, and archaeological analysis.

3.5.1 PIGMENTS

Painting has been used in all types of artworks from antiquity to modern times. Easel and wall paintings, wood and metal polychromy, illuminated manuscripts, and pottery represent art or craft forms in which pigments have been used [101–104]. Identifying the pigments used is important for several reasons. For example, in the case of painted works of art it can help art historians characterize the materials present and understand techniques and effects used by the artist in achieving the final result. Similarly, in the case of ancient painted pottery or frescoes, pigment characterization may lead to improved understanding of the materials and technology available to the craftsmen of that time. This knowledge may even relate to the origin of the materials, suggesting local or remote sources that might show communication and trade between sites. In addition, pigment identification in painted works of art can often be significant in providing dating information on the basis of the known history of the pigment manufacture or in assessing the state of preservation of a painted work of art in preparation for proper restoration.

Physically the pigments are tiny insoluble particles or grains (with diameters in the sub-micrometer to a few μm range) dispersed and suspended in a matrix normally called the binding medium or the carrier; together they form a relatively viscous paste called paint [101]. The matrix can be a drying oil (oil paint), an acrylic polymer (acrylic paint), or lime (fresco paint). After the paint is applied on the appropriate substrate, it gradually solidifies because of chemical reactions (e.g., cross-polymerization of the fatty acid chains of linseed oil in the case of oil paintings or transformation of calcium hydroxide to calcium carbonate in the case of wall paintings) that trap and stabilize the pigment grains. The artist's techniques vary widely depending upon preferences and style, support types used, pigments and oils available, and other considerations. A schematic illustration of a painted structure composed of two paint layers on top of a ground layer on canvas is shown in Figure 3.5. In this case, the ground layer is a thick (around 0.5 to 2 mm) layer of gypsum or chalk (gesso) in animal glue, which may be additionally covered by a thin preparation layer, or primer, consisting mainly of lead white in oil. The primer acts as an additional barrier between paint and support (e.g., canvas, wood, wall) and often adds luminosity to the painting. The paint layers are then applied in various ways. Finally, on top of the paint, the varnish functions as a transparent protective layer, which also improves the tonality and gloss due to improved refractive index matching. Varnishes are natural or synthetic organic resins. They are applied in the form of solution in a volatile organic solvent, yielding a hard, glassy film on the painting after solvent evaporation.

With the exception of a limited number of organic color substances, most pigments used in paintings from antiquity to modern times are inorganic compounds. The main reasons have been the higher chemical and photochemical stability of inorganic pigments compared to organic ones and the easy availability of most inorganic pigments in the form of natural minerals or as products of relatively simple chemistry. In contrast, lengthy and delicate procedures were often involved in extracting organic pigments from insects or plants. Table 3.3 shows, arranged by color, several pigments used in paintings. Their chemical structure and common names are given along with chronological information on their use.

The elemental composition of various inorganic pigments is reflected in the corresponding LIBS spectra, and this enables their discrimination based on the characteristic atomic emission peaks recorded. Major analytical emission lines, used to identify the presence of specific elements in pigments and other materials through their corresponding LIBS spectrum, are shown in Table 3.4.

The positive identification of a certain pigment results from correlating the spectral data with the color of the paint analyzed. In this respect, information from an art historian or conservator about the possible pigments anticipated is essential in the analysis of the work. In many cases, the characteristic elements can easily lead to determination of the pigment used. For example, the presence of mercury undoubtedly suggests the use of the red pigment vermilion HgS (also known as cinnabar in its natural mineral form). The presence of lead in a white paint is definitive proof of the use of lead white ($Pb(OH)_2 \cdot 2PbCO_3$). But mixtures of pigments are often used by an artist to achieve a desired color and shade. For example, it is not uncommon for a green paint to be the result of mixing a green pigment with a white one or even a yellow pigment with a blue one. The situation of pigment mixtures gives rise to more complex LIBS spectra, but the presence of individual pigments can usually be determined from the elements found. In such a case, microscopic examination, when possible, is of great help in guiding the analysis of the LIBS spectral data, if pigment grains can be optically discriminated.

Typical spectra from model samples of two common pigments are shown in Figure 3.6 indicating how the different spectral emission lines can lead to identification of each pigment on the basis of the specific elements detected. The use of single- or multicomponent pigment model samples is essential in establishing proper measurement conditions before carrying out an analysis of a real object.

An actual pigment analysis case is represented by the study of the miniature (19th century France) shown in Figure 3.7a [51]. This miniature is an example in which the size of the object and the sensitivity of the painted surface (thin and weakly adhering paint) dictated the examination of a minimal number of spots and use of not more than three pulses per spot. Examination of the green paint used extensively on the miniature shows intense emission due to copper and weak emission due to arsenic (Figure 3.8a). This spectrum suggests the use of a green pigment based on a copper-arsenic compound, either emerald green ($Cu(CH_3COO)_2 \cdot 3Cu(AsO_2)_2$) or Scheele's green ($Cu(AsO_2)_2$).

An alternate possibility is the use of a mixture of the blue pigment azurite (a copper compound) and the yellow pigment orpiment (an arsenic compound).

TABLE 3.3
List of Selected Inorganic Pigments Analyzed by LIBS and Elements Identified

Pigment name	Chemical composition/formula	Identified elements	Origin and chronological information[*]
Lead white	$Pb(OH)_2 \cdot 2PbCO_3$	Pb	Synthetic, pre-500 B.C.
Titanium white	TiO_2	Ti	Synthetic, 1920
Zinc white	ZnO	Zn	Synthetic, 1834
Lithopone	$ZnS \cdot BaSO_4$	Ba, Zn (Ca)	Synthetic, 1874
Chalk	$CaCO_3$	Ca	Mineral (calcite)
Barytes, barium sulfate	$BaSO_4$	Ba	Synthetic, early 19th C
Gypsum	$CaSO_4 \cdot 2H_2O$	Ca	Mineral
Cadmium yellow	CdS or CdS ($ZnS \cdot BaSO_4$)	Cd, Zn (Ba)	Synthetic, 1829
Chrome yellow	$PbCrO_4$	Cr, Pb (Ca)	Synthetic, 1818
Cobalt yellow	$2K_3(Co(NO_2)_6) \cdot 3H_2O$	Co	Synthetic, 1861
Orpiment	As_2S_3	As	Mineral
Naples yellow	$Pb_2Sb_2O_7$	Pb, Sb	Synthetic, Egypt, ca. 1500 B.C
Lead tin yellow	Pb_2SnO_4, (Type I)	Pb, Sn	Synthetic, ca. 1300
Strontium yellow	$SrCrO_4$	Sr, Cr	Synthetic, early 19th C
Barium yellow	$BaCrO_4$	Ba, Cr	Synthetic, early 19th C
Yellow ochre	$Fe_2O_3 \cdot nH_2O$, SiO_2, Al_2O_3	Fe, Si, Al	Mineral
Cadmium red	$CdS_xSe_{(1-x)}$	Cd	Synthetic, ca. 1910
Cinnabar/ vermilion	HgS	Hg	Mineral/synthetic, 8th C
Red ochre	Fe_2O_3 (Al_2O_3)	Fe (Al)	Mineral
Realgar	As_2S_2	As	Mineral
Mars red (hematite)	Fe_2O_3	Fe	Synthetic, middle 19th C
Red lead (minium)	Pb_3O_4 ($2PbO \cdot PbO_2$)	Pb	Synthetic
Lapis lazuli/ultramarine	$Na_8Al_6Si_6O_{24}S_3$	Al, Si, Na	Mineral/synthetic, 1828
Egyptian blue	$CaCuSi_4O_{10}$	Cu, Si, Ca	Synthetic, Egypt, ca. 3100 B.C.
Cobalt blue	$CoO \cdot Al_2O_3$	Co, Al, Na	Synthetic, 1802
Cerulean blue	$CoO \cdot nSnO_2$	Co, Sn	Synthetic, 1860
Prussian blue	$Fe_4[Fe(CN)_6]_3 \cdot nH_2O$	Fe (Ca)	Synthetic, 1704
Azurite	$2CuCO_3 \cdot Cu(OH)_2$	Cu (Si)	Mineral
Malachite	$CuCO_3 \cdot Cu(OH)_2$	Cu (Si)	Mineral
Viridian green	$Cr_2O_3 \cdot 2H_2O$	Cr	Synthetic, 1838
Emerald green	$Cu(CH_3COO)_2 \cdot 3Cu(AsO_2)_2$	Cu, As	Synthetic, 1814
Verdigris	$Cu(CH_3COO)_2$	Cu	Synthetic, antiquity
Ivory black (bone black)	$C + Ca_3(PO_4)_2$	Ca, P	Charring of ivory, antiquity
Manganese black	MnO	Mn	Mineral
Magnetite/ mars black	Fe_3O_4	Fe	Mineral/synthetic, mid-19th C

[*] Origin information distinguishes between natural (minerals in use since antiquity) and synthetic pigments. Chronological information indicates when pigments were first used (historical sources) or became commercially available (modern pigments).

TABLE 3.4
Major Analytical Emission Lines of Elements

Element	Wavelength (nm)[a]
Ag	272.18 (I), 282.44 (I), 328.07 (I), 338.29 (I), 487.41 (I), 520.91 (I), 546.55 (I)
Al	308.21 (I), 309.27 (I), 394.40 (I), 396.15 (I)
As	228.81 (I), 234.98 (I), 236.97 (I), 245.65 (I), 249.29 (I), 274.50 (I), 278.02 (I), 286.04 (I), 289.87 (I)
Au	267.60 (I), 274.82 (I), 312.28 (I)
Ba	225.47 (II), 230.42 (II), 233.53 (II), 389.18 (II), 413.07 (II), 455.40 (II), 493.41 (II), 553.55 (I), 614.17 (II), 649.69 (II)
C	247.86 (I)
Ca	315.89 (II), 317.93 (II), 318.13 (II), 393.37 (II), 396.85 (II), 422.67 (I), (526.17–527.03 (I))
Cd	228.80 (I), 288.08 (I), 298.06 (I), 340.37 (I), 346.62 (I), 346.77 (I), 361.05 (I), 467.81 (I), 479.99 (I), 508.58 (I), 643.85 (I)
Co	(240.72–241.53 (I)), (242.49–243.90 (I)). Numerous lines in the range 340–360, 389.41 (I), 399.53 (I), 412.13 (I).
Cr	357.87 (I), 359.35 (I), 360.53 (I), 391.92 (I), 396.37 (I), 396.98 (I), 397.66 (I), 425.44 (I), 427.480 (I), 428.972 (I), (520.45, 520.60, 520.84 (I))
Cu	261.84 (I), 276.64 (I), 296.12 (I) 324.754 (I), 327.396 (I), 510.55 (I), 515.32 (I), 521.82 (I)
Fe	Numerous emission lines (from neutral and ionized atoms) appear throughout the spectrum. Features at 273–275 nm, 373–376 nm, 381–384 nm, and 404–407 nm are characteristic of iron.
Hg	253.65 (I), 296.73 (I), (312.57–313.18 (I)), 365.02 (I), 404.66 (I), 435.83 (I), 546.07 (I), 614.95 (II)
K	766.49 (I), 769.90 (I)
Mg	279.553 (II), 280.27 (II), 285.21 (I), (382.93–383.83 (I)), 517.27 (I), 518.36 (I)
Mn	257.61 (II), 259.37 (II), (279.48–280.11 (I)), 293.31 (II), 293.93 (II), 294.92 (II), 380.67 (I), 382.35 (I), (403.08–403.57 (I)), 404.14 (I)
Na	330.13 (II), 589.00 (I), 589.59 (I)
Ni	301.20 (I), 310.61 (I), 310.18 (I), 336.96 (I), 341.48 (I), 351.50 (I), 352.45 (I), 361.94 (I)
P	(253.40–253.56 (I)), (255.33–255.49 (I))
Pb	280.20 (I), 282.32 (I), 283.31 (I), 287.33 (I), 357.27 (I), 363.96 (I), 367.15 (I), 368.35 (I), 373.99 (I), 405.78 (I), 406.21(I), 416.80 (I), 500.54 (I), 722.90 (I)
Sb	206.83 (I), 217.58 (I), 231.15 (I), 252.85 (I), 259.80 (I), 265.26 (I), 277.00 (I), 287.79 (I)
Si	250.69 (I), (251.43–252.85 (I)), 263.13 (I), 288.16 (I), 390.55 (I)
Sn	235.48 (I), 242.17 (I), 242.95 (I), 284.00 (I), 300.91 (I), 303.41 (I), 317.50 (I), 326.23 (I)
Sr	407.77 (II), 421.55 (II), 460.73 (I)
Ti	Numerous emission lines (from neutral and ionized atoms) appear throughout the spectrum. Features at 323–325 nm, 332–339 nm, 372–376 nm, 390–400 nm, 428–432 nm, 444–446 nm, 451–454 nm, and 498–502 nm are characteristic of titanium.
Zn	213.86 (I), 328.23 (I), 330.26 (I), 330.29 (I), (334.50–334.59 (I)), 468.01 (I), 472.21 (I), 481.05 (I), 636.23 (I)

[a] Wavelengths in parentheses indicate two or more lines not resolved in low-resolution spectra but still characteristic of the elements. The wavelengths refer to emission from neutral atoms when followed by (I) and to emission from singly charged ions when followed by (II).

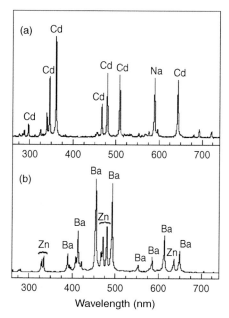

FIGURE 3.6 LIBS spectra from model paint samples of (a) cadmium red and (b) lithopone.

Additional examination of the paint under the microscope verified the identification as emerald green on the basis of its characteristic appearance (small spherulites).

In a similar case, several painted daguerreotypes were examined in order to identify the pigments used (Figure 3.7b). Daguerreotypes were the first form of photograph and were prevalent between 1839 and 1860, before photography based on paper became practical [105]. The images consist of silver particles photochemically produced on silver-coated copper plates. The plates were sometimes colored afterwards to highlight certain features by applying a very thin paint layer. As an example, the white paint used on a daguerreotype (19th century, U.S.A.) was analyzed, and the LIBS spectrum showed the presence of emission lines due to barium, suggesting the use of barium white (Figure 3.8b). Given the delicate nature and thickness of the paint layer (ca. 1 μm), a single laser pulse was used for the analysis, showing that reliable results can be obtained with careful selection of working parameters.

In the context of pigment analysis, LIBS can also be used to identify the type of prior restoration performed on paintings by discriminating between the original paint and that used in the restoration. In certain cases, when a painting was done or when an intervention took place can be indirectly estimated on the basis of known dates that synthetic pigments first became available. In one case, the restoration carried out on several points of an oil painting (Figure 3.7c) was examined and found to contain mainly titanium white, a modern pigment, in contrast to the original paint, which was composed mainly of lead white (Figure 3.9) [42–50].

This result suggests some retouching, done on the original painting after 1920, since titanium white became commercially available at that time. In a similar case

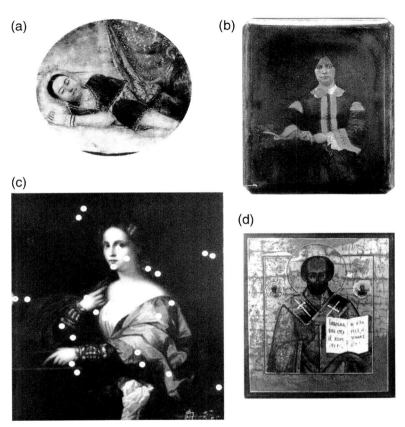

FIGURE 3.7 (See color insert following page 144.) Paintings analyzed by LIBS. (a) 19th century miniature on ivory, (b) 19th century colored daguerreotype, (c) 18th century oil painting, (d) 19th century Byzantine icon.

the "gold" background used on a Byzantine icon was examined by carrying out several spot analyses, which revealed dramatic local differences apparently resulting from a past restoration. The original supposedly gold foil was actually shown to be a silver foil on the basis of characteristic emission peaks due to Ag in the LIBS spectrum. The golden appearance was achieved by coating the silver foil with a yellow varnish (Figure 3.10a). On the other hand, the LIBS spectrum of the restoration suggests that a copper foil has been used instead (Figure 3.10b).

In the context of archaeological objects, pigment analysis is also important. For example, the study of pigments and painting techniques reveals information on technological progress in ancient cities. Such data can even be used to discriminate objects on the basis of their production site and reveal possible links between such sites. Examples of the application of LIBS in ancient paint analysis are shown in Figure 3.11.

The black paint on a fragment of an ancient ceramic shard (12th century B.C., eastern Crete) was analyzed and found to contain manganese in addition to the usual clay components (Al, Si, and Mg). Detection of Mn suggests the use of MnO_2, a black pigment used commonly in antiquity. In a different case, the blue paint on a

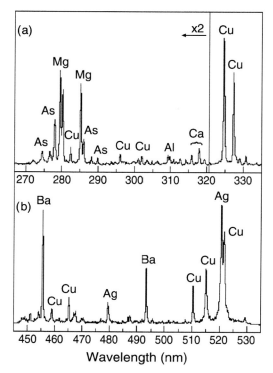

FIGURE 3.8 LIBS spectra from (a) green paint on miniature (Figure 3.7a) and (b) white paint on a daguerreotype (Figure 3.7b).

fresco fragment from Egypt was examined, and the LIBS spectrum showed the presence of Mg, Si, Ca, Cu, and Al. The spectral data suggest the use of Egyptian blue ($CaCuSi_4O_{10}$), which is further supported by comparison with a known sample of the pigment.

Having the ability to quantify the spectral information is obviously important, especially in the case of compositionally similar paints. An example is shown in Figure 3.11c from the LIBS analysis of two different Roman fresco samples (2nd century, St. Albans, U.K.) performed on an échelle spectrograph system [48]. The wide spectral range covered and the high resolution achieved are obvious. By applying the CF-LIBS analysis, it was possible to determine the elemental content of red and yellow paints, detecting elements with concentrations as low as 0.1%.

In several cases, LIBS has been successful in revealing stratigraphic information on paint layers by means of depth profile analysis [42,46,50,55,59]. In the case of a Byzantine icon examined (19th century, Russia, Saint Nicholas, egg tempera on silver foil, Figure 3.7d), depth profile analysis mapped the different paint layers and the silver foil between the brown paint (iron-based pigment, possibly red ochre) and the gypsum ground layer (Figure 3.12). Similar analysis on a different spot on the same icon revealed a thin zinc white overpaint layer on the original lead white paint [46]. Examination of the red paint on another icon (18th century, Greece, Three Saints,

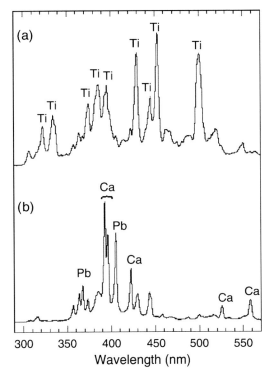

FIGURE 3.9 LIBS spectra of (a) titanium white paint on restoration and (b) original lead white paint on 18th century oil painting (Figure 3.7c).

egg tempera) showed the use of the red pigment vermilion (HgS) on a dark background containing iron, possibly a red ochre or a sienna (Figure 3.13).

The ground layer is rich in calcium, suggesting the use of gypsum or chalk. These results demonstrate the ability of LIBS to extract information on successive paint layers by means of depth profile analysis. This approach is not limited to painted structures but is general for surfaces that have a multilayer structure or elemental concentration gradients.

3.5.2 POTTERY

Ceramic objects are the most common remnants of ancient life, uncovered in numerous excavations, having been used as storage containers, serving dishes, and votive figurines, among other things. They are made of clay, which is often decorated by paint depending on the use and quality of the object. In the analysis of ceramic shards, questions are related to the determination of the elemental composition of pigments, inclusions, and encrustations, and the clay itself. Analysis of clay inclusions can be of importance if they are characteristic of the clay origin or the technique used to make the object. An indicative example is shown in Figure 3.14a and Figure 3.14b, where LIBS spectra from two different types of microscopic inclusions are shown.

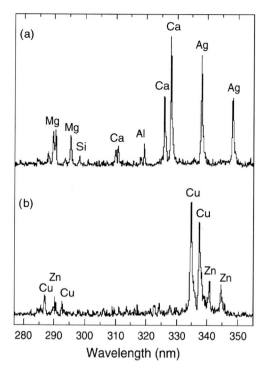

FIGURE 3.10 LIBS spectra of (a) silver foil and (b) restoration on silver foil used in Byzantine icon (St. Catherine 17th century A.D.).

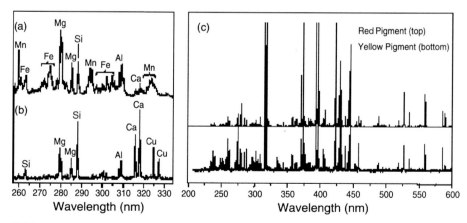

FIGURE 3.11 LIBS spectra from (a) black paint on pottery shard, (b) blue fresco paint sample, and (c) red and yellow fresco paint samples. (c is reproduced from Borgia et al., *J. Cult. Heritage,* **1**, S281, 2000.)

The dark inclusion appears to be based on clay with high iron content (possibly from magnetite, Fe_3O_4, a black mineral), while the white inclusion shows a high calcium content, most likely in the form of calcite. Quantitative elemental analysis

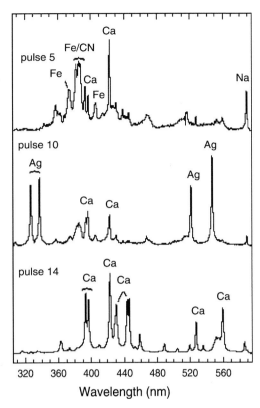

FIGURE 3.12 Depth profiling studies using multiple laser pulses on Byzantine icon of Saint Nicholas (Figure 3.7d).

of pottery shards is important in differentiating between various types of clay from the same or different excavation sites, aiding archaeologists in classifying objects in order to draw conclusions about materials and techniques, which relate to the socioeconomic status of cities and populations.

3.5.3 Marble, Stone, Glass, and Geological Samples

Marble and stone materials have been used extensively in monuments, sculpture, and tools. Their elemental analysis related to geological data can be helpful in discriminating among different types of marble or stone and determining the material source [62]. An important application of LIBS to the analysis of marble or stone monuments has to do with the investigation of environmental pollution effects. It is known that atmospheric pollution through physical and chemical interactions leads to the formation of encrustation on exposed stone. Different types of encrustation arise depending on prevailing environmental conditions and deposition processes, their thickness being in the range of one half to several millimeters [44,53,62]. These

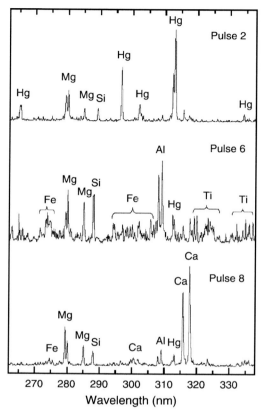

FIGURE 3.13 Depth profiling studies on Byzantine icon (Three Saints, 18th century)

processes involve partial or even total transformation of the chemical structure of the surface with additional deposition and inclusion of airborne particles. Biological crusts are formed if microorganisms develop on the surface. Furthermore, protective or aesthetic coatings applied during past conservation treatments may also constitute major obstacles to conservation. Most conventional approaches for characterizing stone encrustation involve sampling from the monument/sculpture followed by laboratory microscopic examination and spectroscopic or chemical analysis. In this respect LIBS, in the form of a transportable unit, offers a tool for the conservator or the restorer that is appropriate for the *in situ* characterization of different types of crust. The major question is the determination of the crust thickness and the distribution of pollutants as a function of depth from the surface. Obviously a depth profiling study is needed for addressing such questions as has been described for marble, stone, and glass [44,45,54,62]. In such a study, LIBS spectra of marble encrustation were recorded, and an example is shown in Figure 3.14c. Emission from several elements that originate from environmental pollutants including Fe, Al, Si, and Ti is detected in the encrustation but not in the unweathered marble [54]. The concentration changes of these elements can be used to depth profile the encrustation and understand its stratigraphy. For example, iron and titanium are the

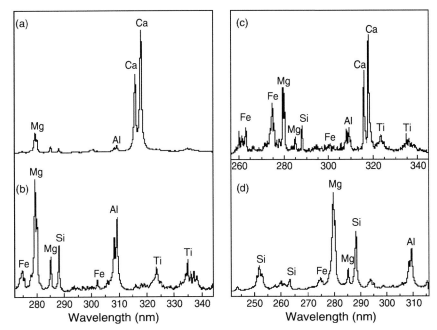

FIGURE 3.14 LIBS spectra from (a) white and (b) dark pottery inclusions, (c) marble encrustation (dendritic crust), and (d) ancient glass.

main transition metals involved in the marble sulphation process and therefore can be used as indirect indicators for the gypsum layer.

Analysis by LIBS of old glass objects can be a quick way to differentiate between various types of glass such as sodium (Na), potassium (K), or lead (Pb) glass or characterize surface corrosion effects [62]. LIBS analysis has been carried out to quantify the composition of glaze on ancient pottery [57], while analysis of glass has been performed in the context of investigating the process of laser cleaning of stained glass [45].

Geological materials such as obsidian have been used since Neolithic times to fabricate simple tools and utensils and to make sculpture. Their elemental composition, particularly of their minor and trace elements, has been related to the object provenance. Such analysis can be of importance in uncovering, for example, the sources of raw materials used to make certain objects by comparison of compositional patterns between the object in question and the original raw materials. Quick screening of geological samples can be done on the basis of qualitative analysis, if differences in elemental content are significant [106,107]. For more detailed studies, a systematic quantitative analysis of major and minor constituents is obviously needed.

3.5.4 Metals

Objects made of metal or alloys, including sculpture, tools, weapons, home utensils, and jewelry, have been widely used for different purposes since metallurgy was invented. The main materials used include copper and bronze, that is, copper–tin alloys, in the Bronze Age and later iron. Other metals used include lead and tin, while precious metals such as silver or gold alloys have been used in jewelry and for decorating objects. Metal alloys have been used extensively in coinage. The first analytical aim concerning a metal object is to identify the metal or metals used in an alloy. This information could be enough for an initial classification of an object, for example, to distinguish between copper and bronze. Furthermore, accurate quantitative analysis of various metals and trace elements can lead to a more complete characterization of materials, yielding information about objects regarding the period or technology of manufacture and possibly the provenance of raw materials.

Examples of spectra collected from various metal objects including an ancient bronze tool, a piece of ancient gold jewelry, and a 20th century coin are shown in Figure 3.15.

Quick screening of ancient metal objects on the basis of elemental contents can be done efficiently by LIBS. For example, discriminating between copper and bronze objects is important for the classification of archaeological findings. Quantitative

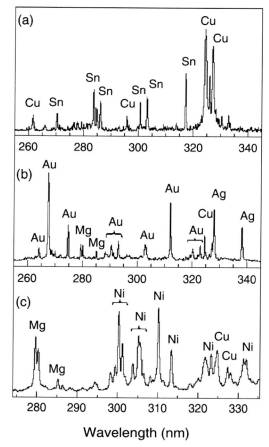

FIGURE 3.15 LIBS spectra from (a) Minoan bronze sample, (b) Minoan golden jewelry, and (c) 20th century coin.

determination of the various metals present and particularly of minor and trace elements can reveal valuable information about metallurgical technology, period of manufacture and possibly provenance of the raw materials. In a recent study, archeological bronze artifacts from a burial site in southern Tuscany, Italy, dated in the period 2500–2600 BC, were examined using micro-LIBS analysis and following the calibration-free LIBS method in order to obtain the concentration of the elements in the alloy [64]. Furthermore, quantitative analysis can be important in assessing the gold content of the alloy, and this can be done by means of appropriate calibration curves using standards or following the CF-LIBS approach [56].

In a recent study, LIBS analysis of daguerreotypes [105] carried out in the depth-profiling mode was used to map the different material layers on the plate [59]. The basis of making a daguerreotype is a highly polished, silver coated copper plate, which upon exposure to iodine along with bromine or chlorine forms a light-sensitive silver halide surface. The sensitized plate is then placed in the focal plane of a camera and exposed to light reflected from the subject photographed. Exposure to light forms the latent image, which is composed of small silver clusters. After removal from the camera, the image is developed with mercury vapor, which leads to the growth of 0.1 to 100 μm particles of silver–mercury amalgam. The image is fixed with sodium thiosulfate and gilded with a warm solution of gold chloride. The gilding improves the durability of the surface and adds richness to the aesthetic quality of the image. Through the successive LIBS spectra recorded after each laser pulse as a function of depth, it was possible to distinguish the gilded upper part of the silver layer (measured to be ca. 1 to 2 μm thick) on the basis of the characteristic gold emission (Figure 3.16).

In fact, evidence for a very thin superficial layer of environmental origin is also provided by the Al, Mg, and Si emissions observed in the spectrum obtained with the first laser pulse. The transition from the silver layer to the bulk copper plate is also clearly seen by the abrupt relative increase of the copper emission relative to that of silver.

3.5.5 BIOMATERIALS

Biological samples are an area of potential interest to archaeological research on human remains to identify certain diseases, nutritional habits, or deficiencies, or the presence of potentially toxic elements [108–111]. The characterization of calcified tissues using LIBS is an example of such a quantitative analysis of traces of aluminum (Al), strontium (Sr), and lead (Pb) in human bones or teeth. It was demonstrated in this study [112] that the above trace elements can be semi-quantitatively detected and in fact surface mapped in calcified tissue samples. These results indicate strongly the potential advantages of LIBS in the analysis of bioarchaeological samples.

3.5.6 CONTROL OF LASER CLEANING

A major breakthrough in art conservation has been brought about in the last two decades by the introduction of laser-based techniques for cleaning art objects ranging from marble, stone, metal, and stained glass to paintings, icons, and paper [113].

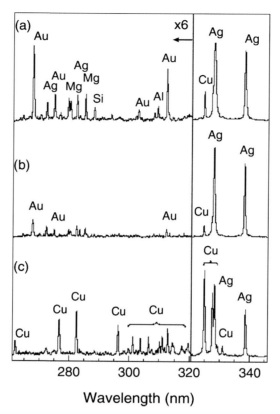

FIGURE 3.16 Depth profile analysis of 19th century daguerreotype after irradiation with (a) one, (b) three, and (c) ten laser pulses at 1054 nm.

The cleaning process relies on the controlled removal of contaminants or other unwanted materials from the object surface. This cleaning is accomplished by laser ablation through the interaction of focused short laser pulses with the material (see Chapter 8). The process depends strongly on the material properties (absorptivity, surface roughness, physical characteristics, mechanical stability) and the irradiation parameters (wavelength, fluence, repetition rate, pulse duration). There have been extensive studies on laser ablation cleaning that are beyond the scope of this chapter, but one critical question regarding the success of the cleaning process relates to LIBS. It is very important, when carrying out any type of cleaning method, to know where (and when) to stop the process, which means to be able to assess reliably to what extent the undesirable material has been removed. Such control of laser cleaning can, in certain cases, be achieved by monitoring the optical emission resulting from material ablation. In essence, LIBS measurement is carried out simultaneously with the laser cleaning. As discussed in Section 6.3, successive laser pulses on the same area can provide compositional information about the surface and into the bulk material. If distinct differences exist between the LIBS spectra of the contamination layer to be removed and the original surface, then it is, in principle, possible to

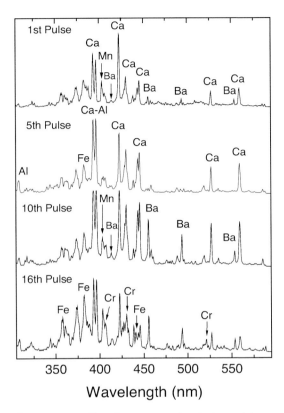

FIGURE 3.17 LIBS spectra revealing the cross-sectional elemental composition of several paint layers as these are probed during the laser removal or overpaint. The different spectral features in the successive layers are used to monitor the progress of laser cleaning.

control the process of cleaning by an algorithm implemented on a computer. This is a delicate approach, and careful preliminary tests of working parameters, with respect to the specific case in hand, have to be done beforehand in order to define properly the end point of cleaning. Such control of laser cleaning has been demonstrated in several cases including in the removal of overpaint from frescoes [43], of encrustation from marble [44,54,114], of glass [45], and so forth. An example is shown in Figure 3.17 indicating evident spectral changes as the overpaint layer is removed, which allows effective, on-line control of the process. Cleaning is then stopped before exposing the original paint layer to the laser irradiation in order to avoid any pigment modification.

3.6 LIBS IN COMBINATION WITH OTHER TECHNIQUES

A combination of analytical techniques can be advantageous, since more complete information on materials can be found on an object. Strengths of one technique could perhaps complement weaknesses of another. The use of LIBS in combination with other laser analytical techniques such as Raman microscopy, laser-induced fluorescence (LIF) spectroscopy, and laser time-of-flight mass spectrometry is described below. Examples of combining results from LIBS analysis with other analytical techniques such as diffuse reflectance spectroscopy and hyperspectral imaging [53] have also been reported but are not discussed here.

3.6.1 LIBS AND RAMAN MICROSCOPY

Raman spectroscopy provides molecular species information that is complementary to elemental data obtained by LIBS. The Raman effect relies on inelastic scattering of light from molecules or materials and is sensitive to vibrational transitions, which in many cases are characteristic of the substrates analyzed (see also Chapter 4). In particular, pigment studies have shown that Raman spectra constitute a unique *spectral fingerprint* for each pigment [115–120]. Raman microscopy is an excellent technique for the scientific investigation of materials used on artworks because it has high sensitivity and is nondestructive. In addition, it can be performed *in situ*, thus obviating the need for sampling. The use of a microscope is essential to this technique as it permits probing of isolated grains of pigment (down to 1 μm). Moreover, the very high spatial resolution thus achieved gives rise to spectra reasonably free from interference from adjacent materials.

The combined use of Raman microscopy and LIBS in the analysis of objects of cultural heritage has been reported in the literature [46,47,51,52,55]. For example, the LIBS analysis of the yellow paint on a Byzantine icon (*The Annunciation,* 18th century A.D., Greece) [51] revealed the presence of lead and chromium, suggesting the use of chrome yellow, which was indeed verified by the paint analysis using Raman microscopy (Figure 3.18a,b). Similarly, the blue paint on the French miniature (of Figure 3.7a) was analyzed and found to contain iron using LIBS. Raman analysis proved the presence of Prussian blue (Figure 3.18c,d) based on characteristic Raman vibrational frequencies in the Raman spectra.

The combined application of LIBS and Raman microscopy involves the use of two separate instrumentation units. The simultaneous application of the two techniques on a unified instrument has been demonstrated recently [121, 122], and this result is quite promising for achieving parallel LIBS and Raman analysis optimizing the combined analytical approach.

3.6.2 LIBS AND FLUORESCENCE SPECTROSCOPY

Fluorescence (photoluminescence) spectroscopy is a sensitive, nondestructive technique, shown in several cases to be capable of identifying pigments used in painting on the basis of their characteristic molecular emission bands. The use of laser excitation leads obviously to increased sensitivity and simpler instrumentation because no monochromator is needed for selecting the excitation wavelength from a conventional white light source, which is commonly used. The term LIF (laser-induced fluorescence) refers to fluorescence or photoluminescence emission spectroscopy using pulsed or cw lasers as the excitation source.

Several examples of the application of LIF spectroscopy to the analysis of selected pigments that are photoluminescent have been reported in the literature

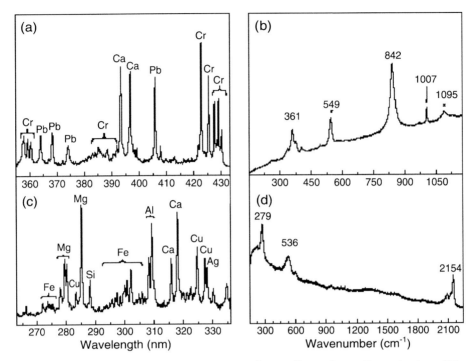

FIGURE 3.18 LIBS (a) and Raman (b) spectra from yellow paint on Byzantine icon (*The Annunciation*, 18th century A.D., Greece). LIBS (c) and Raman (d) spectra from blue paint on French miniature (Figure 3.7a).

[121–127]. In this respect LIF analysis can be complementary to LIBS since it can provide molecular information on materials. In fact the combined use of LIBS and LIF was critical in fully characterizing the white retouching of the Byzantine icon shown in Figure 3.7d [46]. The LIBS spectrum showed that zinc was the main component of the retouching, while the LIF spectrum confirmed the presence of zinc white, based on the characteristic intense narrow-band photoluminescence of ZnO at 383 nm (Figure 3.19).

A particularly interesting example of a practically simultaneous application of LIBS and LIF, which makes use of a single optical setup, is described in [50]. It involves the analysis of a model sample consisting of a cadmium yellow ($CdS\cdot ZnS\cdot BaSO_4$) paint layer applied on a white background made of gypsum ($CaSO_4\cdot 2H_2O$) containing a small amount of zinc white (ZnO), a highly luminescent material. The application exploits the depth profile and elemental analysis capabilities of LIBS in combination with the spectroscopic analytical features of LIF. A standard LIBS setup is used based on laser pulses at 355 nm (third harmonic of the Nd:YAG laser) focused on the sample surface by a 50-mm focal length lens. First, the time-integrated fluorescence/luminescence emission of the sample is collected. It is important to stress that for the fluorescence measurement, the laser energy per pulse employed is much lower than that corresponding to the ablation threshold of

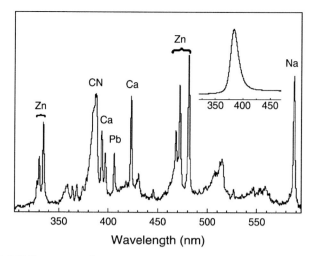

FIGURE 3.19 LIBS spectrum from white overpaint layer analyzed on the Saint Nicholas Byzantine icon (Figure 3.7d). Inset: Fluorescence emission spectrum from the same point on the icon, proving the presence of ZnO.

the material examined. Following the fluorescence experiment, a single pulse of higher energy is delivered on the same spot, and a time-resolved emission spectrum using LIBS is recorded. This completes the first cycle of the measurement, where two spectroscopic tools are used in the analysis of the sample. Additional cycles are repeated, in which each additional LIF measurement probes a newly exposed layer of the sample as a result of the ablation of material that took place in the LIBS measurement done in the previous cycle. In the example described, LIF-LIBS spectra are shown (Figure 3.20) that were collected while performing a depth profiling study of the sample. A transition from the characteristic luminescence emission of CdS at ca. 488 nm from cadmium yellow to that of ZnO at ca. 383 nm from the ground layer (here, the primer) is seen in the LIF spectra. Similarly, atomic emission lines due to Cd, Zn, and Ba are seen in the LIBS spectra corresponding to the upper paint layer, while mainly Ca and Zn are seen in the ground layer of the sample.

3.6.3 LIBS AND MASS SPECTROMETRY

Laser ablation mass analysis (LAMA), a long-established technique, relies on measuring the masses of species ejected from solid surfaces as a result of the laser ablation process. Ionized species or even neutral species (postionized) can be detected and their masses can be related to the material analyzed. Combining analytical LIBS data from the optical emission of the plasma to the mass analysis of the ejected species (LAMA), complementary information about surface composition can be provided. Two modes of analysis are possible. Simultaneous LIBS and LAMA analysis of the ejected plume are carried out for each ablating laser pulse. Alternatively, mass analysis is carried out first, with the laser operating in the desorption regime, followed by optical emission analysis of the plume produced with a subsequent single pulse above the ablation threshold.

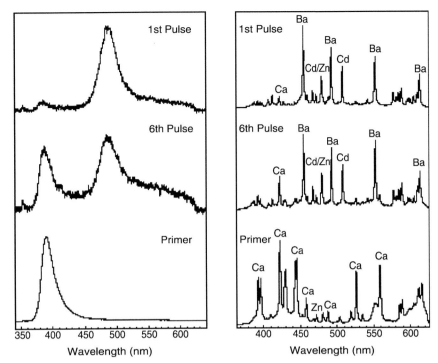

FIGURE 3.20 LIF (left) and LIBS (right) spectra obtained in the depth profile analysis of a model sample composed of a yellow paint layer (cadmium yellow: $CdS \cdot ZnS \cdot BaSO_4$) on top of a white background (gypsum: $CaSO_4 \cdot 2H_2O$ containing a small amount of ZnO). Top spectra: The LIF spectrum obtained is characteristic of CdS luminescence (490 nm), while the LIBS spectrum shows the presence of Ba, Cd, and Zn. Middle spectra: Emission from ZnO in the white background increases at 383 nm (after 10 pulses), while cadmium and barium emission diminish relatively to that of Ca. Bottom spectra: Intense emission from ZnO while very weak CdS emission is observed (after 30 pulses). The LIBS spectrum is dominated by emission due to Ca, while only low intensity features due to Ba or Cd are observed. Characteristic emission due to Zn is also present. For both LIF and LIBS spectra, the excitation wavelength was 355 nm.

The combined use of LIBS with laser ablation mass spectrometry has been known to scientists for some time and was recently demonstrated in the field of artwork analysis with a study on daguerreotypes [128].

Mass spectrometric analysis has identified the tarnish layer (the thin dark corrosion layer appearing on the daguerreotype) as consisting of silver sulfide (Ag_2S) and shown it to be efficiently removed by laser cleaning with the second harmonic (532 nm) of a picosecond Nd:YAG laser (Figure 3.21).

It was also sensitive in detecting mercury used in the development process and gold used in the gilding process of the developed silver plate. LIBS, as already shown, was used in mapping the depth profile of the silver plate and for detecting pigment used in several colored daguerreotypes. The complementarity of the two techniques is obvious upon examination of the spectra [58,128]. Both approaches

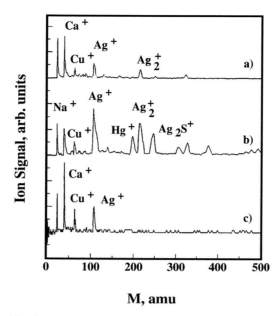

FIGURE 3.21 Positive ion mass spectra taken from a daguerreotype (a) on untarnished area, (b) before laser cleaning (tarnished area), and (c) after laser cleaning. (Reproduced from Hogan et al., *Appl. Spectrosc.*, **53**, 1161, 1999.)

can detect Ag, Cu, and Au quite well. However, mass spectrometry can detect sulphur and some organic contaminants, which LIBS either cannot detect or detects only with difficulty. This is because the expected sulphur lines (lying mainly in the vacuum ultraviolet) are difficult to detect when the analysis is performed in ambient air, while molecular or organic species are easily discriminated by using LIBS.

The use of optical emission and mass spectrometric detection performed as a single unit is very attractive and in principle straightforward. In practical terms, however, the vacuum required for the operation of the mass spectrometer adds complexity, reduces the overall speed of analysis, and limits the type of objects that can be examined. Fortunately, recent developments have enabled laser ablation sampling of solids under atmospheric pressure and introduction of the ejected material through a capillary to the mass spectrometry analyzer [129]. Based on this technology, the development of a dual analysis system (optical emission and mass spectrometry) having no requirement for placing samples or objects in a vacuum chamber appears feasible.

3.7   CONCLUDING REMARKS

As discussed in the previous sections and illustrated with the examples presented, LIBS features several advantages that make its use in archaeological and artwork analysis attractive. First of all, it requires no sample removal from the object. The analysis can be performed *in situ* and requires only optical contact with the object. Material loss in a typical LIBS measurement is minimal, and any damage to the sample surface is practically invisible to the naked eye. Thus LIBS can be considered as a nearly nondestructive technique. The absence of sampling and sample preparation, and the need for only a single laser pulse measurement, complete in less than a second, offer unparalleled speed to the technique. The spatial resolution achieved by LIBS across a surface is nearly microscopic. In addition, the technique can provide depth profiling information if spectra from successive laser pulses delivered at the same point are recorded individually. Finally, the equipment used is compact and can be packaged in a man-portable or transportable unit.

REFERENCES

1. Brech, F. and Cross, L., Optical microemission stimulated by a ruby maser, *Appl. Spectrosc.*, **16**, 59, 1962.
2. Runge, E.R., Minck, R.W., and Bryan, F.R., *Spectrochim. Acta,* **20**, 733, 1964.
3. Runge, E.R., Bonfiglio, S., and Bryan, F.R., *Spectrochim. Acta,* **22**, 1678, 1965.
4. Cremers, D.A. and Radziemski, L., *Handbook of Laser-Induced Breakdown Spectroscopy*, Wiley, New York, 2006.
5. Adrain, R.S. and Watson, J., Laser microspectral analysis: a review of principles and applications, *J. Phys. D, Appl. Phys.*, **17**, 1915, 1984.
6. Majidi, V. and Joseph, M.R., Spectroscopic applications of laser-induced plasmas, *Crit. Rev. Anal. Chem.*, **23**, 143, 1992.
7. Ciliberto, E. and Spoto, G., *Modern Analytical Methods in Art and Archaeology*, Chemical Analysis, Monographs on Analytical Chemistry and Its Applications, Vol. 155, John Wiley, New York, 2000.
8. Ferreti, M., *Scientific Investigations of Works of Art*, ICCROM-International Centre for the Study of Preservation and the Restoration of Cultural Property, Rome, 1993.
9. Pollard, A.M. and Heron, C., *Archaeological Chemistry*, Royal Society of Chemistry, Cambridge, U.K., 1996.
10. Mirti, P., Analytical techniques in art and archaeology, *Ann. Chim.*, **79**, 455, 1989.
11. Spoto, G., Torrisi, A., and Contino, A., Probing archaelogical and artistic soil materials by spatially resolved analytical techniques, *Chem. Soc. Rev.* **29**, 429, 2000.
12. Moens, L., von Bohlen, A., and Vandenabeele P., X-ray fluorescence, In *Modern Analytical Methods in Art and Archaeology* Ciliberto, E. and Spoto, G., Eds., John Wiley, New York, 2000, 55–79.
13. Hanson, V.F., Quantitative analysis of art objects by energy dispersive x-ray fluorescence analysis, *Appl. Spectrosc.*, **27**, 309, 1973.
14. Mantler, M. and Schreiner, M., X-ray fluorescence spectrometry in art and archaeology, *X-Ray Spectrometry*, **29**, 3, 2000.
15. Janssens, K., Vittiglio, G., Deraedt, I., Aerts, A., Vekenmans, B., Vincze, L., Wei, F., Deryck, I., Schalm, O., Adams, F., Rindby, A., Knochel, A., Simionovici, A., and Snigirev, A., Use of microscopic XRF for non-destructive analysis in art and archaeometry, *X-Ray Spectrometry*, **29**, 73, 2000.
16. Karydas, A.G., Kotzamani, D., Bernard, R., Barrandon, J.N., and Zarkadas, C., A compositional study of a museum jewellery collection (7th–1st BC) by means of a portable XRF spectrometer, *Nucl. Instr. Meth. Phys. Res. B*, **226**, 15, 2004.
17. Ferretti, M., Miazzo, L., and Moioli, P., The application of a non-destructive XRF method to identify different alloys in the bronze statue of the Capitoline horse, *Stud. Conserv.*, **42**, 241, 1997.
18. Philippakis, S.E., Perdikatsis, B., and Assimenos, K., X-ray analysis of pigments from Vergina Greece (second tomb), *Stud. Conserv.*, **24**, 54, 1979.

19. Klockenkamper, R., van Bohlen, A., and Moens, L., Analysis of pigments and inks on oil paintings and historical manuscripts using total reflection x-ray fluorescence spectrometry, *X-Ray Spectrometry,* **29**, 119, 2000.

20. Dran, J.-C., Calligaro, T., and Salomon, J., Particle-induced x-ray emission, In *Modern Analytical Methods in Art and Archaeology* Ciliberto, E. and Spoto, G., Eds., John Wiley, New York, 2000, 135–166.

21. Calligaro, T., Dran, J.-C., Salomon, J., and Walter, P., Review of accelerator gadgets for art and archaeology, *Nucl. Instr. Meth. Phys. Res. B,* **226**, 29, 2004.

22. Dran, J.-C., Salomon, J., Calligaro, T., and Walter, P., Ion beam analysis of art works: 14 years of use in the Louvre, *Nucl. Instr. Meth. Phys. Res. B,* **219–220**, 7, 2004.

23. Swann, C.P., Ferrence, S., and Betancourt, P.P., Analysis of minoan white pigments used on pottery from Palaikastro, *Nucl. Instr. Meth. Phys. Res. B,* **161–163**, 714, 2000.

24. Kallithrakas-Kontos, N., Katsanos, A.A., Potiriadis, C., Oeconomidou, M., and Touratsoglou, J., PIXE analysis of ancient Greek copper coins minted in Epirus, Illyria, Macedonia and Thessaly, *Nucl. Instr. Meth. Phys. Res. B,* **109–110**, 662, 1996.

25. Olsson, A.-M.B., Calligaro, T., Colinart, S., Dran, J.-C., Lovestam, N.E.G., Moignard, B., and Salomon, J., Micro-PIXE analysis of an ancient Egyptian papyrus; identification of pigments used for the "Book of the Dead," *Nucl. Instr. Meth. Phys. Res. B,* **181**, 707, 2001.

26. Gratuze, B., Blet-Lemarquand, M., and Barrandon, J.N., Mass spectrometry with laser sampling: a new tool to characterize archaeological materials, *J. Radioanalytical Nucl. Chem.,* **247**, 645, 2001.

27. Young, S.M.M., Budd, P., Haggerty, R., and Pollard, A.M., Inductively coupled plasma-mass spectrometry for the analysis of ancient metals, *Archaeometry,* **39**, 379, 1997.

28. Gratuze, B., Obsidian characterization by laser ablation ICP-MS and its application to prehistoric trade in the Mediterranean and the Near East: sources and distribution of obsidian within the Aegean and Anatolia, *J. Archaeol. Sci.,* **26**, 869, 1999.

29. Garcia Alonso, J.I., Ruiz Enhinar, J., Martinez, J.A., and Criado, A.J., Origin of El Cid's sword revealed by ICP-MS metal analysis, *Spectroscopy Europe,* **11(4)**, 10, 1999.

30. Hughes, M.J., Cowell, M.R., and Craddock, P.T., Atomic absorption techniques in archaeology, *Archaeometry,* **18**, 19, 1976.

31. Rauret, G., Casassas, E., and Baucells, M., Spectrochemical analysis of some medieval glass fragments from Catalan gothic churches, *Archaeometry,* **27**, 195, 1985.

32. Casoli, A. and Mirti, P., The analysis of archaeological glass by inductively coupled plasma optical emission spectroscopy, *Fresenius' J. Anal. Chem.,* **334**, 104, 1992.

33. Mirti, P., Casoli, A., and Appolonia, L., Scientific analysis of Roman glass from Augusta-Praetoria, *Archaeometry,* **35**, 225, 1993.

34. Spoto, G., Secondary ionization mass spectrometry in art and archaeology, *Thermochim. Acta,* **365**, 157, 2000.

35. Noll, W., Holm, R., and Born, L., Painting of ancient ceramics, *Angew. Chem. Int. Ed.,* **14**, 602, 1975.

36. Mirti, P., X-ray microanalysis discloses the secrets of ancient Greek and Roman pottery, *X-Ray Spectrometry,* **29**, 63, 2000.

37. Philippakis, S.E., Perdikatsis, B., and Paradelis, T., An analysis of blue pigments from the Greek bronze age, *Stud. Conserv.,* **21**, 143, 1976.

38. Neff, H., Neutron activation analysis for provenance determination in archaeology, In *Modern Analytical Methods in Art and Archaeology* Ciliberto, E. and Spoto, G., Eds., John Wiley, New York, 2000, 81–134.

39. Kilikoglou, V., Bassiakos, Y., Grimanis, A.P., Souvatzis, K., Pilali-Papasteriou, A., and Papanthimou-Papaefthimiou, A., *J. Archaeol. Sci.*, **23**, 343, 1996.

40. Wagner, G.A., Isotope analysis, dating, and provenance methods, In *Modern Analytical Methods in Art and Archaeology* Ciliberto, E. and Spoto, G., Eds., John Wiley, New York, 2000, 445–464.

41. Gale, N.H. and Stos-Gale, Z., Lead isotope analysis applied to provenance studies, In *Modern Analytical Methods in Art and Archaeology* Ciliberto, E. and Spoto, G., Eds., John Wiley, New York, 2000, 503–584.

42. Anglos, D., Couris, S., and Fotakis, C., Laser diagnostics of painted artworks: laser induced breakdown spectroscopy of pigments, *Appl. Spectrosc.*, **51**, 1025, 1997.

43. Gobernado-Mitre, I., Prieto, A.C., Zafiropulos, V., Spetsidou, Y., and Fotakis, C., On-line monitoring of laser cleaning of limestone by laser induced breakdown spectroscopy, *Appl. Spectrosc.*, **51**, 1125, 1997.

44. Maravelaki, P.V., Zafiropulos, V., Kilikoglou, V., Kalaitzaki, M., and Fotakis, C., Laser induced breakdown spectroscopy as a diagnostic technique for the laser cleaning of marble, *Spectrochim. Acta B*, **52**, 41, 1997.

45. Klein, S., Stratoudaki, T., Zafiropulos, V., Hildenhagen, J., Dickmann, K., and Lehmkuhl, T., Laser-induced breakdown spectroscopy for on-line control of laser cleaning of sandstone and stained glass, *Appl. Phys. A*, **69**, 441, 1999.

46. Burgio, L., Clark, R.J.H., Stratoudaki, T., Doulgeridis, M., and Anglos, D., Pigment identification. A dual analytical approach employing laser induced breakdown spectroscopy (LIBS) and Raman microscopy, *Appl. Spectrosc.*, **54**, 463, 2000.

47. Castillejo, M., Martin, M., Silva, D., Stratoudaki, T., Anglos, D., Burgio, L., and Clark, R.J.H., Analysis of pigments in polychromes by use of laser induced breakdown spectroscopy and Raman microscopy, *J. Molec. Struct.*, **550–551**, 191, 2000.

48. Borgia, I., Burgio, L.M.F., Corsi, M., Fantoni, R., Palleschi, V., Salvetti, A., Squarcialupi, M.C., and Tognoni, E., Self-calibrated quantitative elemental analysis by laser-induced plasma spectroscopy: application to pigment analysis, *J. Cult. Heritage*, **1**, S281, 2000.

49. Corsi, M., Palleschi, V., Salvetti, A., and Tognoni, E., Making LIBS quantitative: a critical review of the current approaches to the problem.and references therein, *Res. Adv. Appl. Spectrosc.*, **1**, 41, 2000.

50. Anglos, D., Laser-induced breakdown spectroscopy in art and archaeology, *Appl. Spectrosc.*, **55**, 186A, 2001.

51. Burgio, L., Melessanaki, K., Doulgeridis, M., Clark, R.J.H., and Anglos, D., Pigment identification in paintings employing laser induced breakdown spectroscopy (LIBS) and Raman microscopy, *Spectrochim. Acta Part B*, **56**, 905, 2001.

52. Castillejo, M., Martin, M., Oujja, M., Silva, D., Torres, R., Domingo, C., Garcia-Ramos, J.V., and Sanchez-Cortes, S., Spectroscopic analysis of pigments and binding media of polychromes by the combination of optical laser-based and vibrational techniques, *Appl. Spectrosc.*, **55**, 992, 2001.

53. Melessanaki, K., Papadakis, V., Balas, C., and Anglos, D., Laser induced breakdown spectroscopy (LIBS) and hyper-spectral imaging analysis of pigments on illuminated manuscripts, *Spectrochim. Acta Part B*, **56**, 2337, 2001.

54. Maravelaki-Kalaitzaki, P., Anglos, D., Kilikoglou, V., and Zafiropulos, V., Compositional characterization of encrustation on marble with laser induced breakdown spectroscopy, *Spectrochim. Acta Part B*, **56**, 887, 2001.

55. Bicchieri, M., Nardone, M., Russo, P.A., Sodo, A., Corsi, M., Cristoforetti, G., Palleschi, V., Salvetti, A., and Tognoni, E., Characterization of azurite and lazurite by laser induced breakdown spectroscopy and Raman microscopy, *Spectrochim. Acta Part B*, **56**, 915, 2001.

56. Corsi, M., Cristoforetti, G., Palleschi, V., Salvetti, A., and Tognoni, E., A fast and accurate method for the determination of precious alloys caratage by laser induced plasma spectroscopy, *Eur. Phys. J. D.,* **13**, 373, 2001.

57. Yoon, Y., Kim, T., Yang, M., Lee, K., and Lee, G., Quantitative analysis of pottery glaze by laser induced breakdown spectroscopy, *Microchem. J.,* **8**, 251, 2001.

58. Anglos, D., Melessanaki, K., Zafiropulos, V., Gresalfi, M.J., and Miller, J.C., Laser-induced breakdown spectroscopy for the analyses of 150-year old daguerreotypes, *Appl. Spectrosc.,* **56**, 423, 2002.

59. Colao, F., Fantoni, R., Lazic, V., and Spizzichino, V., Laser-induced breakdown spectroscopy for semi-quantitative and quantitative analyses of artworks — application on multilayered ceramics and copper based alloys, *Spectrochim. Acta Part B,* **57**, 1219–1234, 2002.

60. Anzano, J.M., Villoria, M.A., Gornushkin, I.B., Smith, B.W., and Winefordner, J.D., Laser-induced plasma spectroscopy for characterization of archaeological material, *Canad. J. Anal. Sci. Spectrosc.,* **47**, 134, 2002.

61. Muller, K. and Stege, H., Evaluation of the analytical potential of laser-induced breakdown spectrometry (LIBS) for the analysis of historical glasses, *Archaeometry,* **45**, 421, 2003.

62. Lazic, V., Fantoni, R., Colao, F., Santagata, A., Morona, A., and Spizzichino, V., Quantitative laser-induced breakdown spectroscopy analysis of ancient marbles and corrections for the variability of plasma parameters and of ablation rate "Q," *J. Anal. At. Spectrom.,* **19**, 429, 2004.

63. Colao, F., Fantoni, R., Lazic, V., Caneve, L., Giardini, A., and Spizzichino, V., LIBS as a diagnostic tool during the laser cleaning of copper based alloys: experimental results, *J. Anal. At. Spectrom.,* **19**, 502, 2004.

64. Corsi, M., Cristoforetti, G., Giuffrida, M., Hidalgo, M., Legnaioli, S., Maotti, L., Palleschi, V., Salvetti, A., Tognoni, E., Vallebona, C., and Zanini, A., Archaeometric analysis of ancient copper artifacts by laser-induced breakdown spectroscopy technique, *Microchim. Acta,* **152**, 105–111, 2005.

65. Moenke-Blackenburg, L., Laser micro analysis, *Prog. Anal. Spectrosc.,* **9**, 335, 1986 and references therein.

66. Roy, A., The laser micro spectral analysis of paint, *National Gallery Technical Bulletin,* **3**, 43, 1979.

67. Rusak, D.A., Castle, B.C., Smith, B.W., and Winefordner, J.D., Fundamentals and applications of laser-induced breakdown spectroscopy, *Crit. Revs. Anal. Chem.,* **27**, 257, 1997.

68. Rusak, D.A., Castle, B.C., Smith, B.W., and Winefordner, J.D., Recent trends and the future of laser-induced plasma spectroscopy, *Trends Anal. Chem.,* **17**, 453, 1998.

69. Song, K., Lee, Y.I., and Sneddon, J., Recent developments in instrumentation for laser induced breakdown spectroscopy, *Appl. Spectrosc. Rev.,* **37**, 89, 2002.

70. Lee, W.B., Yu, J.Y., and Sneddon, J., Recent applications of laser-induced breakdown spectrometry: a review of material approaches, *Appl. Spectrosc. Rev.,* **39**, 27, 2004.

71. Schechter, I., Laser induced plasma spectroscopy. A review of recent advances, *Revs. Anal. Chem.,* **16**, 173, 1997.

72. Niemax, K., Laser ablation — reflections on a very complex technique for solid sampling, *Fresenius' J. Anal. Chem.,* **370**, 332, 2001.

73. Palanco, S. and Laserna, J.J., Design considerations, development and performance of a remote sensing instrument based on open-path atomic emission spectrometry, *Rev. Sci. Inst.,* **75**, 2068, 2004.

74. Lopez-Moreno, C., Palanco, S., and Laserna, J.J., Remote laser-Induced plasma spectroscopy for elemental analysis of samples of environmental interest, *J. Anal. At. Spectrom.,* **19**, 479, 2004.

75. Menut, D., Fichet, P., Lacour, J.-L., Rivoallan, A., and Mauchien, P., Micro-laser-induced breakdown spectroscopy technique: a powerful method for performing quantitative surface mapping on conductive and nonconductive samples, *Appl. Optics,* **42**, 6063, 2003.

76. Kossakovski, D. and Beauchamp, J.L., Topographical and chemical microanalysis of surfaces with a scanning probe microscope and laser-induced breakdown spectroscopy, *Anal. Chem.,* **72**, 4731, 2000.

77. Body, D. and Chadwick, B.L., Optimization of the spectral data processing in a LIBS simultaneous elemental analysis system, *Rev. Sci. Instrum.,* **72**, 1625, 2000.

78. Body, D. and Chadwick, B.L., A simultaneous elemental analysis system using laser induced breakdown spectroscopy, *Spectrochim. Acta Part B,* **56**, 725, 2001.

79. Bauer, H.E., Leis, F., and Niemax, K., Laser induced breakdown spectrometry with an échelle spectrometer and intensified charge-coupled device, *Spectrochim. Acta Part B,* **53**, 1815, 1998.

80. Haisch, C., Panne, U., and Niessner, R., Combination of an intensified charge-coupled device with an échelle spectrograph for analysis of colloidal material by laser-induced plasma spectroscopy, *Spectrochim. Acta Part B,* **53**, 1657, 1998.

81. Uhl, A., Loebe, K., and Kreuchwig, L., Fast analysis of wood preservers using laser induced breakdown spectroscopy, *Spectrochim. Acta Part B,* **56**, 795, 2001.

82. Sabsabi, M., Detalle, V., Harith, M.A., Tawfik, W., and Imam, H., Comparative study of two new commercial échelle spectrometers equipped with intensified CCD for analysis of laser-induced breakdown spectroscopy, *Appl. Optics,* **42**, 6094, 2003.

83. Yamamoto, K.Y., Cremers, D.A., Ferris, M.J., and Foster, L.E., Detection of metals in the environment using a portable laser-induced breakdown spectroscopy instrument, *Appl. Spectrosc.,* **50**, 222, 1996.

84. Castle, B.C., Night, A.K., Visser, K., Smith, B.W., and Winefordner, J.D., Battery powered laser-induced plasma spectrometer for elemental determinations, *J. Anal. At. Spectrom.,* **13**, 589, 1998.

85. Rosenwasser, S., Assimielis, G., Bromley, B., Hazlett, R., Martin, J., Pearce, T., and Zigler, A., Development of a method for automated analysis of ores using LIBS, *Spectrochim. Acta Part B,* **56**, 707, 2001.

86. Gornushkin, I.B., Smith, B.W., Nasajpour, H., and Winefordner, J.D., Identification of solid materials by correlation analysis using a microscopic laser-induced plasma spectrometer, *Anal. Chem.,* **71**, 5157, 1999.

87. Stratoudaki, T., Manousaki, A., Melesanaki, K., Orial, G., and Zafiropulos, V., Discoloration of pigments induced by laser irradiation, *Surface Eng. J.,* **17**, 249, 2001.

88. Pouli, P., Emmony, D.C., Madden, C.E., and Sutherland, I., Analysis of the laser-induced reduction mechanisms of medieval pigments, *Appl. Surf. Sci.,* **173**, 252, 2001.

89. Castillejo, M., Martin, M., Oujja, M., Santamaria, J., Silva, D., Torres, R., Zafiropulos, V., van den Brink, O.F., Heeren, R.M.A., Teule, R., and Silva, A., Evaluation of the chemical and physical changes induced by KrF laser irradiation of tempera paint, *J. Cult. Heritage,* **4**, S83–S91, 2003.

90. Kim, T., Lin, C.T., and Yoon, Y., Compositional mapping by laser-induced breakdown spectroscopy, *J. Phys. Chem. B,* **102**, 4284, 1998.

91. Romero, D. and Laserna, J.J., Surface and tomographic distribution of carbon impurities in photonic grade silicon using laser-induced breakdown spectrometry, *J. Anal. At. Spectrom.,* **13**, 557, 1998.

92. Assion, A., Wollenhaupt, M., Haag, L., Mayorov, F., Sarpe-Tudoran, C., Winter, M., Kutschera, U., and Baumert, T., Femtosecond laser-induced-breakdown spectrometry for Ca^{2+} analysis of biological samples with high spatial resolution, *Appl. Phys. B*, **77**, 391–397, 2003.

93. Anderson, D.R., McLeod, C.W., English, T., and Smith, A.T., Depth-profile studies using laser-induced plasma emission spectrometry, *Appl. Spectrosc.*, **49**, 691, 1995.

94. Vadillo, J.M., Garcia, C.G., Palanco, S., and Laserna, J.J., Nanometric depth-resolved analysis of coated steels using laser-induced breakdown spectrometry with a 308 nm collimated beam, *J. Anal. At. Spectrom.*, **13**, 793, 1998.

95. Mateo, M.P., Cabalin, L.M., and Laserna, J.J., Depth-resolved analysis of multilayered samples by laser induced breakdown spectrometry, *Appl. Opt.*, **42**, 6057, 2003.

96. Margetic, V., Bolshov, M., Stockhaus, A., Niemax, K., and Hergenroder, R., Depth-profile analysis of multilayer samples using femtosecond laser ablation, *J. Anal. At. Spectrom.*, **16**, 616, 2001.

97. Tognoni, E., Palleschi, V., Corsi, M., and Cristoforetti, G., Quantitative microanalysis by laser induced breakdown spectroscopy: a review of experimental approaches, *Spectrochim. Acta Part B*, **57**, 1115, 2002.

98. Wisbrun, R., Schechter, I., Niessner, R., Schroder, H., and Lompa, K., Detector for trace elemental analysis of solid environmental samples by laser plasma spectroscopy, *Anal. Chem.*, **66**, 2964, 1994.

99. Bulatov, V., Krasniker, R., and Schechter, I., Study of matrix effects in laser plasma spectroscopy by combined multifiber spatial and temporal resolutions, *Anal. Chem.*, **70**, 5302, 1998.

100. Ciucci, A., Corsi, M., Palleschi, V., Rastelli, S., Salvetti, A., and Tognoni, E., New procedure for quantitative elemental analysis by laser-induced plasma spectroscopy, *Appl. Spectrosc.*, **53**, 960, 1999.

101. Gettens, R.J. and Stout, G.L., *Painting Materials*, Dover, New York, 1956.

102. Friedstein, H.G., A short history of the chemistry of painting, *J. Chem. Educ.*, **58**, 291, 1981.

103. Roy, A., *Artists' Pigments*, Vol. 2, National Gallery of Art, Washington D.C., 1993.

104. West Fitzhugh, E., *Artists' Pigments*, Vol. 3, National Gallery of Art, Washington D.C., 1997.

105. Barger, M.S. and White, W.B., *The Daguerreotype*, The Johns Hopkins University Press, Baltimore, MD, 1991.

106. Vadillo, J.M. and Laserna, J.J., Laser-induced breakdown spectrometry of silicate, vanadate and sulfide rocks, *Talanta*, **43**, 1149, 1996.

107. Vadillo, J.M., Vadillo, I., Carrasco, F., and Laserna, J.J., Spatial distribution profiles of magnesium and strontium in spelecthems using laser-induced breakdown spectrometry, *Fresenius' J. Anal. Chem.*, **361**, 119, 1998.

108. Sanford, M.K., Repke, D.B., and Earle, A.L., Elemental analysis of human bone from Carthage: a pilot study, in *The Circus and a Byzantine Cemetery at Carthage*, Vol. 1, J.H. Humphrey, Ed., University of Michigan Press, Ann Arbor, 1988, chap. 9.

109. Kyle, J.H., Effect of post-burial contamination on the concentrations of major and minor elements in human bones and teeth-the implications for paleodietary research, *J. Archaeol. Sci.*, **13**, 403, 1984.

110. Klepinger, L.L., Kuhn, J.K., and Williams, W.S., An elemental analysis of archaeological bone from Sicily as a test of predictability of diagenetic change, *Am. J. Phys. Anthropol.*, **70**, 325, 1986.

111. Carvalho, M.L., Casaca, C., Marques, J.P., Pinheiro, T., and Cunha, A.S., Human teeth elemental profiles measured by synchrotron x-ray fluoresce: dietary habits and environmental influence, *X-Ray Spectrom.*, **30**, 190, 2001.

112. Samek, O., Beddows, D.C.S., Telle, H.H., Kaiser, J., Liska, M., Caseres, J.O., and Gonzales Urena, A., Quantitative laser-induced breakdown spectroscopy analysis of calcified tissue samples, *Spectrochim. Acta Part B*, **56**, 865, 2001.

113. Cooper, M., *Laser Cleaning*, Butterworth-Heinemann, Oxford, 1998.

114. Salimbeni, R., Pini, R., and Siano, S., Achievement of optimum laser cleaning in the restoration of artworks: expected improvements by on-line optical diagnostics, *Spectrochim. Acta Part B*, **56**, 877, 2001.

115. Clark, R.J.H., Raman microscopy: application to the identification of pigments on medieval manuscripts, *Chem. Soc. Rev.*, **24**, 187, 1995.

116. Clark, R.J.H., Pigment identification by spectroscopic means: an arts/science interface: medieval manuscripts, pigments and spectroscopy, *C.R. Chimie*, **5**, 7, 2002.

117. Vandenabeele, P., Wheling, B., Moens, L., Edwards, H., De Reu, M., and Van Hooydonk, G., Analysis with micro-Raman spectroscopy of natural organic binding media and varnishes used in art, *Anal. Chim. Acta*, **407**, 261, 2000.

118. Vandenabeele, P., Raman spectroscopy in art and archaeology, *J. Raman Spectrosc.*, **35**, 607, 2003.

119. Bell, I.M., Clark, R.J.H., and Gibbs, P.J., Raman spectroscopic library of natural and synthetic pigments (pre-~1850 A.D.), *Spectrochim. Acta Part A*, **53**, 2159, 1997.

120. Burgio, L. and Clark, R.J.H., Library of FT-Raman spectra of pigments, minerals, pigment media and varnishes, and supplement to existing library of Raman spectra of pigments with visible excitation, *Spectrochim. Acta Part B*, **57**, 1491, 2001.

121. Marquardt, B.J., Stratis, D.N., Cremers, D.A., and Michael Angel, S., Novel probe for laser-induced breakdown spectroscopy and Raman measurements using an imaging optical fiber, *Appl. Spectrosc.*, **52**, 1148, 1998.

122. Giakoumaki, A., Osticioli, I., and Anglos, D., Spectroscopic analysis using a hybrid LIBS-Raman system, *Appl. Phys. A.*, **83**, 537–541, 2006.

123. de la Rie, E.R., Fluorescence of paint and varnish layers (Part 1), *Stud. Conserv.*, **27**, 1, 1982.

124. Miyoshi, T., Ikeya, M., Kinoshita, S., and Kushida, T., Laser-induced fluorescence of oil colors and its applications to the identification of pigments in oil paintings, *Jpn. J. Appl. Phys.*, **21**, 1032, 1982.

125. Miyoshi, T., Fluorescence from varnishes for oil paintings under N_2 excitation, *Jpn. J. Appl. Phys.*, **26**, 780, 1987.

126. Anglos, D., Solomidou, M., Zergioti, I., Zafiropulos, V., Papazoglou, T.G., and Fotakis, C., Laser-induced fluorescence in artwork diagnostics: an application in pigment analysis, *Appl. Spectrosc.*, **50**, 1331, 1996.

127. Borgia, I., Fantoni, R., Flamini, C., Di Palma, T.M., Giardini Guidoni, A., and Mele, A., Luminescence from pigments and resins for oil paintings induced by laser excitation, *Appl. Surf. Sci.*, **127–129**, 95, 1998.

128. Hogan, D.L., Golovlev, V.V., Gresalfi, M.J., Chaney, J.A., Feigerle, C.S., Miller, J.C., Romer, G., and Messier, P., Laser ablation mass spectroscopy of nineteenth century daguerreotypes, *Appl. Spectrosc.*, **53**, 1161, 1999.

129. Stockle, R., Setz, P., Deckert, V., Lippert, T., Wokaun, A., and Zenobi, R., Nanoscale atmospheric pressure laser ablation–mass spectrometry, *Anal. Chem.*, **73**, 1399, 2001.

4 Raman Spectroscopy in Cultural Heritage

ABSTRACT

Raman microscopy has been established as a reliable tool for the noninvasive analysis of a wide spectrum of both inorganic and organic materials in art and archaeological objects presenting unique advantages over other molecular analysis techniques. It has high sensitivity and specificity enabling analysis of a wide variety of materials *in situ*, noninvasively, at relatively short times, and with excellent spatial resolution. This chapter provides a brief introduction to the physical principles of the Raman effect, describes the current state of the art in Raman instrumentation, and presents representative examples from various studies that demonstrate the many applications of Raman spectroscopy in the field of cultural heritage.

4.1 INTRODUCTION

Raman spectroscopy probes vibrational transitions within materials. These transitions represent distinct and well-defined ways (vibrational modes) that atoms oscillate within a molecule or crystal lattice and as such are very specific to chemical bonds, molecular species, and lattice structure [1]. As a result, the Raman spectrum is essentially a fingerprint, which can be used for the identification of the material probed.

The Raman effect was discovered during a series of experiments performed at Calcutta University in 1928 by Sir Chandrasekhara Venkata Raman (1888–1970), who for this discovery was awarded the Nobel Prize in Physics in 1930. The effect relies on the inelastic scattering of light from a molecule. The spectrum of the inelastically scattered radiation, called the Raman spectrum, reveals the molecule's structure and identity based on characteristic spectral bands corresponding to various vibrational modes of the molecule.

Numerous advances in instrumentation, including the use of various types of powerful laser sources for excitation, novel optics arrangements, and sensitive detectors, have enabled Raman spectroscopy to become a reliable tool not only in research but also in routine analysis and process control in industry. In particular the introduction of Raman microscopy [2] by Delhaye and Dhamelincourt in 1975 and the development of portable spectrometers have revolutionized the use of Raman spectroscopy in the field of art diagnostics and archaeometry [3–8]. The paper by Bernard Guineau [9] published in *Studies in Conservation* in 1984 was the first report of the use of Raman microscopy in relation to art materials.

Certain key features make Raman microspectroscopy a very attractive choice for identifying diverse types of materials in objects of art and archaeological findings. Because of the distinct fingerprint nature of the Raman spectrum, the technique is highly specific, enabling one to distinguish among similar materials and even among different crystal phases of the same material such as in the case of minerals. The microscope also offers superb spatial resolution that permits analysis of very small features and, in certain cases, of distinct grains or particulates. This is particularly important when heterogeneous mixtures of materials are examined. Raman spectroscopy is a nondestructive technique and can be performed *in situ* (provided the object can be accommodated by the microscope stage), thus eliminating sampling. Raman microscopes are available commercially; they are quite straightforward to use and offer versatile choices with respect to excitation sources, spectrographs, and CCD (charge-coupled device) detectors that enable sensitive recording of high-resolution spectra in no more than a few seconds. Also, specialized fiber-optic Raman probes are available for analysis of large objects remotely from the spectrometer. Several useful databases are available with Raman spectra of minerals, pigments, and other relevant materials that facilitate the identification of unknown components in objects of cultural heritage.

Despite the advantages outlined above, the Raman microscope remains a rather specialized tool, mainly limited to research laboratories and not yet widely used by the conservation, art history, and archaeology communities. One of the reasons for this is the instrument's cost, which is often prohibitive for a small organization or private conservation laboratory. From the actual experimental perspective, a serious challenge is interference from broadband fluorescence emission, which often dominates the relatively weak Raman scattering signal, for example, in the case of fluorescing organic substances. However, it is noted that with continuous improvements in instrumentation and systematic studies of a wide range of test cases, Raman microscopy is constantly gaining appreciation among users as a reliable diagnostic tool in art conservation and archaeology.

At this point, other molecular or elemental analysis techniques, popular in cultural heritage diagnostics and often used in place of or as complementary to Raman spectroscopy, will be briefly discussed, to allow the reader a broader perspective on alternative options. Relevant references are listed in Table 4.1. It is recognized that a more thorough understanding of each technique's capabilities and advantages or disadvantages is required before a decision is made for the technique of choice to address a specific analytical problem.

Absorption (in diffuse reflectance mode) and fluorescence spectroscopy or microspectroscopy share some common features with Raman spectroscopy from the instrumentation standpoint, although given the considerably stronger signals involved, they are less demanding in detection sensitivity. From the analytical perspective the broad spectral features associated with the electronic absorption or emission transitions of molecular systems limit considerably the specificity of these techniques compared to the Raman fingerprinting capabilities. On the other hand, by employing multispectral imaging approaches, both diffuse reflectance and fluorescence spectroscopy provide important surface information and are capable of mapping the distribution of materials across the surface of an object on the basis of the

TABLE 4.1
Molecular Analysis Techniques in Cultural Heritage [10,11,12]

Analytical method	Applications	Refs.
UV-visible absorbance/reflectance spectroscopy	Analysis of inorganic pigments	[4,12–17]
Fluorescence emission spectroscopy	Pigment analysis	[18–27]
Infrared spectroscopy (FTIR)	Paint analysis (pigments, binders)	[28–35]
Raman spectroscopy/microscopy	Inorganic and organic pigments, binder and varnish analysis	[3–5, 9, 36–39]
X-ray diffraction (XRD)	Pigment analysis	[40–44]
Gas chromatography (GC), gas chromatography–mass spectrometry (GC-MS), pyrolysis GC-MS	Analysis of organic components such as binders, varnishes, etc.	[45–55]
Mass spectrometry (MALDI-TOF, DTMS, SIMS)	Pigments, minerals, organic components such as binders, varnishes, etc.	[34, 56–59]
High-performance liquid chromatography (HPLC)	Analysis of organic components such as binders, varnishes, etc.	[60, 61]
Nuclear magnetic resonance (NMR) spectrometry	Analysis of organic binding media and varnishes	[62]

difference in color or fluorescence emission, respectively. Also, in the case of fluorescence, the use of laser sources for excitation (laser-induced fluorescence, LIF) coupled to optical fiber technology has enabled the analysis of various pigments or binding media *in situ*. Finally, with the use of pulsed lasers, fluorescence-LIDAR (laser-induced detection and ranging) has been employed for remote mapping of large monument structures, while FLIM (fluorescence lifetime imaging) has been shown to provide a complementary tool to fluorescence spectral imaging for analysis and mapping.

Fourier transform infrared (FTIR) microspectrometry probes vibrational transitions (through their corresponding absorptions) in both organic and inorganic materials and often provides similar information to Raman microscopy, even though the infrared absorption bands tend to be broader than those recorded in the Raman spectra. The relatively lower spatial resolution of micro-FTIR spectroscopy (it is around 10 to 20 μm compared to Raman microscopy, which reaches down to 1 to 2 μm) is rather restrictive for the identification of paint stratigraphies or for probing single pigment grains. *In situ* applications of micro-FTIR (in reflectance mode) are quite limited, and reliable results are obtained mainly with analysis of small samples. Much improved performance (primarily because of the high irradiance available) and very encouraging results are obtained with synchrotron radiation FTIR.

Chromatographic methods such as gas chromatography (GC), pyrolysis gas chromatography (Py/GC), and high-performance liquid chromatography (HPLC) are very popular for the analysis of organic materials such as oils, protein media, varnishes, and even organic residues in archaeological findings. Sampling is necessary, but the quantities required are very low (on the order of 1 μg). Before GC or HPLC analysis, the sample, particularly in the case of macromolecular materials, is normally subject to some chemical treatment (usually hydrolysis and derivatization), which is important for obtaining reliable analytical results. In the case of Py/GC, the sample is subject to controlled pyrolytic cleavage before injection into the gas chromatograph. Coupled to mass spectrometric detectors, chromatographic techniques are among the best for detailed organic analysis, offering high sensitivity and specificity and quantitative compositional information.

Other mass spectrometric techniques, such as MALDI-TOF-MS (matrix assisted laser desorption/ionization, time-of-flight mass spectrometry) and DT-MS (direct temperature mass spectrometry), have been employed in the examination of organic substances, with no prior separation, but the applications are rather limited to research studies or to cases in which fast characterization (monitoring) of materials is required. Finally, secondary ionization mass spectrometry (SIMS), offering submicron spatial resolution and being sensitive to both inorganic and organic materials, has been shown recently to be very promising for the analysis of painting cross sections.

Recently, the use of NMR spectrometry was shown to provide an alternative method of organic analysis of art materials, but it is still far from being routine; the instrumentation cost and size are the main drawbacks. An additional obstacle with conventional NMR analysis has been the rather high sample quantity required. Nowadays, thanks to advances in capillary and cryoprobe technology, much smaller sample volumes are required (corresponding to a few μg of analyte). In general, a small quantity of solid is sampled from the object, and the analysis is performed nondestructively, in solution, allowing further use of the material. Different types of organic substances can be identified while aging processes can be probed.

For crystalline substances, x-ray diffractometry (XRD) is a powerful technique that enables not only determination of the molecular structure but also discrimination among different crystal phases, such as for example in minerals. Sampling of a small quantity of powder is normally required, although *in situ* analysis has become possible with modern instruments or at synchrotron radiation facilities. The latter offer high-brilliance x-rays that also permit routine x-ray microbeam diffraction measurements with spatial resolution down to 1 μm (focusing for example on single pigment grains or even on single biopolymer fibers such as cellulose). Quite promising for field applications that require portability are recent developments with compact XRD instruments (intended for planetary explorations).

Elemental analysis techniques such as, for example, x-ray fluorescence (XRF), particle induced x-ray emission (PIXE), scanning electron microscopy energy dispersive x-ray spectrometry (SEM-EDS), and laser-induced breakdown spectroscopy (LIBS) (see Chapter 3 and Table 3.1), which provide qualitative or semiquantitative information on the composition of materials in certain cases, lead to indirect but quite reliable identification of inorganic materials, including metals and metal alloys not detectable through Raman spectroscopy. Also, in certain cases, such as in the analysis and classification of complex matrices (glasses, metals, ceramics), the quantitative elemental composition is preferred over the identification of a multitude of different minerals for the classification of objects or findings.

In the following sections, the physical principles of Raman spectroscopy and basic elements of instrumentation are provided. Representative examples from various studies that demonstrate the wide applications of Raman spectroscopy are also presented.

Several excellent reviews are available in books and scientific journals that provide many more examples of the use of Raman analysis in the field of cultural heritage [3–8].

4.2 PHYSICAL PRINCIPLES

As briefly mentioned, the Raman effect arises from inelastic scattering of light following its interaction with a molecule. When electromagnetic radiation, in the form of light, interacts with a material, several processes are possible, such as absorption, transmission, reflection, and scattering. Absorption occurs when the incident photons are in resonance with transitions between bound states (electronic, vibrational, rotational). Electronic transitions fall in the ultraviolet, visible, and near-infrared region of the electromagnetic spectrum, while vibrational and rotational transitions lie in the infrared and far-infrared, respectively. Transmission of light through clear media indicates weak or no absorption. Reflection of light takes place at interfaces separating media with different indices of refraction. Scattering of light can be elastic or inelastic. The former is termed Rayleigh scattering, in which the energy of scattered radiation is of the same frequency (wavelength) as the incident radiation. The latter is termed Raman scattering, and is distinguished further into Stokes and anti-Stokes Raman when the scattered frequencies are lower or higher, respectively, than that of the incident radiation.

It is important to note that the photon energy loss or gain (in Stokes and anti-Stokes scattering, respectively) relates to well-defined quanta of energy corresponding to vibrational transitions in the molecule (Figure 4.1). According to the scheme in Figure 4.2, incident monochromatic radiation (typically from a laser source) at frequency v_0 (and energy $E = hv_0$) excites the molecule from the ground state into a nonstationary (virtual) excited state. From this state, the molecule relaxes back to the ground state, releasing a photon of equal energy in the case of Rayleigh scattering. In the case of Stokes Raman scattering, the molecule relaxes to an excited vibrational level of the ground electronic state, releasing a photon whose energy (frequency: $v_S = v_0 - v_v$) is lower by the corresponding vibrational quantum. Similarly in the case of anti-Stokes Raman scattering, the molecule initiates a transition from an excited vibrational level of the ground electronic state to a virtual state and relaxes to the ground level, releasing a photon of energy increased by one vibrational quantum (frequency: $v_{AS} = v_0 + v_v$).

Spectrally resolving the scattered light produces the Raman spectrum. This is the plot of the intensity of the scattered light as a function of the frequency difference between the incident and scattered radiation (Figure 4.3). The frequency difference Δv is called the Raman shift, and within the spectroscopy community it is commonly expressed in units of wavenumber (cm^{-1}, inverse wavelength).

The Raman bands correspond to certain vibrational modes that constitute a characteristic fingerprint of the material examined and as such can be used for identification and discrimination. The three modes of vibration for water, a bent triatomic molecule, are shown in Figure 4.4.

It is known that vibrational transitions are probed directly through their absorption of radiation in the infrared (IR spectroscopy). However, it is stressed that because of the fundamentally different nature of the absorption and scattering phenomena,

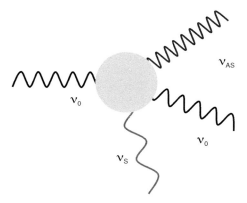

FIGURE 4.1 Schematic diagram indicating the different scattering waves, Rayleigh, Stokes, and anti-Stokes Raman, arising from a molecule irradiated by light at frequency v_0.

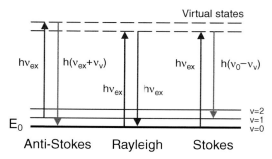

FIGURE 4.2 Energy level diagram indicating the different processes involved in the Rayleigh, Stokes, and anti-Stokes Raman scattering following irradiation of a molecule by light at frequency v_0.

not all vibrational modes are observed in the Raman or the IR spectra. Raman active modes are those that change the polarizability of the molecule while IR active modes are those that change the dipole moment. In Raman spectroscopy, even though vibrational transitions corresponding to infrared wavelengths are probed, the radiation involved (excitation and scattered light) is usually in the visible or near infrared part of the spectrum. This has an important impact on the instrumentation because optics and detectors in the visible are considerably more sensitive than in the IR.

In simple terms, the intensity of Raman scattered radiation I_{ij} corresponding to a vibrational transition between levels i and j is described by the equation

$$I_{ij} = K v_S^4 p_{ij}^2 N_i I_0$$

where K represents instrumental factors including illumination and collection geometry and detector sensitivity, v_S is the frequency of the scattered photon, p_{ij} is the transition probability, N_i is the concentration of scattering molecules, and I_0 is the intensity of the incident radiation.

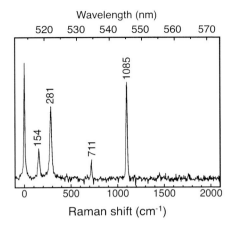

FIGURE 4.3 Raman spectrum of calcite with lattice vibrational bands at 154 cm^{-1} and 281 cm^{-1} and in-plane bending and symmetric stretch modes of the carbonate at 711 cm^{-1} and 1085 cm^{-1} respectively.

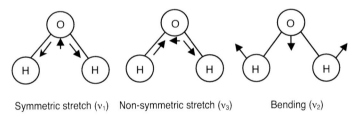

Symmetric stretch (ν_1)   Non-symmetric stretch (ν_3)   Bending (ν_2)

FIGURE 4.4 The vibrational modes of water, H_2O, a bent (not linear) triatomic molecule. As they vibrate, the atoms move about an equilibrium position maintaining at all times the center of mass of the molecule fixed in space.

Because of the virtual states involved, the transition probabilities p_{ij} associated with the Raman processes are very low and correspondingly the scattering signal produced is several orders of magnitude lower than that of fluorescence emission, which involves transition between bound states. Obviously, fluorescent impurities that are excited in a Raman measurement could result in a number of emitted photons comparable to or even much larger than the number of Raman scattered photons. This is a major problem in Raman spectroscopy, and several approaches have been proposed to minimize fluorescence effects. The most common solution is to employ excitation wavelengths in the red or near infrared part of the spectrum, where fluorophores exhibit negligible absorption. The drawback in this case is the dependence of the Raman intensity on ν_S^4, which leads to considerable reduction of the Raman scattered signal when using near-infrared excitation. In some cases, excitation in the UV has been proposed so that the red-shifted Raman scattering is at shorter wavelengths than those of the emitted fluorescence.

4.3 INSTRUMENTATION

A Raman measurement is conceptually simple and involves irradiation of the sample (solid, liquid, or gas) by the excitation source, collection of the scattered light into a spectrograph, and recording of its intensity as a function of the wavelength (frequency). This is what Raman did in his experiments at Calcutta University in 1928, which led to the discovery of the Raman effect. He made use of nearly monochromatic light from a mercury vapor lamp to irradiate various liquids and solids. Using a spectrograph and a photographic plate, he recorded spectra of the scattered radiation and discovered that besides light at frequency equal to that of excitation (Rayleigh scattering), a much weaker but consistent emission was present at lower frequencies as well. The displacement of these frequencies with respect to the frequency of excitation was characteristic for each liquid and independent of the excitation frequency. This observation demonstrated that the spectra of the scattered light carried information about the molecular structure of the liquids, opening a vast field of opportunities in the study of the structure of matter.

Since its discovery and until the 1970s, the use of Raman spectroscopy was restricted to research laboratories. Its application to practical analytical problems was prohibitive, mainly because of the inherently weak signal of the Raman effect, which required very long recording times (several hours) for collecting spectra. The conventional excitation sources employed (commonly mercury discharge lamps with filters isolating selected emission lines) did not provide adequate excitation power. Also, in order to eliminate the strong Rayleigh scattering that overwhelms the weak Raman signal, double or even triple monochromators were used that resulted in considerable signal loss.

Dramatic improvements took place with the introduction of lasers as excitation sources. High intensity and monochromaticity in combination with ease of beam focusing led to a vast increase in the number of photons delivered to the sample. Then the introduction of sensitive array detectors eliminated the need of scanning, while interferometric edge or notch filters enabled efficient rejection of the Rayleigh line while allowing high throughput of the Raman scattering corresponding to frequency shifts as low as 100 cm^{-1} and consequently permitting spectra collection with the use of a compact single spectrograph (in place of the large double or triple monochromators). Another big step forward has been the combination of Raman spectroscopy with optical microscopy. Raman microscopes offer superb spatial resolution (down to about 1 to 2 µm) and considerably improved signal collection (through the microscope objective). Several brands of Raman microscopes are commercially available and find increasing use not only in research laboratories but also in chemical analysis and industrial facilities for routine analysis of samples and materials. The components of a typical Raman microscope are shown schematically in Figure 4.5, while relevant technical information is outlined below.

Common laser sources used in Raman spectroscopy include continuous wave (cw) ion lasers such as argon ion, krypton ion, or helium-neon, which offer several convenient excitation lines in the visible, and diode lasers operating in the near infrared (Table 4.2). If excitation further into the NIR is required, the Nd:YAG laser is

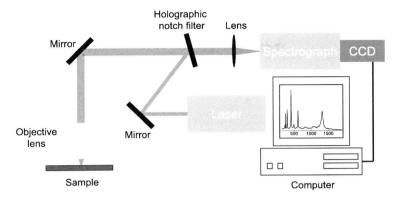

FIGURE 4.5 (See color insert following page 144.) Schematic diagram of a Raman microspectrometer.

TABLE 4.2
Laser Sources Used in Raman Spectroscopy

Laser	Wavelength (λ_{exc}) (nm)
Helium–cadmium	441.6, 325
Argon ion (Ar[+])	457.9, 476.5, 488.0, 496.5, 514.5, 229[a], 244[a], 257[a]
Krypton ion (Kr[+])	530.9, 568.2, 647.1, 676.4
Helium–neon (He–Ne)	632.8
Diode (GaIn$_{0.5}$P$_{0.5}$)	670
Diode (GaAlAs)	750, 785, 830, 850
Neodymium:YAG[b]	1064 (532, 355, 266)[c]
Dye lasers	Tunable (400 to 800)

[a] 2nd harmonics of the 457.9, 488.0, 541.5 nm lines.
[b] Neodymium:yttrium aluminum garnet (Nd:YAG).
[c] 2nd, 3rd, and 4th harmonics, respectively, produced by using special nonlinear crystals.

---

employed. In certain cases, excitation in the ultraviolet is also used. All these lasers offer output powers in the range of a few to hundreds of milliwatts (mW).

The beam is typically directed to the sample by the use of several laser mirrors and focused onto the sample with the microscope objective. Good quality gaussian beams typically produced by cw lasers enable focusing down to about 1 or 2 μm [see also Equation (3.2) in Chapter 3]. In the case of sensitive surfaces such as those of works of art, care is taken to adjust the beam power at levels that sample irradiation does not lead to thermal decomposition or photodegradation of the material. Typical excitation power in the visible is in the range of 0.1 to 5 mW.

The back-scattered light is collected by the objective lens and directed to the entrance slit of the spectrograph. A special laser-line filter (edge or notch filter) is used that minimizes transmission of the intense Rayleigh line into the spectrograph. In several arrangements the notch filter plays a dual role, acting also as a mirror for the incoming laser beam. The scattered radiation is then introduced into a diffraction grating spectrograph and recorded on a charge-coupled device (CCD) array detector.

High spectral resolution, 1 cm⁻¹, can be achieved by use of the proper grating. Typical data collection times are 1 to 10 seconds. In cases of poorly scattering materials, integration up to a minute may be necessary, provided the surface probed can withstand prolonged irradiation.

The use of excitation at 1064 nm produces Raman scattered light at longer wavelengths and is marginally compatible with CCD silicon detectors, which have relatively poor spectral response beyond 900 nm. For these reasons, a Fourier transform Raman spectrometer is used, based on a Michelson interferometer, for the analysis of the scattered radiation into its spectral components (similar to the one used in FT-IR spectrometers) [1].

An important challenge from the instrumentation standpoint is portability. Raman microscopes offer some versatility and can be transported to a museum or a collection for carrying out analyses on a large set of objects that cannot be moved off location. In certain cases, however, for example, the examination of wall paintings in a church or the analysis of large objects, portability and instrument adaptability in the local conditions are important. An answer to this challenge is the use of compact instrumentation coupled to fiber-optic technology for delivering the excitation light to a remote location and for transmitting the signal back to the spectrometer. A probe head that enables accurate positioning of the objective lens in front of the sample is used in this case. An example of such an approach is a mobile Raman instrument (MartA: mobile art analyzer) developed by P. Vandenabeele, E. Grant, and coworkers [63] for the nondestructive, *in situ* investigation of works of art. The instrument has been tested successfully in several campaigns in museums, conservation laboratories, and remote monuments, which represent environmental and measurement conditions far from those achieved in a research laboratory. Several compact, portable Raman instruments are now available commercially, opening up new opportunities for the *in situ* examination of works of art using Raman spectroscopy. An example of the use of such a commercial instrument, a fiber-optic portable Raman spectrometer, in the analysis of pigments in a Venetian icon is shown in Figure 4.6. The flexible fiber-optic probe enables one easily to approach different points on the object's surface to collect spectra.

An alternative approach to remote analysis involves the stand-off monitoring of surfaces and makes use of a telescope both for delivering the excitation beam to the object and for collecting the Raman scattered signal. In this case, the spatial resolution is compromised, because tight focusing is obviously not possible with the telescope optics employed. On the other hand, the probing of large surfaces is possible, and useful information can still be collected. This approach has been employed up to now for analysis of geological materials in relation to planetary mission applications for ground survey [64].

The field/remote analysis includes also underwater objects. In fact, research work is currently underway exploring the technological and analytical challenges of applying Raman microspectrometry for surveying and analyzing cultural heritage objects in subaquatic/submarine environments [65]. Laboratory simulation measurements have been performed in water by immersing a low-power microscope objective that delivers the excitation light (514.5 nm, 100 mW at the output of the objective) and

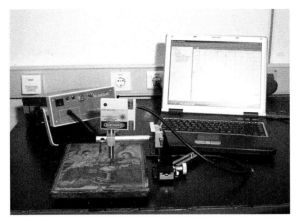

FIGURE 4.6 A fiber-optic Raman spectrometer (R3000, Ocean Optics) used to analyze pigments on a painted icon.

collects the Raman scattered radiation. It is noted that meaningful spectra were collected not only in distilled water but also in sodium chloride and sugar saturated water as well as in red wine and in a mixture of the three (simulating opaque water with dissolved or suspended biogenic substances). These results show the versatility and demonstrate the potential of Raman microspectrometry for *in situ,* mobile, subaquatic archaeometry applications.

Interpretation of Raman spectra, particularly in the case of heterogeneous mixtures or unknown materials, requires special attention and experience. The characteristic vibrational frequencies recorded are the basis for assigning the Raman features to different materials. A great help in this task is the availability of reference spectra of pure materials for cross-reference. Thanks to systematic research by several groups and scientific communities, various databases have been put together (some of them accessible through the internet) that provide reference spectra of materials relevant to art and archaeology. These spectral libraries cover pigments, minerals, and organic materials and are valuable tools for anyone who works in the field of scientific analysis of works of art (Table 4.3). Furthermore, as more and more data are collected from a broad variety of materials, which represent different types of objects and analytical questions, additional Raman spectra libraries are expected to become available.

4.4 EXAMPLES OF RAMAN ANALYSIS IN ART AND ARCHAEOLOGY

A great number of examples involving Raman analysis of a wide variety of art objects can be found in the literature, the majority of papers focusing on pigment analysis. Representative cases are described in this section to illustrate how Raman spectroscopy is being used to address different analytical questions encountered in art history, conservation, and archaeological analysis.

TABLE 4.3
Databases of Raman Spectra

Material	Reference
Pigments, minerals	I.M. Bell, J.H. Clark, P.J. Gibbs, Raman spectroscopic library of natural and synthetic pigments (pre-~1850 A.D.), *Spectrochim. Acta,* A **53**, 2159–2179, 1997. (http://www.chem.ucl.ac.uk/resources/raman/index.html)
Pigments, minerals, painting media, and varnishes	L. Burgio and R.J.H. Clark, Library of FT-Raman spectra of pigments, minerals, pigment media and varnishes, and supplement to existing library of Raman spectra of pigments with visible excitation, *Spectrochim. Acta,* A **53**, 1491–1521, 2001.
Natural organic binding media and varnishes	P. Vandenabeele, B. Wehling, L. Moens, H. Edwards, M. DeReu, and G. Van Hoydonk, Analysis with micro-Raman spectroscopy of natural organic binding media and varnishes used in art, *Anal. Chim. Acta,* **407**, 261–274, 2000.
Modern azo pigments	P. Vandenabeele, L. Moens, H.G.M. Edwards, and R. Dams, Raman spectroscopic database of azo pigments and application to modern art studies, *J. Raman Spectrosc.,* **31**, 509–517, 2000.
Colored glazes	P. Colomban, G. Sagon, and X. Faurel, Differentiation of antique ceramics from the Raman spectra,of their colored glazes and paintings, *J. Raman Spectrosc.,* **32**, 351–360, 2001.
Minerals, metal corrosion products	M. Bouchard and D.C. Smith, Catalogue of 45 Raman spectra of minerals concerning research in art history or archaeology, especially on corroded metals and coloured glass, *Spectrochim. Acta,* A **59**, 2247–2266, 2003.
Minerals	California Institute of Technology, Division of Geological and Planetary Sciences (USA) (http://minerals.gps.caltech.edu/files/raman and links therein)
Pigments	University of Florence, Department of Chemistry (Italy) (http://www.chim.unifi.it/raman/lista_pigmen.html)
Artists' and related materials	Infrared and Raman Users Group (IRUG) Spectral Database© (http://www.irug.org)

4.4.1 PIGMENTS

Raman analysis has been proven quite successful in determining the identity of pigments on many types of painted artifacts, enabling historical and/or artistic characterization of the objects, including dating and authentication, on the basis of the pigments present, and also guiding conservation treatments. Some information about pigments and paints and their chemical and physical structure was already provided in Section 3.5.1 and Table 3.3 (in Chapter 3).

Typical Raman spectra from several pigments are shown in Figure 4.7. With only a few exceptions, both inorganic and organic pigments yield Raman spectra, which essentially carry the fingerprint information necessary to identify these materials. As already mentioned, the identification of a pigment is done, in practice, by comparison with a proper set of spectra from relevant reference samples or with

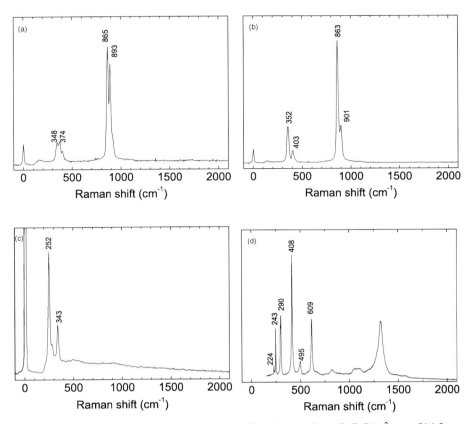

Figure 4.7 Raman spectra of pure pigments. (a) Strontium yellow ($SrCrO_4$), λexc : 514.5 nm (b) Barium yellow ($BaCrO_4$), λexc : 514 nm (c) Vermillion (HgS) λexc : 780 nm and (d) Heamatite (Fe_2O_3) λexc : 780 nm.

the help of spectral libraries (Table 4.3). Analysis of an artifact can be carried out *in situ*, provided the instrument can accommodate this artifact size and shape. In this case, the method is entirely noninvasive. Representative spectra from the *in situ* Raman microscopic analysis of a miniature (Figure 3.7a) are shown in Figure 4.8. The red paint on the woman's lips is found to be a mixture of lead red ($2PbO \cdot PbO_2$) and massicot (PbO). The black paint on the hair is a mixture of Prussian blue and carbon black as verified by the spectra of the pure pigments collected during the analysis [66]. An important advantage when paints are examined under the Raman microscope is that the tight focusing of the laser beam quite often enables probing of a single pigment grain or a small area of uniform pigment composition permitting accurate determination of the pigment analyzed free of spectral interferences from other components present.

Furthermore, in the case of objects that cannot be directly placed on the microscope stage, special fiber-optic probes can be used that facilitate analysis; but the high spatial resolution of the microscope is in some cases compromised because

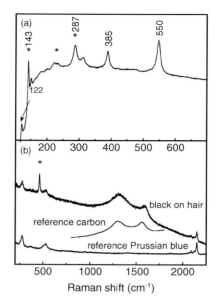

FIGURE 4.8 Raman spectra of (a) red paint (red lead and massicot, marked with asterisks) and (b) black paint (carbon black; quartz band is indicated by asterisk) from 19th century French miniature (shown in Figure 3.7). λexc : 780 nm.

of difficulties experienced with stabilizing the optical probe (particularly if that has to be on scaffold) or the object surface (as for example in large canvas paintings). Several recent reports describe cases of *in situ* Raman investigations, which demanded that a remote analysis approach be followed. For example a fiber-optic probe connected to an FT-Raman spectrometer (λ_{exc} = 1064 nm) was used in the *in situ* analysis of a series of modern paintings, enabling identification of several pigments even under varnish layers as thick as 0.5 mm [67]. A similar probe connected to a dispersive Raman spectrometer was likewise used in the analysis of pigments in several illuminated manuscripts dated from the 16th century A.D. [68]. In other cases, which include the investigation of large or fragile objects, sampling is preferred because it saves the object and/or the instrument from unnecessary manipulations and transportation. Minute samples taken from the object are embedded in a resin and polished and subsequently analyzed. When an appropriate cross section has been sampled, detailed mapping of the paint stratigraphy is feasible.

Interference in the Raman analysis of pigments arises mainly from the coexistence of fluorescence components in the paint (even at the impurity level) that can lead to strong fluorescence emission that masks the Raman scattering signal. Varnishes and organic painting media are common sources of fluorescence. As already mentioned, moving the excitation to longer wavelengths is one way of minimizing the interference from fluorescence.

A particularly interesting application of Raman microscopy is in the examination of illuminated manuscripts. Elaborate drawing and lettering with very fine features created using a host of different pigments and inks can be studied in detail under the Raman microscope. Spectral data are used to characterize the artist's palette to identify possible forgeries and guide conservation. The rather high concentration of pigment in the paint and the absence of fluorescing binding media and varnishes in illuminated manuscripts provide quite favorable conditions for using Raman microscopy. Extensive work by the group of R.J.H. Clark at University College London (U.K.) describes in-depth studies on the identification of pigments in a large variety of books and manuscripts [5,69,70]. In addition, besides the analysis of pigments, studies of degradation products have in certain cases been catalytic in revealing the exposure history of the manuscript but also in helping our understanding of degradation mechanisms and causes. Illuminated manuscripts often show the selective blackening of areas painted with lead white, a widely used white pigment. An example of this pathology is found in a 13th century Byzantine/Syriac gospel lectionary in the British Library [71]. Analysis of the black product has shown it to be lead(II) sulfide (PbS), and the transformation was attributed to a reaction with hydrogen sulfide (H_2S) originating from the polluted atmosphere in 19th-century London. These findings triggered further systematic laboratory studies, which showed that upon exposure to a controlled H_2S atmosphere various lead- and copper-containing pigments react forming black products, confirmed through their Raman spectra to be PbS and CuS, respectively. However, on real paintings, only lead white is observed to blacken, with copper pigments remaining intact. This suggests an alternative and more complicated scenario involving the activity of sulphur metabolizing bacteria capable of converting lead white to PbS. The fact that copper pigments are not suffering from the same pathology is attributed to the known antimicrobial activity of copper ions [72].

Organic pigments and dyes, even though not as widely used as the mineral ones, offer important insight into the making of artifacts and the technology of pigment extraction (from plants or animals) and preparation. Nowadays an increasing number of synthetic organic pigments are available and popular among artists. Raman spectroscopy, being a molecular analysis technique, has obvious advantages over x-ray based or atomic spectrometry methods commonly used in the analysis of inorganic pigments. In this context a database with Raman spectra from 21 azo pigments has been published along with a protocol proposed for discrimination among the pigments [73]. It is noted that the similarities in molecular structure within the different families of synthetic pigments is reflected by the Raman spectra, which share many common features and as such are not straightforward to discriminate; thus the unambiguous identification of a pigment in a modern painting is a challenging task. Spectra correlation or chemometric techniques have been proposed to resolve such problems. For example, in a recent Raman study, indigo samples from three different sources were examined. The different pigments included the synthetic pigment and pigments from the woad (*Isatis tinctoria*) and the indigo plant (*Indigofera tinctoria*). It was shown that Raman spectroscopy in combination with suitable chemometric methods can discriminate between synthetic and natural indigo samples [74].

Pigment analysis is also important in the context of archaeological science as it often provides information on production and painting technology that indirectly reflect cultural and economic aspects of ancient communities. For example, in a recent study, Raman spectroscopy allowed the identification of 6,6'-dibromoindigotin, the major coloring compound of conchylian purple in selected paint details of the ample wall compositions found in the town of Akrotiri of Thera in Greece, dated to the apogee of the settlement in the early Late Bronze Age on the eve of its longstanding interment under the tephra of the eruption around 1650 (±50) B.C. [75]. The spectral identification of Tyrian purple present in the examined samples definitely verified previous analytical indications (by XRF and XRD analysis) for the murex provenance of the pigment. The corresponding Raman spectra also indicated vibrational bands attributed to two allotropic forms of calcium carbonate: calcite and aragonite, the latter being typically found in crystals developed in sea environment. The identification of the conchylian purple dyestuff in the form of a pigment, used on the wall paintings at Akrotiri of Thera, is shown to be the earliest tangible evidence for the development of the vat dyeing technology in the Aegean ever since the early periods of the Late Bronze Age. These valuable findings of the use of conchylian purple pigment at Akrotiri reverse to a certain extent the archaeological setback to the recovery of purple dyeing evidence (up to now only indirectly proven) and help to explore the details of the dyeing procedure used in the Late Bronze Age. Additional evidence for the use of indigo in ancient Greece comes from another Raman study performed on several painted plaster fragments from different sites in mainland Greece that date to the period of 1450 to 1200 B.C. [76]. These findings suggest the need for further investigation of the presence of indigoid or other organic pigments in the Aegean and of the provenance and technology of paint preparation and application to wall paintings and plaster.

In a different study, ancient pottery from China was analyzed *in situ* by Raman microscopy [77]. Shards dating back to the period of 3900 to 3100 B.C. (from the archaeological site of Xishan at Henan, China), were found to contain anatase (TiO_2) as a white background layer, a very surprising finding, since the use of anatase as a pigment (titanium white) has only become popular after its production by chemical synthesis around 1920. The presence of anatase and not rutile (a more stable crystal structure obtained at high temperatures) suggests that firing of the pottery took place at a temperature below 800°C. Likewise, the coexistence of heamatite (Fe_2O_3) and magnetite ($Fe_2O_3 \cdot FeO$) in varying proportions in several of the shards examined suggests a precise control on either the raw materials used or the firing temperature and atmosphere employed. Hematite converts to magnetite at temperatures in the range of 950 to 1250°C. The conclusion drawn through the Raman analysis of pigments is that the knowledge of and the control over the conditions required to achieve the desired color existed 5000 years ago.

In an interesting work, *in situ* Raman analysis was carried out on fragments of Egyptian faience dating from the XVIIIth dynasty (ca. 1570 to 1293 B.C.) excavated at El Amarna in Middle Egypt [78]. The analysis of pigments within the glaze was seriously hindered by the glaze fluorescence. But it was possible to access directly exposed pigment areas at the edges of cracked fragments. While green, blue, and white paint resulted in spectra corresponding to silica, giving no hint about the pigment structure, red and yellow paint were unambiguously identified as red earth (iron oxide, silica, and clay) and lead antimonate ($Pb_2Sn_2O_7$), respectively. The latter was not a common pigment in ancient Egyptian times, and its presence in the yellow glazes has been correlated to the limited availability of a source of antimony around the time of the XVIIIth Dynasty. In general, colored glazes pose a challenge to Raman spectroscopy. The different ways of coloring the glass matrix produce different Raman features. Transition metal ions dissolved in the glass or pigment nanoparticles give rise to very weak Raman scattering. A confocal arrangement in the microscope is important for ensuring that the signal detected originates from the glaze and not the underlying luster layer or glaze–body interface and for excluding fluorescence from random impurities. In a detailed study, various types of colored glazes were analyzed by Raman microscopy and a spectral database of the different materials has been compiled [79]. The technique appears well suited for the analysis of white, green, yellow, brown, and red colors.

4.4.2 Organic and Biological Materials

In addition to determining the identity of pigments, Raman micro-spectroscopy can be used to obtain information about the nature of organic binding media and varnishes used in paintings. According to their chemical structure, natural binders and varnishes can be classified into four major categories: proteinaceous, polysaccharide, fatty acid, and resinous media. Synthetic materials, for example acrylic resins, are quite common in modern paints as both binding media and varnishes. These organic materials can be discriminated on the basis of distinct Raman scattering bands arising from a characteristic chemical group, for example, carbonyl $(C = O)$ and unsaturated bonds $(C = C)$. The combination of certain spectral features serves as a classification tool for such materials, as is demonstrated in recent works in which database spectra of several organic binders and varnishes have been recorded and interpreted [38]. The chemical similarity of different substances belonging to the same class, and furthermore the effect of degradation products, may hinder the complete identification of an organic medium. On the other hand, such degradation products, if identified, could provide insight into the history of the material in relation to exposure to environmental factors. It is further stressed at this point that Raman microscopy offers a noninvasive approach unlike that of chromatographic techniques, which could offer more detailed information but at the expense of the consumed sample. As already pointed out, an issue of concern is fluorescence emission, which is quite common in organic materials. Excitation in the near infrared with dispersive or FT-Raman instruments is the usual approach to reduce the contribution of fluorescence to the overall signal recorded.

An important field of potential application of Raman spectroscopy is the characterization of biomaterials of archaeological significance. These biomaterials may be human or animal remains, plant or animal products, raw or processed, including, for example, foods, wood, textiles, and hair. The chemical structure and composition of such biomaterials or their type and degree of deterioration may reflect their origin, date, use, or burial context. In this respect, very interesting results have been obtained by FT-Raman analysis of human skin from mummies including the 5200-year-old Ice Man found on the border between Italy and Austria [80], the 1000-year-old skin samples from mummified bodies from the Chiribaya culture from the Southern Peruvian desert [81], samples from a cache of 15th century mummies near the abandoned settlement of Qilakitsoq, Northwest Greenland [82], and mummies from the "Tomb of the Two Brothers" (Khnum-Nakht and

Nekht-Ankh) from Rifeh in Middle Egypt dating to about 2000 B.C. [83]. On the basis of the Raman spectra it was possible to characterize the degree of deterioration or preservation of the human skin and determine whether mummification was due to artificial or natural burial causes. It is impressive that in the Egyptian mummies clear bands in the Raman spectra corresponding to the well-known α-helix and β-sheet secondary structures of the skin protein were observed and were indicators of the degree of preservation.

In a similar type of study, Raman spectra from a (rather limited) set of ancient human teeth have provided some evidence of the dating potential that is based on certain spectral indicators. Both enamel and dentine show clean spectral features corresponding to the inorganic (carbonated hydroxylapatite) and the proteinaceous organic components. It is observed that the ratio of the organic to inorganic component in enamel (based on the CH-stretch to $(PO_4)^{3-}$-stretch intensity ratio) exhibits a monotonically decreasing dependence on burial time. This suggests a loss of the organic material to the environment, quite drastic over the first 1000 years, which could in principle be correlated to the age of the specimen examined [84].

4.4.3 POTTERY

Pottery objects have been made and used by people since very ancient times. Their making, style, and context of use constitute indicators of technology, culture, and trade activities. Examples of pigment analysis in painted pottery were mentioned in Section 4.4.1. Also, the analysis of different mineralogical phases and/or inclusions may reveal information about the ceramic production technology, provenance, and dating. Established methods such as x-ray diffraction or polarized light microscopy have been used successfully in these types of investigations. However, a nondestructive, nonsampling approach such as Raman microscopy is a very attractive option [85,86]. Typical mineralogical phases identified in ceramics through their Raman spectra include α-quartz, feldspars, calcite, diopside, heamatite, magnetite, rutile, and so forth. In an example, Raman analysis of ancient pottery samples from Syria (3rd millennium B.C.) provided strong indication about the firing temperature and atmosphere [86]. The presence of magnetite ($FeO \cdot Fe_2O_3$) or heamatite (Fe_2O_3) suggests firing during a reducing or oxidizing atmosphere, respectively. Also, the presence of diopside ($CaSiO_3 \cdot MgSiO_3$) and the absence of calcite ($CaCO_3$) suggest a firing temperature above 900°C, at which diopside forms in calcareous clays because of the decomposition of calcite. In contrast, a different sample showed calcite present with no detectable diopside, suggesting a firing temperature below 900°C. Furthermore, in the case of the reducing firing process, an upper limit in the temperature could be set at around 1000°C because of the absence of certain high-temperature reduced minerals such as hercynite ($FeO \cdot Al_2O_3$) and fayalite ($FeO \cdot FeSiO_3$).

4.4.4 GLASS, STONE, AND MINERALS

The chemical composition and structure of natural or artificial glass and various minerals in the form of stone or jewels can be important in provenance, dating, and technology studies. Raman spectroscopy is among the tools that can be used in such studies, exploiting the noninvasive *in situ* character of the technique as well as the portability of the instruments.

Obsidian, a natural glass of volcanic origin, used extensively by prehistoric people in making tools, utensils, and weapons, is an important indicator for tracing trade communication between communities. Both geochemical and elemental analysis methods have been employed in provenance studies of obsidian objects, but noninvasive techniques also are investigated as alternatives particularly useful for valuable or rare objects. Raman microscopy recently used in a systematic study of obsidian glass from the western Mediterranean has shown that the technique enables discrimination of different types of archaeological obsidian according to their origin, offering valuable insight into the provenance of the objects [87]. The Raman spectrum of obsidian is essentially that of silica glass with a major band in the range of 300 to 500 cm^{-1}, corresponding to bridging oxygen bending vibrations, and two additional bands at 800 cm^{-1}, related to stretching of isolated $[SiO_4]^{4-}$ tetrahedra, and 950 to 1150 cm^{-1}, associated with vibrations of silica units with zero, one, or two nonbridging oxygens. The latter was found to vary according to the source of obsidian. Fitting this band to three gaussians provides six parameters (position and area for each) that if properly correlated lead to a complete discrimination of the different archaeologically significant sources.

Likewise, synthetic glasses as well as glazes and porcelain can be studied by Raman microscopy [88,89]. The Raman spectra of glassy (or crystalline) silicates provide information on the glass composition and structure. For example, the relative intensities of the envelopes corresponding to the bending (500 cm^{-1}) and stretching (1100 cm^{-1}) modes of the amorphous silicates (referred to as Raman index of polymerization) correlate strongly with the glass composition and firing temperature, allowing discrimination among different types of objects and processing. In the case of porcelains, the different crystalline phases can be differentiated in the Raman spectra leading to easy classification between soft- and hard-paste porcelain bodies. The high peak intensity of β-wollastonite ($CaSiO_3$) and/or tricalcium phosphate $[β-Ca_3(PO_4)_2]$ phases is characteristic of the soft pastes. These phases are absent from hard pastes, which are composed of mullite ($3Al_2O_3 \cdot 2SiO_2$) or mullitelike glassy-phase spectra.

In a different study, similar types of artistic glass, dated in the period of 1750 to 1940, were investigated in an effort to deduce appropriate Raman markers in order to determine the age of the objects [90]. No particular resonances or spectral shifts were found to depend on the age of the glass. However, it was observed that the fluorescence emitted from the glass (the excitation wavelength was at 488 nm) showed a clear increase with age. While the origin of this fluorescence is not clear, it appears that the intensity ratio of the Raman band at 1080 cm^{-1} (symmetric stretch of O–Si–O bonds) over that of the fluorescence shows a decreasing trend with time (the logarithm of the ratio has a linear dependence on the glass age). This is an example of dating and authenticity information arising from the spectral data.

In the case of exposed stone monuments, it is well known that atmospheric pollution, through physical and chemical interactions, leads to the formation of various types of encrustation. These polluting crusts involve partial or even total transformation of the chemical structure of the stone surface with additional deposition and inclusion of particles (soil dust, pollution particles). Biological crusts are formed if microorganisms develop on the surface. Furthermore, coatings applied during past conservation treatments may also constitute a major obstacle to conservation. The need of a technique that can work on site to study the different patinas, identify the transformation products, and characterize the encrustation on stone sculpture is obvious. In a recent study, the crust on two different stone samples, Hontoria limestone from the Burgos Museum (Spain) and Pentelic marble from a neoclassic monument in Athens (Greece) was investigated using Raman and FT-Raman microspectrometry in the context of a broader study related to laser cleaning. The hard black dendritic encrustation layer on the limestone sample was found to contain gypsum ($CaSO_4 \cdot 2H_2O$) and carbon. An intermediate yellow/brown patina covers the stone as a protective layer or as the basis on which paint was applied. Raman analysis of the patina shows the presence of gypsum, oxalate ($C_2O_4^{2-}$), and phosphate (PO_4^{3-}) salts. Gypsum originates from the interaction of limestone (calcite) with the atmospheric SO_2, while the oxalate and phosphate salts indicate the decomposition of the organic medium used in the patina layer. The thin soiling crust on the (much younger) marble sample contained only gypsum and carbon. The Raman bands for these materials were used to evaluate the extent and possible side effects of laser cleaning interventions in the two stone samples [91].

Another emerging field of application of Raman microscopy is in the identification and authentication of gemstones. The problem has intensified in recent years, as a large number of gemstones of questionable identity and quality have entered the market. Even experienced jewellers are not always in a position to identify fake or modified gemstones. Visual inspection and possibly refractometry have been the state of the art diagnostic tools in the field of gemology; Raman microspectrometry appears as an ideal method for investigating gemstones. It combines reliable, fingerprint spectral information with applicability *in situ,* and it shows no need for sample preparation. As an example, Raman analysis has clearly differentiated topaz, a common golden brown to yellow gemstone ($AlSiO_3$ $Al(F,OH)_3$) that has been used for centuries in jewelry, from the less valuable citrine (impure quartz), which is often sold under the name topaz or gold topaz [92].

4.4.5 METALS

While Raman spectroscopy is not sensitive to metals, its use in the characterization of corrosion products in objects made of metal or metal alloys can be very important. Identifying corrosion products that may have originated throughout the burial history of an object or through its exposure to the atmosphere or because of its display microenvironment can clarify the underlying corrosion processes and more importantly lead to proper conservation treatments protecting the metal object from further deterioration. In this context a catalogue with 45 Raman spectra of minerals relevant to corroded metals has recently been presented and provides a useful reference databank for studies of metal corrosion [93].

In a recent campaign, copper corrosion products from several bronze sculptures in the park of Mariemont (Morlanweltz, Belgium) were examined by Raman microscopy [94]. Different types of copper(II) sulfate minerals arising from atmospheric sulfur dioxide (SO_2) were identified and easily discriminated from one another on the basis of their characteristic Raman scattering bands. The nature of the corrosion products is sensitive to the level of pollution (for example, urban vs. rural environments) and even to the exposure to rain, wind, and sun. For example, Raman analysis of corrosion samples from the sculptures showed that areas exposed to rain favored the formation of brochantite ($3Cu(OH)_2 \cdot CuSO_4$), while sheltered areas in which more acidic conditions prevailed favored the formation of antlerite ($2Cu(OH)_2 \cdot CuSO_4$). In a similar study, corrosion products originating from the different alloying elements of ancient Chinese bronze trees (1st to 3rd century A.D.) were studied by Raman microscopy, providing insight into the different environments to which the bronze tree branches were exposed [95]. The excellent spatial resolution of the technique allowed mapping of the distribution of the corrosion products within individual corrosion layers and even within individual metallographic phases. For example, it was seen that in one tree only the alpha phase of the Cu–Sn alloy was corroded (presence of cuprite, Cu_2O), while in a different tree the delta (δ) phase was the one showing major corrosion. Also, the detection of lead sulfate ($PbSO_4$) was linked to a burial history involving anaerobic soils with sulfate-reducing bacteria with an alternative possibility of exposure to modern air pollution. Finally, Raman spectra also proved the presence of several modern pigments including phthalocyanine blue and green and Hansa yellow G, which revealed recent interventions applied to cover repairs.

4.5 RAMAN ANALYSIS IN COMBINATION WITH OTHER TECHNIQUES

In several cases, analytical results obtained by combining Raman analysis with other spectroscopic techniques (LIBS, LIF, FT-IR, diffuse reflectance spectrometry), x-ray techniques (x-ray fluorescence, x-ray diffractometry, PIXE), or even chromatographic methods (GC-MS) have led to a more complete investigation of the objects examined. While the examples mentioned above refer to the combined use of techniques but with entirely different instruments, it is noted that recently researchers have focused on developing hybrid units, which combine techniques. These cases are described below.

4.5.1 RAMAN MICROSCOPY AND LIF

Raman and fluorescence spectroscopy are considered incompatible, and the usual approach is to carry out the Raman measurement under conditions that minimize electronic excitation leading to fluorescence emission. However, with a proper approach, the two spectroscopies can provide complementary information using the very same instrument. One needs only to recognize that a Raman spectrometer is, in principle, also a laser fluorescence spectrometer. In an obviously favorable case (Section 4.4), it was shown that Raman scattering and photoluminescence emission, present simultaneously in the same spectrum, could be used for dating artistic glasses. In a more ambitious direction, exploiting recent instrumentation advances that favor modularity and versatility, it is conceivable to employ a Raman spectrometer as a fluorescence spectrometer by introducing an alternative laser for excitation in the UV (or even using light-emitting diodes for fluorescence excitation) and by properly configuring the spectrograph (with the selection of a low resolution grating) to permit the broader spectral coverage required for recording the fluorescence emission spectra. These modifications are rather simple and enable one to switch easily between the two modes of operation. Furthermore, carrying out fluorescence imaging microscopy and Raman microscopy on the same unit is entirely feasible.

4.5.2 RAMAN MICROSCOPY AND LIBS

The combination of Raman microscopy with LIBS has been covered in Section 3.6.1. In this section, emphasis is placed on hybrid instruments that combine in one unit the necessary instrumentation for running both Raman and LIBS analysis. Recently this type of approach has received considerable attention, and research groups have developed hybrid LIBS-Raman units for stand-off analysis of minerals (to be used in future planetary missions) [96] and for the analysis of environmental and biological samples [97]. In another recent study, a novel system was proposed that utilizes a straightforward optical setup for carrying out Raman and LIBS measurements on materials related to cultural heritage [98]. The system is effectively a single spectrometer that permits analysis of a surface on the basis of either laser-induced Raman scattering or laser-induced breakdown optical emission by simply adjusting the energy of the laser pulse(s) either below or above the material's ablation threshold, respectively. This way, exactly the same point can be analyzed by both techniques. Furthermore, with this approach, a LIBS depth-profiling study of layered materials can be easily complemented by Raman analysis.

4.5.3 RAMAN AND XRF MICROANALYSIS

X-ray fluorescence spectroscopy has been among the best choices for obtaining elemental composition information when art and archaeological objects are investigated, and recent advances with micro-XRF [99–101] and portable instrumentation [102] have enhanced such applications considerably. XRF analysis has, in several cases, been combined with Raman analysis in an effort by researchers to obtain complete information on materials composition. For example, in the case of a mural painting by the Italian miniaturist Napoleone Verga [103], XRF mapping with a portable unit was used to carry out a preliminary assessment of pigment distribution. This guided sampling using a Raman microscope, and subsequent analysis, identified 13 different pigments.

Unlike LIBS or LIF, which are both optical spectroscopies (in the UV and visible or near-IR parts of the spectrum), XRF shares no common instrumentation with Raman spectroscopy, and so a hybrid instrument is not feasible. However, the two techniques share the general microprobe approach, and their combination in a dual unit has been quite an attractive option. Indeed, the result of recent work has led to the construction of a new portable instrument that takes advantage simultaneously of micro-XRF and micro-Raman spectrometries [104]. Initial tests on an overpainted model Byzantine icon have been encouraging and have demonstrated the instrument's capabilities for *in situ* and noninvasive identification of pigments. Combining complementary spectral data from the two techniques enables a complete analysis of almost all the pigments present in the painting. Furthermore, in certain cases, in which the overpaint permits adequate penetration of radiation, the XRF analysis is shown to provide valuable information on the underpaint as well. Examination of cross sections is also feasible, and then maximum information on the stratigraphy of the paint is obtained.

4.6 CONCLUDING REMARKS

Continuously gaining popularity as a result of thorough scientific investigations and critical instrumentation advances, Raman spectroscopy shows excellent potential to become a reliable analytical tools in the fields of archaeological science, art history, and conservation. The capability of the technique to analyze almost any material, of mineral, organic, or biological origin, nondestructively and *in situ*, permits diverse studies on objects of cultural heritage such as the identification of pigments, the characterization of pottery, glass, and stone objects, the analysis of metal corrosion products and environmental pollution deposits, and the analysis of organic and biomaterials. Several Raman spectral libraries have been compiled and are available in the literature or through the Internet, facilitating data analysis, while chemometric techniques can aid the interpretation of complex spectral data. Continuous improvements in instrumentation have led to commercially available, versatile, and straightforward-to-use instruments that enable the collection of high-quality spectra in a short time. Most of the modern work is performed on Raman microscopes that provide uniquely high spatial resolution, permitting the investigation of very fine features down to, for example, pigment grains. New developments have led to reliable portable Raman microspectrometers that offer versatility within the museum and also enable campaigns outdoors, at remote locations, or even under water; they are expected to give an increasing contribution to the field of cultural heritage in future. Finally, research on hybrid systems that combine Raman analysis with other techniques, providing complementary molecular or species information, may lead to further advances in the analytical capabilities provided by Raman spectroscopy.

REFERENCES

1. J.J. Laserna, Ed., *Modern Techniques in Raman Spectroscopy,* John Wiley, New York, 1996.
2. M. Delhaye and P. Dhamelincourt, Raman microprobe and microscope with laser excitation, *J. Raman Spectrosc.,* **3**, 33–43, 1975.
3. F. Cariati and S. Bruni, Raman spectroscopy, in *Modern Analytical Methods in Art and Archaeology,* E. Ciliberto and G. Spoto, Eds., Chemical Analysis, A Series of Monographs on Analytical Chemistry and Its Applications, J.D. Winefordner, Ed., John Wiley, New York, **155**, 255–278, 2000, chap. 10.
4. S.P Best, R.J.H. Clark, M.A.M. Daniels, C.A. Porter, and R. Withnall, Identification by Raman microscopy and visible reflectance spectroscopy of pigments on an Icelandic manuscript, *Stud. Conserv.,* **40**, 31–40, 1995.
5. R.J.H. Clark, Raman microscopy: application to the identification of pigments on medieval manuscripts, *Chem. Soc. Rev.,* **24**, 187–196, 1995.

6. G.D. Smith and R.J.H. Clark, Raman microscopy in art history and conservation science, *Reviews in Conservation,* **2**, 92–106, 2001.

7. R.J.H. Clark, Raman microscopy in archaeological science, *J. Archaeological Sci.,* **31**, 1137–1160, 2004.

8. P. Vandenabeele, Raman spectroscopy in art and archaeology, *J. Raman Spectrosc.,* **35**, 607–609, 2004.

9. B. Guineau, Analyse non-destructive des pigments par microsonde Raman laser: examples de l' azurite et de la malachite, *Stud. Conserv.,* **29**, 35–41, 1984.

10. M. Ferreti, Scientific investigations of works of art, ICCROM — International Centre for the Study of Preservation and the Restoration of Cultural Property, Rome, 1993.

11. A.M. Pollard and C. Heron, *Archaeological Chemistry,* Royal Society of Chemistry, Cambridge, U.K., 1996.

12. E. Cilberto and G. Spoto, Modern analytical methods in art and archaeology, *Chemical Analysis, A series of monographs on analytical chemistry and its applications,* Winefordner, J.D., Ed.,Vol. 155, Wiley, New York, 2000.

13. M. Bacci, F. Baldini, R. Carla, and R. Linari, A color analysis of the Brancacci chapel frescoes, *Appl. Spectrosc.,* **45**, 26–31, 1991.

14. A. Casini, F. Lotti, M. Picollo, L. Stefani, and E. Buzzegoli, Image spectroscopy mapping technique for non-invasive analysis of paintings, *Stud. Conserv.,* **44**, 39–48, 1999.

15. M. Bacci, UV-VIS-NIR, FT-IR and FORS spectroscopies, in *Modern Analytical Methods in Art and Archaeology,* E. Ciliberto and G. Spoto, Eds., Chemical Analysis, A Series of Monographs on Analytical Chemistry and Its Applications, J.D. Winefordner, Ed., John Wiley, New York, 2000; **155**, 321–361, chap. 12.

16. J. van der Weerd, M.K. van Veen, R.M.A. Heeren, and J.J. Boon, Identification of pigments in paint cross sections by reflection visible light imaging microspectroscopy, *Anal. Chem.,* **75**, 716–722, 2003.

17. C. Balas, V. Papadakis, N. Papadakis, A. Papadakis, E. Vazgiouraki, and G. Themelis, A novel hyper-spectral imaging apparatus for the non-destructive analysis of objects of artistic and historic value, *J. Cultural Heritage,* **4**, 330s–337s, 2003.

18. E.R. de la Rie, Fluorescence of paint and varnish layers (Part I), *Stud. Conserv.,* **27**, 1–7, 1982.

19. E.R. de la Rie, Fluorescence of paint and varnish layers (Part II), *Stud. Conserv.,* **27**, 65–69, 1982.

20. S. Shimoyama, Y. Noda, and S. Katsuhara, Non-destructive analysis of ukiyo-e prints: determination of plant dyestuffs used for traditional Japanese woodblock prints, employing a three-dimensional fluorescence spectrum technique and quarz fibre optics, *Dyes in History and Archaeology,* **15**, 27–42, 1996.

21. T. Miyoshi, M. Ikeya, S. Kinoshita, and T. Kushida, Laser-induced fluorescence of oil colours and its application to the identification of pigments in oil paintings, *Jpn. J. Appl. Phys.,* **21**, 1032–1036, 1982.

22. T. Miyoshi, Fluorescence from varnishes for oil paintings under N_2 laser excitation, *Jpn. J. Appl. Phys.,* **26**, 780–781, 1987.

23. D. Anglos, M. Solomidou, I. Zergioti, V. Zafiropulos, T.G. Papazoglou, and C. Fotakis, Laser-induced fluorescence in artwork diagnostics: an application in pigment analysis, *Appl. Spectrosc.,* **50**, 1331–1334, 1996.

24. I. Borgia, R. Fantoni, C. Flamini, T.M. Di Palma, A. Giardini Guidoni, and A. Mele, Luminescence from pigments and resins for oil paintings induced by laser excitation, *Appl. Surf. Sci.,* **127–129**, 95–100, 1998.

25. A. Romani, C. Miliani, A. Morresi, N. Forini, and G. Favaro, Surface morphology and composition of some "lustro" decorated fragments of ancient ceramics from Deruta, *Appl. Surf. Sci.,* **157**, 112–122, 2000.

26. P. Weibring, T. Johansson, H. Edner, S. Svanberg, B. Sundner, V. Raimondi, G. Cecchi, and L. Pantani, Fluorescence lidar imaging of historical monuments, *Appl. Opt.,* **40**, 6111–6120, 2001.

27. D. Comelli, C. D'Andrea, G. Valentini, and R. Cubeddu, Fluorescence lifetime imaging and spectroscopy as a tool for non destructive analysis of works of art, *Appl. Opt.,* **43**, 2175–2183, 2004.

28. M.R. Derrick, D.C. Stulik, and J.M. Landry, *Infrared Spectroscopy in Conservation Science,* J. Paul Getty Trust Publications, Los Angeles, 2000.

29. T. Learner, The use of FT-IR in the conservation of twentieth century paintings, *Spectrosc. Eur.,* **8**(4), 14–19, 1996.

30. J.C. Shearer, D.C. Peters, G. Hoepfner, and T. Newton, FTIR in the service of art conservation, *Anal. Chem.,* **55**, 874A–880A, 1985.

31. R.J. Meilunas, J.G. Bentsen, and A. Steinberg, Analysis of aged paint binders by FTIR spectroscopy, *Stud. Conserv.,* **35**, 33–51, 1990.

32. M. Bacci, M. Fabbri, and M. Picollo, Non-invasive fibre optic Fourier transform-infrared reflectance spectroscopy on painted layers — Identification of materials by means of principal component analysis and Mahalanobis distance, *Anal. Chim. Acta,* **446**, 15–21, 2001.

33. J. van der Weerd, H. Brammer, J.J. Boon, and R.M.A. Heeren, Fourier transform infrared microscopic imaging of an embedded paint cross-section, *Appl. Spectrosc.,* **56**, 275–283, 2002.

34. K. Keune and J.J. Boon, Imaging secondary ion mass spectrometry of a paint cross section taken from an early Netherlandish painting by Rogier van der Weyden, *Anal. Chem.,* **76**, 1374–1385, 2004.

35. N. Salvado, S. Buti, M.J. Tobin, E. Pantos, A.J.N.W. Prag, and T. Pradell, Advantages of the use of SR-FT-IR microspectroscopy: applications to cultural heritage, *Anal. Chem.,* **77**, 3444–3451, 2005.

36. R.J.H. Clark, Pigment identification by spectroscopic means: an arts/science interface: medieval manuscripts, pigments and spectroscopy, *C.R. Chimie,* **5**, 7–20, 2002.

37. L. Burgio, R.J.H. Clark, and H. Toftlund, The identification of pigments used on illuminated plates from 'Flora Danica' by Raman microscopy, *Acta Chem. Scandinavica,* **53**, 181–187, 1999.

38. P. Vandenabeele, B. Wheling, L. Moens, H. Edwards, M. DeReu, and G. Van Hooydonk, Analysis with micro-Raman spectroscopy of natural organic binding media and varnishes used in art, *Anal. Chim. Acta,* **407**, 261–274, 2000.

39. H.G.M. Edwards and M.J. Falk, Investigation of the degradation products of archaeological linen by Raman Spectroscopy, *Appl. Spectrosc.,* **51**, 1134, 1997.

40. W. Noll, R. Holm, and L. Born, Painting of ancient ceramics, *Angew. Chem. Int. Ed.,* **14**, 602–613, 1975.

41. S.E. Philippakis, B. Perdikatsis, and T. Paradelis, An analysis of blue pigments from the Greek bronze age, *Stud. Conserv.,* **21**, 143–153, 1976.

42. S.E. Philippakis, B. Perdikatsis, and K. Assimenos, X-ray analysis of pigments from Vergina Greece (second tomb), *Stud. Conserv.,* **24**, 54–58, 1979.

43. M. Muller, B. Murphy, M. Burghammer, C. Riekel, M. Roberts, M. Papiz, D. Clarke, J. Gunneweg, and E. Pantos, Identification of ancient textile fibres from Khirbet Qumran caves using synchrotron radiation microbeam diffraction, *Spectrochim. Acta,* **B59**, 1669–1674, 2004.

44. I. De Ryck, A. Adriaens, E. Pantos, and F. Adams, A comparison of microbeam techniques for the analysis of corroded ancient bronze.objects, *Analyst,* **128**, 1104–1109, 2003.

45. J.S. Mills and R. White, *The Organic Chemistry of Museum Objects,* 2nd ed., Butterworth-Heinemann, Oxford, U.K., 1994.

46. R.P. Evershed, Biomolecular analysis by organic mass spectrometry, in *Modern Analytical Methods in Art and Archaeology,* E. Ciliberto and G. Spoto, Eds., Chemical Analysis, A Series of Monographs on Analytical Chemistry and Its Applications, J.D. Winefordner, Ed., John Wiley, New York, **155**, 177–239, 2000, chapter 8.

47. J.S. Mills and R. White, Natural resins of art and archaeology: their sources, chemistry and identification, *Stud. Conserv.,* **22**, 12–31, 1977.

48. R. White, The characterization of proteinaceous binding media in art objects, *National Gallery Technical Bulletin,* **8**, 5–14, 1984.

49. H.H. Hairfield, Jr. and E.M. Hairfield, Identification of a late Bronze Age resin, *Anal. Chem.,* **62**, 41A–45A, 1990.

50. S.L. Vallance, Applications of chromatography in art conservation: techniques used for the analysis and identification of proteinaceous and gum binding media, *Analyst,* **122**, R75–R81, 1997.

51. G.A. van der Doelen, K.J. van der Berg, and J.J. Boon, Comparative chromatographic and mass spectrometric studies of triterpenoid varnishes: fresh material and aged samples from paintings, *Stud. Conserv.,* **43**, 249, 1998.

52. R.P. Evershed, S.N. Dudd, M.S. Copley, R. Berstan, A.W. Stott, H. Mottram, S.A. Buckley, and Z. Crossman, Chemistry of archaeological animal fats, *Acc. Chem. Res.,* **35**, 660–668, 2002.

53. M.P. Colombini and F. Modugno, Characterisation of proteinaceous binders in artistic paintings by chromatographic techniques, *J. Sep. Sci.,* **27**, 147–160, 2004.

54. M.P. Colombini and H. Bonaduce, Characterisation of beeswax in works of art by gas chromatography-mass spectrometry and pyrolysis-gas chromatography-mass spectrometry procedures, *J. Chromatogr.,* **A1028**, 297–306, 2004.

55. M. Regert, J. Langlois, and S. Colinart, Characterisation of wax works of art by gas chromatographic procedures, *J. Chromatogr.,* **A1091**, 124–136, 2005.

56. R. Hynek, S. Kuckova, J. Hradilova, and M. Kodicek, Matrix-assisted laser desorption/ionization time-of-flight mass spectrometry as a tool for fast identification of protein binders in color layers of paintings, *Rapid Comm. Mass Spectrometry,* **18**, 1896–1900, 2004.

57. O.F. van den Brink, J.J. Boon, P.B. O'Connor, M.C. Duursma, and R.M. Heeren, Matrix-assisted laser desorption/ionization Fourier transform mass spectrometric analysis of oxygenated triglycerides and phosphatidylcholines in egg tempera paint dosimeters used for environmental monitoring of museum display conditions, *J. Mass Spectrom.,* **36**, 479–492, 2001.

58. D. Scalarone, J. Van Der Horst, J.J. Boon, and O. Chiantore, Direct-temperature mass spectrometric detection of volatile terpenoids and natural terpenoid polymers in fresh and artificially aged resins, *J. Mass Spectrom.,* **38**, 607–617, 2003.

59. G. Spoto, Secondary ion mass spectrometry in art and archaeology, *Thermochim. Acta,* **365**, 157–166, 2000.

60. M.R. Guasch-Jane, M. Ibern-Gomez, C. Andres-Lacueva, O. Jauregui, and R.M. Lamuela-Raventos, Liquid chromatography with mass spectrometry in tandem mode applied for the identification of wine markers in residues from ancient Egyptian vessels, *Anal. Chem.,* **76**, 1672–1677, 2004.

61. C.M. Grzywacz, Identification of proteinaceous binding media in paintings by amino acid analysis using 9-fluorenylmethyl chloroformate derivatization and reversed-phase high-performance liquid chromatography, *J. Chromatogr.*, **A676**, 177–183, 1994.

62. A. Spyros and D. Anglos, Study of ageing in oil-paintings by 1D and 2D NMR spectroscopy, *Anal. Chem.*, **76**, 4929–4936, 2004.

63. P. Vandenabeele, T.L. Weis, E.R. Grant, and L.J. Moens, A new instrument adapted to *in situ* Raman analysis of objects of art, *Anal. Bioanal. Chem.*, **379**, 137–142, 2004.

64. S.K. Sharma, S.M. Angel, M. Ghosh, H.W. Hubble, and P.G. Lucey, Remote pulsed laser Raman spectroscopy system for mineral analysis on planetary surfaces to 66 meters, *Appl. Spectrosc.*, **56**, 699–705, 2002.

65. D.C. Smith, *In situ* mobile subaquatic archaeometry evaluated by non-destructive Raman microscopy of gemstones lying under impure waters, *Spectrocim. Acta*, **A59**, 2353–2369, 2003.

66. L. Burgio, K. Melessanaki, M. Doulgeridis, R.J.H. Clark, and D. Anglos, Pigment identification in paintings employing laser induced breakdown spectroscopy and Raman microscopy, *Spectrochim. Acta*, **B56**, 905–913, 2001.

67. P. Vandenabeele, F. Verpoort, and L. Moens, Non-destructive analysis of paintings using Fourier transform Raman spectroscopy with fiber optics, *J. Raman Spectrosc.*, **32**, 263–269, 2001.

68. R.J.H. Clark and P.J. Gibbs, Analysis of 16th century Qwazini manuscripts by Raman microscopy and remote laser Raman microscopy, *J. Archaeolog. Sci.*, **25**, 621–629, 1998.

69. K.L. Brown and R.J.H. Clark, Analysis of key–Anglo-Saxon manuscripts (8–11th centuries) in the British Library: pigment identification by Raman microscopy, *J. Raman Spectrosc.*, **35**, 181–189, 2004.

70. K.L. Brown and R.J.H. Clark, Analysis of pigmentary materials on the Vinland map and Tartar relation by Raman microprobe spectroscopy, *Anal. Chem.*, **74**, 3658–3661, 2002.

71. R.J.H. Clark and P.J. Gibbs, Raman microscopy of a 13th-century text, *Anal. Chem.*, **70**, 99A–104A, 1988.

72. G.D. Smith and R.J.H. Clark, The role of H_2S in pigment blackening, *J. Cult. Heritage*, **3**, 101–105, 2002.

73. P. Vandenabeele, L. Moens, H.G.M. Edwards, and R. Dams, Raman spectroscopic database of azo pigments and application to modern art studies, *J. Raman Spectrosc.*, **31**, 509–517, 2000.

74. P. Vandenabeele and L. Moens, Micro-Raman spectroscopy of natural and synthetic indigo samples, *Analyst*, **128**, 187–193, 2003.

75. I. Karapanagiotis, S. Sotiropoulou, S. Chryssikopoulou, P. Magiatis, K.S. Andriko-poulos, and Y. Chryssoulakis, Investigation of Tyrian purple occurring in historical wall paintings of Thera, *Dyes Hist. Archaeol.*, 23, in press, 2006.

76. A. Brysbaert and P. Vandenabeele, Bronze Age painted plaster in Mycenean Greece: a pilot study on the testing and application of micro-Raman spectroscopy, *J. Raman Spectrosc.*, **35**, 686–693, 2004.

77. J. Zuo, C. Xu, C. Wang, and Z. Yushi, Identification of the pigment in painted pottery from the Xishan site by Raman microscopy, *J. Raman Spectrosc.*, **30**, 1053–1055, 1999.

78. R.J.H. Clark and P.J. Gibbs, Non-destructive *in situ* study of ancient Egyptian faience by Raman microscopy, *J. Raman Spectrosc.*, **28**, 99–103, 1997.

79. P. Colomban, G. Sagon, and X. Faurel, Differentiation of antique ceramics from the Raman spectra, of their colored glazes and paintings, *J. Raman Spectrosc.*, **32**, 351–360, 2001.

80. H.G.M. Edwards, M. Gniadecka, S. Petersen, J.P.H. Hansen, O.F. Nielsen, D.H. Christensen, and H.C. Wulf, NIR-FT Raman spectroscopy as a diagnostic probe for mummified skin and nails, *Vib. Spectrosc.,* **28**, 3–15, 2002.

81. M. Gniadecka, H.G.M. Edwards, J.P.H. Hansen, O.F. Nielsen, D.H. Christensen, S.E. Guillen, and H.C. Wulf, Near infrared Fourier transform Raman spectroscopy of the mummified skin of the alpine iceman, Qilakitsoq Greenland mummies and Chiribaya mummies from Peru, *J. Raman Spectrosc.,* **30**, 147–153, 1999.

82. M. Gniadecka, H.C. Wulf, O.F. Nielsen, D.H. Christensen, and J.P.H. Hansen, Fourier transform Raman spectroscopy of 15th century mummies from Qilakitsoq Greenland, *J. Raman Spectrosc.,* **28**, 179–184, 1997.

83. S. Petersen, O.F. Nielsen, D.H. Christensen, H.G.M. Edwards, D.W. Farwell, R. David, P. Lambert, M. Gniadecka, and H.C. Wulf, Near-infrared Fourier transform Raman spectroscopy of skin samples from the "Tomb of the Two Brothers," Khnum-Nakht and Nekht-Ankh, XIIth Dynasty Egyptian mummies (ca 2000 B.C.), *J. Raman Spectrosc.,* **34**, 375–379, 2003.

84. A. Bertoluzza, P. Brasili, L. Castri, F. Facchini, C. Fagnano, and A. Tinti, Preliminary results in dating human skeletal remains by Raman spectroscopy, *J. Raman Spectrosc.,* **28**, 185–188, 1997.

85. B. Wopenka, R. Popelka, J.D. Pasteris, and S. Rotroff, Understanding the mineralogical composition of ancient Greek pottery through Raman microprobe spectroscopy, *Appl. Spectrosc.,* **56**, 1320–1328, 2002.

86. A. Zoppi, C. Lofrumento, E.M. Castellucci, and M.G. Migliorini, Micro-Raman technique for phase analysis on archaeological ceramics, *Spectrosc. Eur.,* **14(5)**, 16–21, 2002.

87. L. Bellot-Gurlet, F.-X. Le Bourdonnec, G. Poupeau, and S. Doubernet, Raman micro-spectroscopy of western Mediterranean obsidian glass: one step towards provenance studies? *J. Raman Spectrosc.,* **35**, 671–677, 2004.

88. P. Colomban, Raman spectrometry, a unique tool to analyze and classify ancient ceramics and glasses, *Appl. Phys.,* **A79**, 167–170, 2004.

89. P. Colomban and C. Truong, Non-destructive Raman study of the glazing technique in lustre potteries and faience (9–14th centuries): silver ions, nanoclusters, microstructure and processing, *J. Raman Spectrosc,* **35**, 195–207, 2004.

90. A. Bertoluzza, S. Cacciari, G. Cristini, C. Fagnano, and A. Tinti, Non-destructive *in situ* Raman study of artistic glasses, **26**, 751–755, 1995.

91. P. Pouli, G. Totou, V. Zafiropulos, C. Fotakis, M. Ouija, E. Rebollar, S. Gaspard, M. Castillejo, C. Domingo, P.P. Perez, and A. Laborde, The coloration effect associated with laser cleaning of stonework, *Spectrochim. Acta A,* submitted.

92. A.L. Jenkins and R.A. Larsen, Gemstone identification using Raman spectroscopy, *Spectrosc. Eur.,* **19**(4), 20–25, 2004.

93. M. Bouchard and D.C. Smith, Catalogue of 45 Raman spectra of minerals concerning research in art history or archaeology, especially on corroded metals and coloured glass, *Spectrochim. Acta,* **A59**, 2247–2266, 2003.

94. V. Hayez, J. Guillaume, A. Hubin, and H. Terryn, Micro-Raman spectroscopy for the study of corrosion products on coper alloys: setting up a reference database and studying works of art, *J. Raman Spectrosc.,* **35**, 732–738, 2004.

95. L.I. McCann, K. Trentelman, T. Possley, and B. Golding, Corrosion of ancient Chinese bronze money trees studied by Raman microscopy, *J. Raman Spectrosc.,* **30**, 121–132, 1999.

96. R.C. Wiens, S.K. Sharma, J. Thompson, A. Misra, and P.G. Lucey, Joint analyses by laser-induced breakdown spectroscopy (LIBS) and Raman spectroscopy at stand-off distances, *Spectrochim. Acta,* **A61**, 2324–2334, 2005.

97. M.Z. Martin, S.D. Wullschleger, C.T. Garten, Jr., A.V. Palumbo, and J.G. Smith, Elemental analysis of environmental and biological samples using laser-induced breakdown spectrosocpy and pulsed Raman spectroscopy, *J. Dispersion Sci. Tech.,* **25**, 687–694, 2004.

98. A. Giakoumaki, I. Osticioli, and D. Anglos, Spectroscopic analysis using a hybrid LIBS-Raman system, *Appl. Phys. A,* **83**, 537–541, 2006.

99. K. Janssens, G. Vittiglio, I. Deraedt, A. Aerts, B. Vekemans, L. Vincze, F. Wei, I. Deryck, O. Schalm, F. Adams, A, Rindby, A. Knoechel, A. Simionovici, and A. Snigirev, Use of microscopic XRF for non-destructive analysis in art and archaeology, *X-Ray Spectrom.,* **29**, 73–91, 2000.

100. B. Kanngiesser, W. Malzer, and I. Reiche, A new 3D micro x-ray fluorescence analysis set-up — first archaeometric applications, *Nucl. Instrum. Methods Phys. Res.,* **211**, 259–264, 2003.

101. B. Kanngiesser, W. Malzer, A. Fuentes Rodriguez, and I. Reiche, Three-dimensional micro-XRF investigations of paint layers with a tabletop setup, *Spectrochim. Acta,* Part B: **60**, 41–47, 2005.

102. A.G. Karydas, D. Kotzamani, R. Bernard, J.N. Barrandon, and Ch. Zarkadas, A compositional study of a museum jewellery collection (7th–1st B.C.) by means of a portable XRF spectrometer, *Nucl. Instr. Methods Phys. Res.,* **B226**, 15–28, 2004.

103. F. Rosi, C. Miliani, I. Borgia, B. Brunetti, and A. Sgamelotti, Identification of 19th century blue and green pigments by *in situ* x-ray fluorescence and micro-Raman spectroscopy, *J. Raman Spectrosc.,* **35**, 610–615, 2004.

104. K.S. Andrikopoulos, Sister Daniilia, B. Roussel, and K. Janssens, In vitro validation of a mobile Raman–XRF micro analytical instrument's capabilities on the diagnosis of Byzantine icons, *J. Raman Spectrosc.,* in press, 2006.

5 Laser Interferometry for Direct Structural Diagnostics

5.1 INTRODUCTION: LASER INTERFERENCE AS DIAGNOSTIC TECHNIQUE

Structural diagnosis refers to the integrity state of the work of art. Laser technology developed in recent years in a number of laboratories around Europe can provide the means to assess it safely and accurately. The knowledge about the condition of structural integration is related to problems of invisible deterioration with cumulative effects on the structure due to continuous environmental and seasonal alterations and the unavoidable aging causing the growing of discontinuities in the constructing materials. Nowadays, increased demands on artworks' mobility, continuous exposure to the public, transportation to exhibitions, periodic conservation intervention, and restoration actions provoke research on new methods and alternative practices. The aim is to develop tools to facilitate retrieval and evaluation of the cause–effect relation in artwork deterioration for upgrading preservation and maintenance policies. Laser technology by a number of means and techniques may assist in the effort of cultural preservation. Laser interference techniques having achieved a mature state of nondestructive testing applications in industry rise as an interesting alternative to existing conservation practices and can be specially adjusted and optimized for the application requirements in cultural heritage. An effort is demanded to summarize the potential and offered experience, particularly in a field where interdisciplinary expertise is involved.

This chapter portrays the fundamental principles underlying the interaction of coherent waves in interferometry and holography. The basic properties of coherent laser light, wave optics and the physics of interference and diffraction, the speckle effect, and principles of holography, which allow recording and reconstruction of optical waves, as basic information needed for the understanding of the application of the analytical techniques involved are briefly presented. Further insight includes the unique properties of coherent laser light, wave optics, and the physics of interference and diffraction, the physics of holography, and the speckle effect. These fundamentals are not presented here, since they can be found in textbooks of physics, optics, and engineering usually available at university libraries.

Thus emphasis in this chapter is on presentation of laser structural diagnosis in art conservation demonstrated in a number of representative studies performed mainly with holographic interferometry nondestructive testing (HINDT) and its digital holographic counterpart commonly termed electronic speckle pattern interferometry (ESPI). Some fundamentals on concepts and experimental guidance pave the way. Examples from recent developments with digital speckle shearography and laser Doppler vibrometry are also included. Relevant references are suggested to the reader for more detailed study. Figure 5.1 outlines the fundamentals of laser interferometry dominating this section.

5.1.1 FUNDAMENTAL CONCEPTS: DEFINITION OF TERMS

In the vast majority of interference applications for structural diagnosis of works of art a laser is used as the light source. Continuous wave (cw) lasers emit light waves having output parameters quite constant with time and can be measured with high precision. This text concentrates only on interference with visible light. Most cw lasers emit linearly polarized light in a very narrow spectral range (highly monochromatic) with low divergence (essentially plane output wavefront). Laser light has unique properties, many interrelated, including *monochromaticity, directionality, brightness,* and *coherence,* which have stimulated many different measuring techniques and applications [1–5].

A typical low-cost, low-power laser found in almost all optical laboratories and providing output properties well suited for interference measurements is the linearly polarized cw He–Ne laser with a wavelength $\lambda = 632.8$ nm, a beam of about 2 mm diameter, and divergence angle less than 0.7 mrad. This well collimated beam can be made to form either a spherical or a plane wavefront used in interferometry by using a combination of lenses for beam conditioning. For example, a short focal

---

Laser Interferometry for Structural Diagnosis

Keywords Outline

1. Fundamentals of Laser Interferometry
- Laser Properties
- Coherent phenomena
- Holography principles

2. Structural Diagnostic Techniques
- Optical Holographic Interferometry NDT (common term HINDT)
- Speckle Pattern Interferometry (Electronic or Digital or TV Holography, common term ESPI)
- Speckle Shearography (Digital, common term DSS)
- Doppler Vibrometry (Laser Scanning. common term LDV or SLDV)

FIGURE 5.1 The basic elements constituting the needed information.

length positive lens (f_1) may be placed in the beam to produce a spherical (focusing) wavefront, and that same wavefront can then be collimated by using a second lens of a longer focal length (f_2) placed at a distance f_2 from the origin of the spherical wave, thus forming a collimated (plane) wave uniquely suited as the input beam for interferometry.

Interferometry is brought about by the separation (beamsplitting) of an incident plane wavefront into two beams, one made to produce a reference signal (unperturbed wavefront), the other made to produce a sample signal (modified wavefront) characteristic of the object or surface of interest under examination. A typical optical interference setup is shown in Figure 5.2. The two beams then recombine through one or more optics to construct an interference pattern consisting of alternate bright and dark bands (or fringes) with each pair separated by a half-wavelength ($\lambda/2$) displacement. An interference (or fringe) pattern is produced that may characterize the condition of any object inserted into one of the two beam paths and forms the measuring component of interference. The most accurate measurements in science are made possible by exploiting such interference properties of waves [6].

The phenomenon and application of interference depends on the *coherence* property of laser light, in other words, the ability of light waves to maintain a constant phase difference in space (spatial coherence) and over time (temporal coherence). Thus two light waves that are capable of interfering with each other are termed *coherent,* and the related study of optical phenomena has established a new field called coherent optics. This refers to the laser phenomena of *coherence, interference,* and *diffraction,* which are the fundamental characteristics of the interference techniques involved. *Diffraction* and *interference* are interrelated. The fringe pattern produced by interference of the light beams in Figure 5.2 can be used as a structural diagnostic tool provided a photosensitive medium has captured the information.

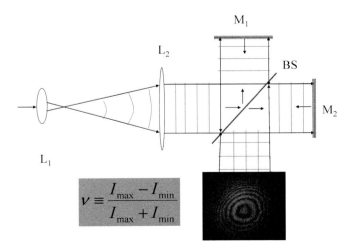

$$v \equiv \frac{I_{max} - I_{min}}{I_{max} + I_{min}}$$

FIGURE 5.2 Schematic of interference: after L, collimating lens, the laser beam is divided and reflected/transmitted at mirrors M to overlap and produce an interference pattern with visibility v.

One type of optical component that can usefully convey the interference pattern is called a diffraction grating because if a beam of laser light illuminates it from one side, then the transmitted light emerges as multiple diffracted beams from the other side. The spatial frequency of the grating ($f_y = \sin\theta/\lambda$, line/mm, and the grating spacing $\Lambda = \lambda/(2\sin\theta/2)$) is a measure of information density content, dominates the separation of the beams, and depends on the wavelength (λ) of the laser light, the angle (θ) between the recording beams, and Λ, the grating spacing produced. In linear records of interfering beams, by keeping within the exposure characteristics of the absorption of interference medium, the coefficient of second and higher order beams becomes negligible, and from the three emerging beams when the original beam illuminates the grating, the reconstructed wave emerges with amplitude linearly proportional to the original subject wave amplitude. The three waves can be expressed at the plane of the grating as a wave with real and imagery parts with an attenuation dependent on the absorption (t) of the photosensitive medium.

To collect and record an interference pattern from an ordinary object, one proceeds as follows: (1) split the beam into two paths of equal optical beam path length, (2) place an object of interest in one of the paths, (3) allow this throughput object beam to rejoin the reference beam from the other path so that they interfere with each other (creating fringes), and (4) capture the resultant interference pattern (interferogram) in a medium that generates a diffraction grating from its interfering beams capable of reconstructing by diffraction to the incident reference wave the original object wavefront. The test setup does not need to use a lens or any other optical element to image the light to the imaging plane, just the use of simple driving and diverging optics for the beams. For reconstruction only the original reference beam is required while the object beam is blocked. This is a brief description of *wavefield recording* and *reconstruction*. Holography is a technique used to record and reconstruct wavefields, and holographic interferometry is a technique used to measure any optical phase change between the recorded wavefields.

The superposition of two interfering wavefronts from the same coherent laser source (as shown in Figure 5.2) is produced by an optical arrangement called an interferometer. Interferometers in general assume division of an incoming laser beam into two equal paths, one to carry the signal of the object under study (termed either the signal beam or the object beam) while the other path carries an unperturbed reference beam. When an interferometer is based on wavefront or amplitude division, as in the Michelson interferometer shown in the figure, strict boundary conditions should be followed, satisfying coherence requirements on which interference patterns and fringe visibility v are based [1,2,7–9].

The range of applications of laser interference is remarkable, a principal application being holographic interferometry, which serves as a scientific and engineering tool for which it is uniquely suited and recognized [3–5,7–13]. The development of holographic interferometry has also influenced the development of several closely related measurement techniques based also on the use of laser light, in which the theory, practice, and application are very close and often complementary to holographic interferometry.

It was holographic interferometry that was first applied to detect subsurface damage in a Donatello statue in Venice and in a fifteenth-century panel painting [14,15], introducing the optical coherent interference measurement as a novel and alternative information source in analysis of structural integrity. Subsurface structural information, in terms of visually exhibited fringe patterns qualitatively and quantitatively evident, can be produced by complex surfaces and three-dimensional shapes. Defective regions became apparent by isolated irregularities or discontinuities in an otherwise continuous distribution of uniform interference fringes and can be visually located, sized at one-to-one scale, and restored.

Diagnosis is an important term signifying the indirect contribution of interference pattern information in the evaluation of the condition of the object under study. Diagnosis indicates the correct use of a source of information on interpretation of which the significant results derive; a lot of research has been conducted to interpret and establish the findings [14–17]. Diagnosis ($\delta\iota\alpha\gamma\nu\omega\sigma\iota\varsigma$) nowadays is used as a medical term. It derives from $\delta\iota\alpha$ = *through* and $\gamma\nu\omega\sigma\iota\varsigma$ = *knowledge*, signifying the ability to synthesize *through* the analysis of all the provided information the *knowledge* wherein the reasoning done to understand the cause is found. Structural analysis of artworks through laser interferometric techniques can provide evidence of an underlying condition or structural change, defect, or weakness. The interference fringe pattern analysis belongs to the category of inverse engineering problems [18]. Thus informative revealed interference patterns form the visual display of a structural condition that may lead us to diagnose a variety of root causes such as distribution of stress, thermal effects, or processes of moisture content equilibrium, from whole field fringe patterns to hidden defects such as cracking, loss, or separation (delamination) of materials and surfaces, or expanding voids and tunnelings from localized (isolated) fringe patterns.

In this context, the title of Chapter 5 is self-explanatory, signifying the use of *laser* sources producing a coherent metrology tool to generate the phenomenon of *interference,* upon evaluation of which *diagnostic* results about the condition of the artwork under inspection are extracted.

5.1.2 Interferometry with Diffuse Artwork Surfaces

The concept of laser interferometry for artwork structural analysis is founded on the inherent property of holographic interferometry, which allows a light wave scattered by an arbitrary object with random surface heights to be holographically recorded and reconstructed with such precision that it can be interferometrically compared with light scattered by the same object at a later time, as in real-time techniques. Alternatively, it can be compared interferometrically with a second holographic reconstruction of light scattered by the object, termed *double exposure* or *double pulse,* when a pulsed laser is used. Accordingly, holographic interferometry is defined as the interferometric comparison of two or more wavefronts at least one of which is holographically reconstructed, and the composite of these wavefronts is referred to as a holographic interferogram [9].

Holography is the recording of all light properties of amplitude, phase, and polarization and so it differs radically from photography, which is only the optical density as time-average of light intensity recorded. Before the invention of holography by Dennis Gabor [19,20], in one of his attempts to increase the resolving capabilities of electron microscopy, interferometry was only possible with optically flat surfaces. Soon after the invention of lasers and their use for holography by Leith and Upatnieks [21], this type of interferometry became known in a variety of investigations such as vibration analysis by Powell and Stetson [22], the development of double exposure and real-time holographic interferometry to study deformation and displacement of diffusely reflecting objects [23–26], and aerodynamic measurements, industrial and medical applications, and particle size analysis [27–32]. It was in that same period that holography was being introduced in the examination of artworks [12,13]. Speckle as an inherent property of lasers produced by photon interference due to the high coherence of emitted frequency tracked the path of development after its measurement properties were recognized [7,11]. For speckle interferometry, the speckle field decorrelation was put into consideration when examination of rough surfaces was required.

It was long after the invention of the holography principle by Dennis Gabor (1947) (Figure 5.3 shows the Gabor on-line holography setup) and the invention of the laser by T. Maiman (1962) that the first holographic recording of an artwork (a painted 15th-century wood carving by Donatello) took place in Venice in 1972 [15]. The potential of optical coherent techniques for art conservation, preservation, and production of 3D images was demonstrated, providing possibilities for unique applications. A significant advance in interferometric metrology applications came about with the advent of higher power, longer-coherence-length pulsed lasers used for holography, typically ruby- and Nd-doped lasers, which introduced inherent improvements and reduced the need for vibration isolation for hologram recording [33].

A subsequent series of systematic investigations on artwork diagnostics by a group of researchers at the University of Aquila [14,34–37] and the University of Firenze [38] brought about the application of holographic interferometry nondestructive testing (HINDT). The technique was inaugurated as a brand new conservation practice for structural evaluation and also as a potential alternative to existing x-ray methods, and it received significant interest from scientists, engineers, and conservators. Initially, detection of subsurface defects was the main aim of investigators trying to prove its feasibility to conservators. Painted artworks and sculptures were examined to visualize hidden defects without the need for physical contact examination.

Although most NDT applications require only a qualitative interpretation of holographic interference patterns to assess visually the condition of the tested object or component, later developments in the industrial field for the sake of graphical representation and quantitative evaluation of fringe patterns have also proven useful to artwork conservation [39–42].

By the end of the 1980s, the use of electronic image acquisition to record holographic and speckle interferograms was spreading, and new terms such as video holography and TV holography were introduced in structural inspection tools in the submicrometer range [43–51]. As the interferograms are now collected with a video camera and analyzed by computer algorithms, the electrooptical components involved are also playing a major role in data capture and assessment. The transition from optical to optoelectronic hologram acquisition involved various compromises, as the tolerances in higher resolution and measurement accuracy demanded. An additional strong point is that the video capture decreased the exposure time required and reduced consumables by avoiding wet processing of commonly used silver halide materials or expensive photothermoplastic films. During the 1990s, digital versions expanded the range of techniques possible and enriched various applications involving monitoring and measurement, reconstruction, digital 3D recording, laser archiving, replication and reconstruction, and so forth. The fact is that during the 1990s, laser use in art conservation became a well-known alternative for structural and analytical as well as laser cleaning procedures. There was also success in numerous other laser diagnostic applications in the field, such as laser spectroscopic techniques for accurate compositional definition [52–72] as described in Chapter 3. The implementation of laser technology research reinforced the expectation for novel alternative tools, methods, and practices in artwork analysis, preservation, and restoration.

From an historical point of view, although the rise of laser applications in artwork conservation started with optical holographic structural diagnosis as early as the 1970s, the more widespread use of lasers in artwork conservation has been established as competitive and complementary tools in a number of applications only during the digital decade of 1990s.

5.1.3 SUITABILITY PROVISIONS

Artwork inspection, more than other fields concerned with industrial applications, requires well-satisfied and specific suitability provisions to qualify a technique for routine implementation or application on real art objects.

Holography and related nondestructive interferometry techniques do not penetrate into surface volume and instead make use of diffuse or scattered surface illumination[1] and can be used only if the presence of a subsurface flaw under minor stress results in anomalous deformation of the tested surface. Laser irradiation does not interact with the surface, and induced stress is kept within the safe range of conservation standards. Common advantages and features of interferometry are shown in the table.

In the following paragraphs, some of these benefits are further highlighted:

(a) *Safe, not penetrating, irradiation*: Laser interference techniques use object surface illumination to acquire the signal of interest. For whole field techniques such as holographic and speckle interferometry, the beam diverges to illuminate either part of or the entire object surface, and thus the intensity due to the directionality property is diminished in favor of whole field and safety limitations important for sensitive surfaces.

The maximum divergence of the beam depends on the sensitivity of the recording medium (specified in $\mu J/cm^2$) — one of the recording medium characteristics crucial for holography — and the laser power output. The required energy density per unit

---

[1] See scheme of laser–matter interactions in introduction.

BOX 1: COMMON CHARACTERISTICS
Main Advantages

- Non-destructive, non-invasive, non-contacting
- No penetrating radiation, only surface illumination
- Safe radiation level for artwork and operator
- No surface preparation and no sample removal
- High information content with high spatial resolution
- Fast, repeatable and objective subsurface assessment

Complementary Functions and Advantages

- Remote on-field inspection, then laboratory-based analysis
- Spatially resolved whole field imaging with scanning matrix optional
- Qualitative assessment and quantitative evaluation

---

area of light to which the photosensitive medium is subjected is about 10 to 70 $\mu J/cm^2$ for most emulsions according to transmittance–exposure characteristics (t-E) of most films and about 150 $\mu J/cm^2$ for glass substrate, and relevant lower values are approximated for CCDs. The divergent beam spreads even further after diffusely reflecting from an artwork surface, further decreasing the signal level and the collected energy density.[2] Morphological and chromatographic monitoring studies have been performed[3] and have confirmed the safety of laser illumination. In particular, collimated cw laser beams at λ = 628, 532, and 450 nm have been directed for 30 to 60 minutes onto colored surfaces without effect (maximum laser output power implemented up to 30 mW). The analysis with multispectral imaging and scanning electron microscopy showed no evidence of any chromatic or morphological alterations [73]. Even in techniques that implement an unexpanded collimated beam, such as laser vibrometry, a low-power laser is used as a source obeying also the Hurter–Driffield curve satisfying linear recording and is as safe as the diverging beam arrangements used with higher-power pulsed lasers.

(b) *Nondestructive*: The term nondestructive implies the detection of structural imperfections by procedures that do not require destruction, sample removal, or any significant alteration of the object being tested. Holographic speckle interferometry, shearography, and vibrometry are typical nondestructive testing techniques. Conventional nondestructive testing uses a variety of techniques including transmission imaging using x-rays, gamma rays, ultrasound, acoustic emission, visual observation by microscopes and stereoscopes, thermal imaging cameras, or spectroscopic imaging with relevant advantages and disadvantages with which laser techniques compete.

---

[2] In optical holography with films, amplitude transmittance defines transmittance–exposure characteristics to set exposure time t_e and reference to object irradiance I_R/I_O.

[3] Undertaken under the EC R&D project LASERART.

(c) *Resolution and sensitivity*: The methods of optical coherent interferometry provide the highest available spatial resolution when compared to other optical means (since laser wavelength and recording beam angle define the spatial frequency $f_y = \sin \theta/\lambda$ line/mm, e.g., in typical hologram recording with $\lambda = 633$ nm and $\theta = 30°$, mean fringe separation is $\Lambda = 1.2$ µm and $f_y = 790$ line/mm with the offset angle between interfering beams being a factor to increase or decrease it at will depending on the application, e.g., small angles for glass inspection or CCD recording). There is though also a dependence on optoelectronic parameters, since the detection sensitivity varies according to (a) the spatial resolution capabilities of the recording medium (number and size of pixel vs. speckle size), (b) the aperture size A (inversely proportional to resolution, the smaller the A, the bigger the speckle size), (c) the field of view of the system (in regard to the illumination area determined by the expanded beam), and (d) the distance from the object (inversely proportional, as the sensor furthest from the object FOV increases, the resolution decreases). One should choose the technique that best suits the particular circumstances and object under examination (e.g., for an icon of typical dimensions and good conservation condition one can expand fully the beam over the surface, whereas in larger artworks or advanced deterioration a beam of smaller diameter can ensure higher resolution of each illuminated part of the object of interest). The high spatial resolution and sensitivity[4] also minimize and self-limit the amount of induced stress required to generate useful displacement and warrant the response of the technique to minimal, transient, and reversible range of nondestructive excitation. Sensitivity affects experimental settings such as exposure time, since the longer experimental duration may introduce external disturbances into the record if proper isolation is not provided, and the self-limiting feature is extracted by the separation of fringes (primal and secondary).[5] Sensitivity indirectly affects also the recording of hygroscopic materials, such as paper and canvas, for the defect detection of which, owing to high responsivity to the surrounding environment, fast recording procedures (of the order of a few tenths of msec) with acquisition grabbers or, ideally, Q-switched double pulse recording, are suggested.

(d) *Complex surface texture and shape*: Several advantages for the investigation of complex objects are provided and, in particular for HINDT techniques, include unobstructed inspection of objects having fairly complicated shapes and rough surfaces. Recording of complicated shapes may not be equally successful with other interferometric techniques, as numerical processing or interference principles assume that the object beam is normal to the hologram or recording plane. Also, rough surface texture may generate speckle decorrelation in some geometries of speckle interferometry, which can merely be minimized with holographic speckle recording [11]. As a general rule, shape, texture, and artwork dimensions do not imply any theoretical or practical unsolved limitation.

---

[4] Sensitivity of the recording emulsion is termed after the exposure E at which maximum diffraction efficiency occurs, $n_{diff} = I_{total}/I_{recon}$, with I_{total} the average irradiance of the reconstructed object wave and $I_{reconstruction}$ the irradiance of the reconstructed wave.
[5] Primal refers to fringe separation in holographic recording and secondary to reconstructed fringes of the interferogram. Further explanations cannot be employed in the context of this chapter.

(e) *No surface preparation*: The lack of special requirements for surface preparation is another strong advantage of laser interference techniques for artwork conservation. However, special attention to black background or encrustation is necessary, since they may reduce the contrast that lowers visibility, and they can make it difficult to detect, or easy to misinterpret, fringes; in some extreme cases they make it impossible to get useful data (in such cases, a superficial first cleaning prior to investigation or proper selection on object illumination angle should be chosen as first step).

(f) *Optional optical 3D display*: An equally important characteristic, particularly of optical holography, is the appearance of results in a visual 3D field mode. The processing of the film is similar to the photographic process and is familiar to many conservators. The 3D representation of the image of the objects provides a good understanding and appreciation of space coordinates with regard to localization of overlapping fringe patterns. Even so, the faster digital recording and optoelectronic reconstruction are often preferred to the optical reconstruction practices for field applications.[6]

(g) *Broad applicability*: Laser interferometry can be suited to a wide variety of inspection requirements in artwork diagnosis (as shown in the box) with most applied to

- State of conservation and defect detection
- Preventive deterioration on simulating environmental conditions
- Comparative long-term assessment or interventions studies

The broad applicability for diagnostics and preventive conservation can be remarkably flexible in the hands of an experienced and properly trained conservator to suit

---

BOX 2: APPLICATIONS
DIAGNOSTICS

- Defects detection and identification – location, size, structure morphology (detachments, cracks, inhomogeneities, stress concentration, and so forth in scale and fidelity)
- Monitoring of conservation actions and interventions
- Environmental studies and materials kinetics
- Estimation of conservation condition - existing defects, their interconnection and their propagation

PREVENTIVE CONSERVATION

- Long term alteration / aging, environmental effects
- Sudden alteration / accident in shipping and handling

---

[6] Combined software development with reconstruction algorithms and fringe projection may replace the originally offered three-dimensionality by depth coordinates on monitor (e.g., edge enhancement algorithms with superimposed speckle data).

uniquely developed methods to various conservation demands as well as for research in conservation science.

5.2 HOLOGRAPHY PRINCIPLES

The term holography signifies the writing of the whole (entire) information of the captured light (Greek όλο = *holo,* whole; γραφή = *graphy,* writing). This term, in contrast to photography (Greek φωτο = *photo,* light; γραφή = *graphy,* writing), indicates the additional recording of the wavefront phase, which for holography signifies the property of light carrying the depth information about the object. The object wavefront is recorded by interference with the reference beam and is reconstructed by diffraction of the reference beam, Figure 5.3. The reproduced hologram is the exact replica of the light wavefront at the instant of recording. Thus, by holography, the frequency, polarization, phase, and amplitude of the light used are recorded.

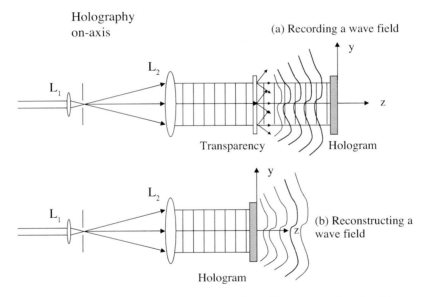

FIGURE 5.3 Gabor on-line holography: (a) Recording; (b) reconstructing the object wave.

BOX 3: SUMMARIZED HOLOGRAPHIC PROPERTIES
Holographic Properties

- *Lensless imaging*
- *3-dimensional imaging*
- *Paraxial viewing*
- *Multiple point interference*
- *Multiple image interference*
- *1-to-1 scale reconstruction*
- *Interconnection of information*

Holography has unique straightforward properties summarized in Box 3, and relevant bibliography references are provided [1,8,10].

5.2.1 OPTICAL APPARATUS FOR HOLOGRAPHIC INTERFEROMETRY

A typical optical system for off-axis holography is shown in Figure 5.4. It allows hologram viewing off the laser beam axis. A more detailed layout for recording (off-axis laser transmission holograms) is shown in Figure 5.5. The beam is divided into an object beam OB and a reference beam RB by an optical element, the beamsplitter BS. The beamsplitter is also used to select the amplitude ratio of the two beams, selecting the more intense (usually transmitted) beam for the object path, since after reflection from the object only a small fraction of the laser light will arrive to the hologram plane H for recording. A microscope objective and pinhole assembly, called a spatial filter collimator SFC, is used to ensure a high-quality reference beam at the recording plane. Typically in this application and as shown in the schematic, a spatial filter should be the last optical element used on the reference beam path before the photosensitive medium, to minimize diffraction effects due to dust particles on optical components. A second SFC may also be used before the beamsplitter to filter spatially both the object and the reference beam and act as beam expander and collimator. To increase the object beam diameter (to illuminate a part or the whole extent of the object), a diverging concave (negative) lens L is often used.

The holographic apparatus must be mounted so that relative motion of the components is eliminated. This is reasonably easy to accomplish if all components, including the laser, are mounted stably on the same rigid surface. If the laser is mounted separately, at least the rest of the optical components, the object and the

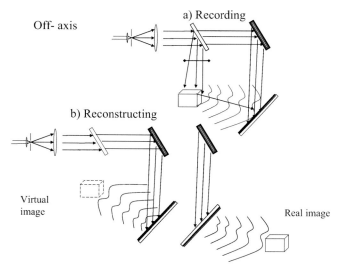

FIGURE 5.4 Off-axis holography: (a) recording the hologram; (b) reconstructing the real and virtual image.

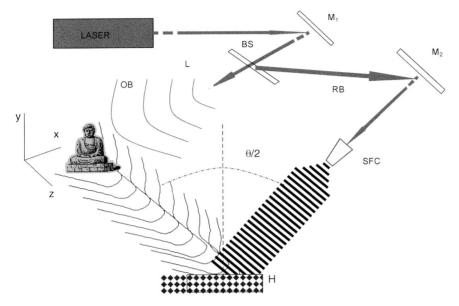

FIGURE 5.5 Schematic setup for recording of holographic interferograms (based on laser transmission holograms). M: mirror; BS: beam splitter; L: concave lens; OB: object beam; RB: reference beam; SFC: spatial filter collimation; H: hologram plane.

plate holder for the photosensitive medium, should be mounted on a common surface. Vibration isolation provided by pneumatically isolated optical tables achieves high rigidity and is best suited for this kind of experiment. Nevertheless, even low-cost custom-made supports made of sand tables or even granite tables mounted on inflated innertubes can be used with success, provided a long settling time for such supports is used prior to first exposure and a short interval is suited for the excitation and the artwork involved.

The precise geometric layout of the holographic system is not critical and can be adjusted to a variety of optical table restrictions provided the difference in the path length of the two beams (from the beamsplitter to the hologram plane) respects the coherence length of the laser (with the upper limit of the delay paths being the coherence length). The visibility of the interference fringes decreases as the optical path length imbalance increases, affecting the fringe contrast, quality, and brightness of the final image. An ideal hologram should be linear to constitute a set of carrier interference fringes of spatial frequency f_y that are modulated in amplitude by $a_0(x, y)$ and in phase by $\phi(x, y)$, have a high signal/noise ratio, and have a high diffraction efficiency in order to result in a bright and clear reconstruction. Thus equal beam paths are better by design. Chemical development concerned in optical holography affects the diffraction efficiency,[7] and for good quality holograms quantitative

---

[7] Diffraction efficiency n of the hologram, termed the efficiency of the hologram to reconstruct a bright image, is measured from the reconstructed beam and the reconstructed intensity of the first-order diffracted beam, $n_{diff} = I_{total}/I_{recon}$.

knowledge about film characteristics for exposure and development procedures is essential. Considerable practical information regarding holographic systems, and details for all types of display holograms, ranging from simple single-beam reflection to white light transmission and rainbow holograms, can be found in the publication of Saxby [12].

Once the hologram is formed by exposing and developing the plate, the object wave can be reconstructed by replacing the plate in the holder and blocking the object wave. An observer located opposite and looking through the hologram will see a realistic three-dimensional virtual image of the object, Figure 5.4.

For more insight on the holography principle and for thorough mathematical and physical appreciation of the phenomena involved, the interested reader should refer to the books referenced at the end of this chapter.

5.2.2 Recording Principle

Holographic interference techniques share a common measuring approach, namely, the need to generate two slightly different images of (or wavefronts from) the same surface, which will overlap to produce differential fringes of displacement (optical path length), revealing the change in surface shape.

The comparison is achieved by recording the object at two different times, namely, before and after an induced excitation (or perturbation) has been purposely applied to the object. The excitation generates a slight displacement of the object from its initial position (see Figure 5.6), forming an optical wavefront whose complex amplitude in the hologram (recording) plane, $U_2(x,y)$, is different from the original one, $U_1(x,y)$. If a coherent laser source illuminates the surface before and after this induced displacement, an irradiance pattern due to interference between the two waves is formed at the observation plane. If during the procedure a photo-sensitive medium (such as a fine silver halide grain photographic plate) is placed at the interference plane to capture first the initial state of the object wave $U_1(x,y)$, adding a reference wave $U_R(x,y)$, and then the displaced state of the object $U_2(x,y)$ together with $U_R(x,y)$, and if it is developed appropriately and illuminated, then, with the $U_R(x,y)$, from diffraction an image of the object will be reconstructed with complex amplitude proportional to $U_1(x,y)+U_2(x,y)$ and irradiance proportional to

$$I(x,y) = |U_1(x,y)+U_2(x,y)|^2 \qquad (5.1)$$

The minute change of the object at the second recording state though does not affect any light property other than the phase of the reflected light. It undergoes a slight optical path length change adding, in the second recording, an additional phase change $\Delta\phi(x,y)$, so that

$$\begin{array}{ll} U_1 = U_0(x,y) & U_0(x,y) = a(x,y)\exp[-i\phi(x,y)] \\ & \text{and} \\ U_2 = U_0'(x,y) & U_0'(x,y) = a(x,y)\exp\{-i[\phi(x,y)+\Delta\phi(x,y)]\} \end{array} \qquad (5.2)$$

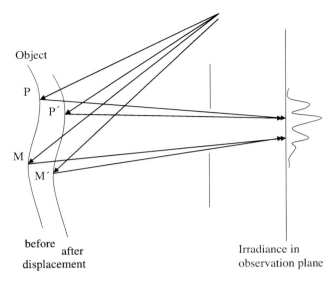

Object

P

P′

M

M′

before
after
displacement

Irradiance in
observation plane

FIGURE 5.6 Schematic of the irradiance pattern generation due to surface induced displacement.

Therefore the irradiance of Equation (5.1) now becomes

$$I(x, y) = \left| a(x, y) \exp[-i\phi(x, y)] + a(x, y)] \exp\left\{-i[\phi(x, y) + \Delta\phi(x, y)]\right\} \right|^2$$

$$= 2a^2(x, y)\left\{1 + \cos[\Delta\phi(x, y)]\right\} \tag{5.3}$$

Thus Equation (5.3) represents the irradiance of the object, $a^2(x,y)$, modulated with a cosine-dependent fringe pattern of $2\{1 + \cos[\Delta\phi(x,y)]\}$ as shown in Figure 5.7. The interference pattern will be permanently stored for reconstruction and photography. The density and distribution of the fringes will determine the reaction of the object to the excitation used. Dark fringes correspond to contours of odd integer multiples of π phase change in $\Delta\phi$, whereas bright fringes correspond to even integer multiples of π as is evident by Equation (5.3). Depending on the application and the excitation load, $\Delta\phi$ is related to physical quantities such as object surface displacement due to induced dimensional change of the artwork. The approach to quantify phase changes by visually examining the fringe pattern is given by the generalized formulation of Vest [7]. It is done by first counting the fringes N to a chosen location from a zero order fringe ($N = 0$) located at a convenient starting point in the interferogram. Then by the number of fringes counted and recognizing that $\Delta\phi = 2\pi N = N\lambda/2$ with $N = 1$, 2, 3 ... consecutively, the total displacement (in units of a half-wavelength) between the starting point and a chosen location is calculated and the resultant displacement is expressed in microns. For flat surfaces with system geometry sensitive to normal displacement, the fringe interpretation is as above. In order

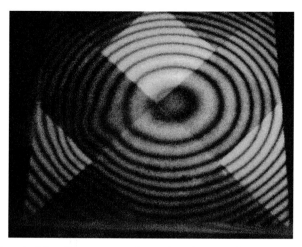

FIGURE 5.7 A typical irradiance pattern due to interference of object displaced and initial position, surface covered with bright maxima and dark minima of light intensity.

to relate the overall vector displacement of points on the object surface, the surface is treated as a collection of point scatterers with propagation vectors in the plane and toward the observer. The illumination direction and phase differences are as related in consideration to system propagation ($k1,k2$) and displacement sensitivity (L) vectors. Thus the phase shift $\delta = \Delta\varphi(x,y)$ of light traveling to ($k1$) and scattered by a particular object point (P) to a particular direction with propagation vector ($k2$) is the result of object point displacement $\varphi_2 - \varphi_1$ and signifies a sensitivity vector (L) of displacement. The approach may use *a priori* knowledge of the direction of the sensitivity vector to determine the magnitude of displacement. For most setups, the sensitivity vector will have the direction close to the normal to the surface, hence the fringe order is most sensitive to normal displacements with sign ambiguities involved (since the cosine, $1 + \cos\delta$, is an even function and the same fringe pattern arises from positive $+z$ and negative $-z$ deformations). The vector approach for fringe interpretation, in brief, is based on vector analysis of system geometry and concludes in a coefficient matrix determined entirely by the geometry of the holographic system and the wavelength. Nevertheless, even in this vector approach, using either multiple angle or multiple hologram analysis, the phases are calculated by counting fringes from different viewing points; implying that a valid zero motion fringe for assignment of the zero-order fringe is a relative easy and accurate quantitative measurement tool for dimensional deformations delivered to the hands of the conservation expert.

Therefore the effectiveness and uniqueness of holographic interferometry (HI) for the application to artwork conservation is based on critical properties that are summarized in the following box.

BOX 4: CRITICAL HI PROPERTIES

- The information content is sufficiently high to permit the recording and reconstruction of fine details of complicated wavefronts with great fidelity and it is this property that allows the study of 3-D diffusely reflecting objects with complicated shape and texture with HI.
- In double-exposure holographic interferograms made from the same object, the amplitude is divided temporally in two. This means the wavefronts that interfere traverse essentially the same length path in space at two different times, and since the superposition of light waves is a linear process, this fundamental property allows almost any objects from rough scattered surfaces even to low optical quality glass to be studied (cancellation of phase errors occurs).[8]
- Optical visualization of the reconstructed holographic object wavefront allows an observer to study the response and allocate the disturbances of the interference pattern to specific areas on the artwork without ambiguities regarding the scale.
- Responses as exhibited in fringe patterns are correlated directly to object condition since the optical wavefronts are modulated directly by the object. This property can be useful for direct estimation of environmental influences, for example.
- Archiving and comparing transient optical wavefronts or external effects and accumulative influence of interventions is possible, and such studies can be important in conservation sciences.

In conclusion, although the formation of fringes of holographic interferometry is a rather complicated phenomenon collected under strict boundary conditions, simple geometric optical arrangements and experimental conditions can be used to generate a unique record of deformation effects provoked by differential displacement of the artwork under investigation. From the rather straightforward observation of fringe patterns of optical or even digital speckle holographic interferograms, one can conduct a critical task for conservation, namely, the diagnosis of the structural conditions of a work of art.

5.3 HOLOGRAPHIC INTERFEROMETRY NDT FOR DIRECT ASSESSMENT OF STRUCTURAL CONDITION OF ARTWORKS

The differential response generated from localized imperfections is often small when compared to the object size and the region over which the induced excitation is made. But the interferometric system configuration can still be chosen to accomplish structural assessment down to the wavelength in magnitude, whether on a small scale or over an extended area. For artwork structural diagnosis, two kinds of

---

[8] Not the case in the refraction by curved surfaces.

interference fringe pattern formation could be used as the information source to assess the actual condition of the artwork under investigation. These are:

(a) *Localized interference fringe patterns* are characterized by a systematic fringe[9] feature limited over a relatively small area of the object.

(b) *Whole-body interference fringe patterns* are characterized by a regular distribution of fringes over an extensive object area. Whole-body fringe patterns may be interrupted in part by localized fringe patterns, whereas the opposite is not usually observed.

5.3.1 LOCALIZED FRINGE PATTERNS

In the investigation of works of art and antiquities, a primary aim is accurately to detect and assess hidden bulk defects. Interferograms accomplish this aim by the generation of localized fringe patterns defined by distinct area limits as compared to the whole-body patterns induced by the excitation displacement. The localized fringe patterns can be used to detect a variety of defect types categorized in Box 5.

BOX 5: DEFECT CLASSIFICATION		
Defect Description	**Category**	**Fringe Pattern Description**
• Detachments between layers (e.g. wood/canvas)	*Cat. A:*	*one or more concentric fringes*
• Material voids in bulk (e.g. stone porosity) • Inclusions near surface or in surface contaminated layers (oxidised varnish layer) • Unevenly applied restoration layers • Inclusions between layers		
• Cracks in the support or ground	*Cat. B:*	*one or more curvature changes*
• Materials separation (delamination)		
OTHER[10] • Surface disruption	*Cat. C:*	*one or more fringes with sudden stop*
• Stress concentration	*Cat. D:*	*increased fringe density*
• Defect propagation	*Cat. E:*	*external fringes continuation to a specific direction*

---

[9] Small local defects often result in one fringe pair displacement or are visualized by causing in the fringe pattern irregularity in just one fringe pair, e.g., inborn cracks under the surface cause dead end fringes over a minute region.

[10] Categories C–E can be summarized as a general category of other defects.

The localized fringe characteristics in the defect cases, which allow one to distinguish separate defect categories (A–E), are evidenced through a distinct fringe shape that typifies a visual fringe pattern provoked by a corresponding defect. Each visual appearance forms a distinct category of localized fringe patterns on which potential defects are correlated. This classification is independent of artwork and material, since the visual appearance of localized fringe patterns due to mechanical subsurface discontinuity is similar, depending only on the defect type [74].

This statement signifies the importance of *a priori* knowledge of the technique or of the construction of a work of art, since a void inside a solid statuette may be represented with the same concentric fringe pattern as a detachment in a multilayered panel or fresco painting. Thus only knowledge about the artwork allows us to declare the defect as a void or a detachment. Despite the visual convenience offered for the analysis of interferograms, one should keep in mind that even minor differences in defect structure affect the fringe pattern's geometrical configuration in terms of curvature, thickness, fringe separation, etc. However, this sensitivity should not be a confusing factor, since *the topological features of each fringe system category corresponding to a defect category are kept similar,* regardless of the variety in construction and materials, e.g., concentric fringes with different curvature are still concentric fringes, and curved fringes with different curvatures are still curved fringes, etc. Furthermore and very important, if thorough analysis is performed and there exists the *a priori* knowledge,[11] the variety in details of a fringe pattern presents only additional information, which is useful toward detailed interpretation of the subsurface condition.

Therefore, by integrating the localized fringe pattern categories, we can derive a table of five basic defect indicative fringe patterns that allow direct structural diagnostic analysis in qualitative terms [42,75,76]. These systems will be further described in a section to follow and highlight the direct diagnosis that may be offered to the diagnosis operation with qualitative mapping of subsurface defects.

5.3.2 WHOLE-BODY FRINGE PATTERNS

The distribution of a whole-body fringe pattern (as shown in Figure 5.7), representing the whole-body displacement[12] due to the provoked dimensional change induced by the excitation mechanism, provides a source for visualizing and assessing a variety of structural details useful in conservation applications, as is described below.

5.3.2.1 Sensitivity Indicator

The overall alteration of a work of art evidenced by optical path changes which are induced by a load is a product of the transient energy exchanged toward its equilibrium. The phenomenon is liable for the generation of a fringe pattern covering the surface

---

[11] In artwork inspection, the restorer is provided with a condition report allowing correlation of results.
[12] Not the rigid-body displacement assigned to undesirable external disturbance in recording system components but the overall reaction of the artwork due to the induced load.

with a distinct fringe density witnessing the structural sensitivity of the artwork to the induced load [49,53,77,78]. The fringe density[13] of the whole-body displacement derived from uniformly induced excitation can be estimated provided there are regions without defect. Changes in the fringe density can be induced by altering excitation settings. Monitoring of fringe density gives a measure of sensitivity.

This information may be used for the estimation of:

- Sensitivity to induced influences (T, RH, pressure, etc.)
- Sensitivity to interventions (before and after, or sequential)

Or it may be used in combination with local defect identified categories to assess

- Interrelation of defect categories
- Visualization of defect propagation toward specific directions

Extracting responses as indications of sensitivity may be done to lead to the redesigning of conservation and preventive deterioration treatments or strategies.

5.3.2.2 Preventive Conservation

Experimental results have further extended procedures and methods to include even preventive conservation with new technological approaches. Thus for preventive conservation schemes, environmental influence has been studied and monitoring schemes have been suggested [37,53,79–81]. The critical advantage over traditional methods such as strain gauges, data loggers, etc., is that information on the influence on the structural condition is assessed directly from the interferometric response of the artwork itself, not indirectly through readings of any externally employed instrument. Similarly, one can monitor conservation interventions by comparing the responses before, during, and after the treatment, as in exemplary studies that monitored laser cleaning by means of a holographic interferometry workstation [82]. The distinct feasibility is also proven through the fact that the same system can be used after treatment for assessment of long-term effects [62,63,83–90]. Integration of complementary optical coherent interferometry techniques, adjusted for conservation demands, to output a flexible multitask structural diagnostic sensor incorporating an advantageous combination of method, instrument, and procedure in a user-friendly environment is under study in an FP5 EC funded project, LASERACT[14] [88]. The flexible applicability of the system aims to allow a wider spread of the use of the techniques and familiarize the conservation community with this type of instrumentation and analysis by being used either for defect direct diagnosis or for preventive and sensitivity studies.

In general, holographic interferometry has demonstrated an ever-widening range of applicability with correspondingly modifiable methods depending on the artworks' diagnostic challenge to be addressed.

---

[13] Spatial resolved fringe number over a pre-defined finite object area.
[14] LASERACT contract number EVK4-CT-2002-00096, www.iesl.forth.gr/programs/laseract.

5.3.3 CHOICE OF INVESTIGATION PROCEDURE

(i) Aim-Dependent

Three main recording procedures stand out depending on the recording allocation with respect to the excitation:

1. The interferogram is formed by projection between the actual object and its prerecorded and repositioned hologram while the object reacts to transient effects. The scheme is termed real-time recording since it allows experiments to be performed in real time and, specifically in digital version, the cumbersome exact hologram reallocation is avoided. The excitation should be kept in its lower possible limits for successful fringe pattern formation, and a fast recording device is to be used for image capture.
2. The interferogram is formed with only one rather long exposure during the excitation time. This scheme is termed time-averaged recording, and the suitable excitation used for detection of vibration nodes is acoustical in nature.
3. The interferogram is formed by double exposing the object to the laser light with the excitation taking place between the exposures. This scheme is termed double exposure or double pulse holographic interferometry, and the thermal excitation is rather effective for allowing longer time duration for differential defect detection.

The double exposure recording with thermal excitation uniquely allows

- Choice of rather large and versatile experimental configurations and settings
- Broader selection of excitation duration and intensity
- Variable time interval between exposures
- Variable settling time after first exposure and excitation
- Exposure duration by alternate use of cw and pulsed lasers

The choice of the investigation procedure is primarily dependent on the aim of the investigation, e.g., in examination of musical instruments the time-average method can be used while for environmental studies the real-time method is more suitable. However, owing to the flexibility of the double exposure technique, several methods using multiplexed sequential or comparative recordings have been introduced to expand the range of double-exposure applications [53,90].

(ii) Standardized Investigation Procedure

Depending on the measurement challenge to be addressed, appropriate methods may be developed. Any methodology employing the double exposure procedure consists of a few basic steps as shown in Table 5.1.

It is appropriate then to summarize some notes and observations regarding the double exposure procedure.

(1) *Thermal monitoring*: After the artwork has been set on the object holder, a thermal monitoring prior to investigation is performed as summarized in Table 5.2.

TABLE 5.1
Key Steps in Double Exposure Procedure

1st	Recording of the artwork in its initial condition
2nd	Excitation induced to the surface (thermal, mechanical, acoustical load)
3rd	Recording of the artwork in its displaced position
4th	Developing[15] or computer processing the interferogram
5th	Visual or image processed evaluation of the fringe patterns

Thermocouples on the surfaces provide the initial temperature. Thermal loading by IR lamps is then applied with different time durations and at different locations and in relation distances to exposed artwork surface. The last two columns show the time and temperature limits in which the complete experiment should be performed. So, for example, when the temperature difference ΔT increases by 3°C with a thermal load directed from the front, the operator has 25 seconds in which to collect the second exposure. As shown in Table 5.2, the reaction of the artwork to the induced load determines the experimental settings for the exposure.

According to the steps of Table 5.1, the first exposure (recording) is done, then the excitation load is applied. The operator then divides the available time for the second exposure: half of the time allocated is made the interval between exposures for the thermal load to affect the whole artwork, and the other half is for the second exposure recording depending on the film exposure time. The exposure time for a film is calculated by dividing the medium sensitivity by the irradiance measured on the recording plane (in $\mu J/cm^2$) [7,12]; in the case of cw lasers, it is better if the exposure time is kept in the order of from seconds to a few tenths of a second. It is recommended that a 1:1 reference-to-object-beam ratio is kept for holographic interferometry, but this ratio can be increased if it is necessary to reduce exposure time. The operator may also choose to use a variable interval time before the second exposure to record surface displacement sequentially in different instances and positions [87,89–90].

Despite the exposure dependence on laser power and object beam divergence, which is diffusely reflected to the recording plane with considerable energy loss, the total experimental time after excitation should not exceed the monitored time of alteration of the object studied. This is of rather critical experimental importance, since if the second exposure is made only after the artwork has returned close to its initial position, no relative point displacement will be present, and a useful holographic interferogram will not be recorded. This is because the induced dimensional change is minute and transient. The same argument holds true for the time interval after the excitation, which allows time for the disturbance to spread over the entire object volume, to excite all defects, and to stabilize the air temperature between the surface and the recording plane. If the second exposure is taken immediately (or too

---

[15] For silver halide photosensitive mediums, there are special procedures similar to photographic process extensively described in literature or in the film packs of distributors. For optoelectronic means, special packages for numerical reconstruction of holograms should be purchased or developed mostly based on Fresnel diffraction theory and Fourier transforms. See references at the end of the section.

TABLE 5.2
Exemplary Definition: Excitation and Time Interval between Exposures

Initial temp. (at 1st exposure) °C	Duration of thermal load (sec)	Direction of thermal load	Final temperature °C	ΔT °C	Remaining experimental time (interval and 2nd exposure) (sec)
23	5	front	26	3	25
23	5	back	26	3	10
23	10	front	29	6	90
23	10	back	29	6	50
23	15	front	30	7	50
23	15	back	30	7	50
23	20	front	31	8	110
23	20	back	31	8	120

soon) after the excitation, the whole artwork will not yet be affected, and except for the relative displacement of some surface areas, the interferogram will again not record useful defect information. The importance of the relation between the applied excitation and the cooling down process to return to equilibrium, which dominates the available time of experimental completion, leads to the consideration of room temperature and the dimensions (mass) of the artwork during the experimental preparations. Special attention needs to be given to the representative graph of the Newton law of cooling (exponential time dependence). Experimental observations show that the mechanical work provoked by the thermal energy induced during the excitation period — whatever the initial temperature settings are — leads to the most resolvable fringe patterns at the middle of the cooling-down curve dependent on the resolving capabilities of the recording medium [77]. An *a priori* estimation of the cooling down curve for each artworkt is thus suggested to specify for the time interval before the second exposure (see ± of exposure in Figure 5.8).

(2) *Direction of excitation load*: It may be obvious that the direction of the induced load should be onto the illuminated (front) surface, and this is the common practice. However, this method is not always sufficient to allow visualization of deeper defects. This is often the case in multilayered artworks mainly because of the position of defects throughout the painted volume and loose structural adhesion between multiple complex layers. The monitoring graphs in Figure 5.8 demonstrate the difference of thermal distribution for various time durations in regard to the direction of excitation and the measurement surface, termed front and back as the two surfaces, with the front being the one facing the object beam illumination.

Despite the effectiveness shown for both excitation directions, the risk is hidden in the lack of adhesion between the surfaces as shown in Figure 5.9, where the thermal excitation cannot be transported to the opposite surface. This case of adhesion loss between surfaces will not allow formation of a useful interferogram if excitation is applied only from any of the opposite surfaces, regardless of the duration of excitation.

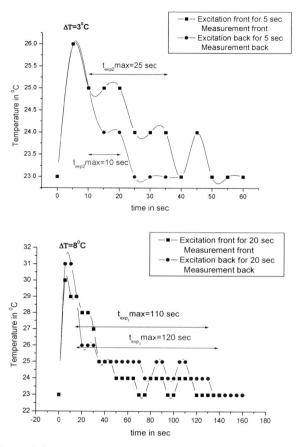

FIGURE 5.8 Thermal alteration monitored through probe reading for different durations and directions of induced alteration.

The above effect can also be demonstrated with relevant unsuccessful formation of holographic interferograms shown in Figure 5.10. The excited back surface did not affect the illuminated front surface, and no detectable defect fringes were formed.

Besides some critical aspects in the excitation method of the opposite direction of excitation to illumination that may prevent the collection of useful holographic interferogram recording (in terms of defect indicative fringe pattern formation), there should also be taken into account problematic situations in using the same direction of excitation and illumination. An example is found in the case of multilayered structures where the surface being excited and illuminated for recording suffers from partial or extended detachment from the supporting medium. In fact, for such cases, the surface layer is attached only at the edges of the frame, thereby visualizing only the relative rigid-body displacement of the surface similar to membrane-type displacement obscuring visualization of smaller localized defects, as is shown in the example of Figure 5.11.

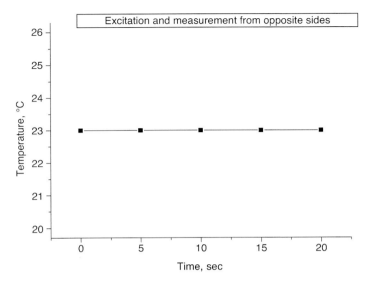

FIGURE 5.9 Case with cohesion failure.

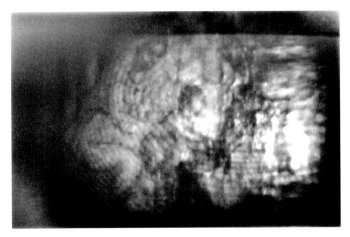

FIGURE 5.10 Holographic interferogram recorded with thermal excitation direction opposite to illumination surface. No evidence of fringes due to local defects is apparent. The fringes seen visualize only the effect from the well-adhered edges toward the center.

Thus total excitation divided in half and directed simultaneously to both front and back surfaces has been experimentally proven to produce optimal results in terms of overall fringe distribution as well as in overall defect visualization. The example shown in Figure 5.12 shows a multilayered panel painting[16] excited by dividing the total excitation time in half for two excitation directions to allow the thermal load to be directed to both surfaces.

---

[16] El Greco "Adoration of Magi" Benaki Museum of Athens, conservators Stergios Stassinopoulos and Nikos Smyrnakis.

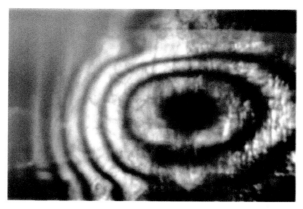

FIGURE 5.11 The same excitation to illumination direction may fail to provide information on defects if the illuminated surface suffers from membrane-type detachment from the support. The overall displacement of the whole surface does not allow smaller localized defects to be exhibited.

FIGURE 5.12 Overall applied excitation reveals higher information defect map than single sided. Local detachments in a variety of depth are visualized through interaction of both sides.

The optimum results obtained with this type of excitation load and direction are attributed to the thermal and mechanical interaction of surfaces that effects the detachments and defects over depth. The same results have been confirmed in a number of cases of artworks of similar construction and the "overall" procedure is suggested here as the optimum method by far for visualizing defects. The excitation method has been applied with the same success even in solid materials and works of art. Thus the uniform temperature change (thermal excitation) induced homogeneously to all environment exhibited surfaces produces volume effects (displacement due to mechanical work product) capable of being detected from the illuminated surface. Thus Table 5.2 can be expanded to Table 5.3.

The spatial frequency shown in the table is a measurement of fringes across the horizontal or vertical axis from the object interferogram used to measure and extract normal vs. defected fringe distributions. The approach was proposed during the EC

TABLE 5.3
Extraction of Defect Localization

Overall heating duration, sec	Initial temperature, T_0, °C	Final temperature, T_1, °C	Temperature difference, $DT = T_1 - T_0$	Spatial frequency on defect, fringes/cm	Spatial frequency on undefected area, fringes/cm	Spatial frequency difference, fringes/cm	
1	24,9	25,7	0,8	0,13	0,17	−0,03	
2	25,2	26,4	1,2	0,67	0,25	0,42	WELL
4	25,6	27,7	2,1	0,80	0,50	0,30	DETECTED
5	26,2	28,6	2,4	1,07	0,58	0,48	DEFECT
7	27,4	30,8	3,4	1,47	0,83	0,63	
9	26,2	30	3,8	2,00	0,92	1,08	WELL DEFINED,
12	26,5	31,3	4,8	2,93	1,33	1,60	LOCALIZED,
15	26,3	32,1	5,8	3,33	1,58	1,75	DEFECT

funded project LASERACT[17] to allow automotive procedure for computer driven defect localization, particularly useful for the fast qualitative analysis of interferograms in routine conservation applications (see example in Figure 5.20).

(3) *Exposing*: Independent of the recording media, which carry their own distinctive requirements and optical geometries [11], the operator should take care to exclude any extraneous perturbations such as vibrations, air flow and temperature drafts, acoustic sources, relative optics and holder motion, and so forth from influencing the object during or between exposures.

(4) *Evaluation*: Optical or digital reconstruction of holographic or speckle interferograms provides 3D and 2D distribution of one or multiple fringe patterns, respectively, covering the object under test. The operator should distinguish the defective regions among the fringe intensity distributions. This may be performed visually by evaluating the continuity of a fringe or the systematic formation of fringe patterns from the edges or the fixed sides toward the center. The continuity is destroyed or altered when the fringe stops entirely or changes direction abruptly or starts a new fringe with a different direction. Fringes of similar inclination or direction form patterns representing a collection of object points with equal or uniform displacement. Uniform patterns — independent of their extent — indicate the uniform displacement of object areas from which evaluation of sensitivity can be delivered. Instead, in each point that these patterns are interrupted to give place to another pattern formation or an abrupt discontinuity, then a defect visualize its presence, e.g., in the example in Figure 5.7, the whole-body fringe pattern, (exhibited as consequent concentric fringes covering the whole image) includes a systematic fringe feature across the diagonal at the left bottom quarter of the image, influencing

[17] Consortium EU patent in preparation.

three consequent fringes with a bendlike feature. If the presented discontinuity has the same effects on lateral neighboring fringes of the uniform pattern as in the example of Figure 5.7, then a new fringe pattern indicating the extent of the defects is traced with the boundary being the extent of the uniform lateral fringe discontinuity (in this case an unknown hidden inborn crack of 3 cm length witnesses its presence). Particularly in evaluation of art, many fringe patterns may be extracted on the basis of only one interferogram, depending on and depicting the extent of deterioration.

5.3.3.1   Defect Visualization: Characteristics, Parameters, Features

General statements can be made about defect magnitude, defect depth, and defect interconnection, in terms of fringe density along an axis (fringe pair/cm),[18] from a series of characterized defects on realistically modeled samples[19] suited to the purpose of conservation research.

In the graphs in Figure 5.13, the density of fringes exhibited from the induced defects with known size and depth inside the artwork sample shows an exponential increase and decrease, respectively.[20] The problem is treated from the optically reconstructed secondary macroscopic fringe formation as is recorded for one-to-one scale representation [7,8]. Another implication concerned is that each defect type generates a distinct fringe pattern shape. According to the fringe pattern shape, various methods have been employed to assess these relations. (e.g., for circular fringe patterns the diameter square over finite area, for fringe density changes the fringe number over a finite length, etc.) Thus the graphs can summarize the defect influence by generalizing the defect type. A mean excitation effect for various defect sizes, illustrating the relation of defect size to surface, is achieved by generating a moderate number of secondary fringes (uniform moderated excitation). Thus the graphs (1) demonstrate suitable standardized experimental procedure of mean excitation delivered from preliminary investigations of sample responses and (2) allow visualization of the modeled defect influences on the surface.

The raw data are obtained from normalized intensity profiles drawn across different directions (x–y or diagonal) of fringe systems and normalized by averaging over a set number of pixels/cm that may change depending on the choice of resolution of the primal holographic record and on the imaging system used to digitize them. An important observation is that the graphs can be generalized because they follow the same slope, even with other fringe numbering approaches (e.g., Δd/whole object from different viewpoints).

The term *size of defect* in the graph is also an indication of degrees of freedom, thus de-bonding from the rest of the structure and is not only size of defect in a

---

[18] The square of the unit is not used because the calculation is along one histogram axis.

[19] All varieties of samples required to draw the general statements of this section were constructed by conservators from Benaki Museum of Athens during the EC project LaserArt (1996–1999).

[20] In treating the problem involved in the expression of the graphs we chose for our purposes not to treat general diffraction problems of nonperiodic objects according to the parabolic approximation valid with imaging systems for high spatial frequencies (upper limit 113 cycles/mm, for d = z = 10 cm, λ = 0.5 μm) since here notations for the general defect visualization are concerned.

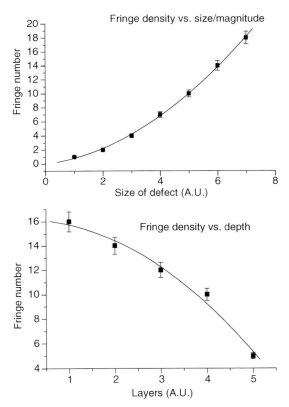

FIGURE 5.13 By systematic inspection of known samples, the relation between fringe density for the same excitation to defect magnitude and depth can be as a rule described by fitting in exponential increase and decrease accordingly.

dimensional extent (which in holography representation is visual in one-to-one directly estimable scale). The size is therefore better described as the magnitude of a defect. As a rule, the size of the defect can be generalized, since the resolution of a system may differ according to the recording parameters and medium without affecting the general character of the graph. The same generalized notation is used for the term layers, since one layer may be 30 microns thick whereas another may be one or even two orders of magnitude thicker or thinner. Nevertheless, deeper within the structure, a defect is less able to affect the surface shape, and in general the probability of producing a fringe pattern is decreased. Therefore the values of the graphs are interdependent, i.e., a low magnitude or a small defect may affect the surface as much as or more than a larger magnitude defect if the former is close to the surface and the latter is buried deeper within the bulk. Hence knowledge about the technique of sample construction and its tendency to one or another type of defect generation should be equally important when defect magnitudes and depth positioning are estimated.

Another important general rule may be made from studies on variable deterioration conditions of simulated artwork samples. Samples with increased complexity

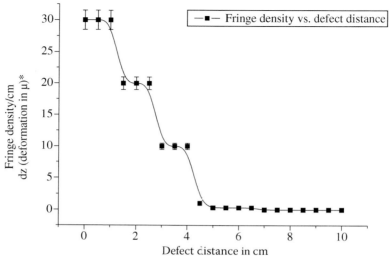

*Data from measurement of fringe visibility.

FIGURE 5.14 The distance between defects contributes to overall artwork deformative effects. In lesser distances, defects show interconnected magnitude, whereas in increased distances, defects uniquely affect the surface and are localized.

of known defects[21] were constructed and underwent the same experimental procedures with fixed excitation load and recording means. The tests concluded that:

1. A whole-body fringe pattern is not likely to be observed in highly deteriorated cases.
2. Localized fringe patterns witnessing the presence of defects dominate the interference field.
3. The categories of defect become less apparent since the effects of underlying defects are combined.

The above were observed with increased defect density, independently of defect magnitude or depth. In order to represent graphically these observations, a graph with the term fringe pattern deformation is introduced to indicate the degree of fringe pattern separation and distinct appearance. The distance between defects makes a considerable contribution to overall deterioration or future risk due to the increase in mechanical instability of the structure. The findings shown in Figure 5.14 indicate critical defect distances[22] such that, with less separation, defects generate high deformative effects on the surface and are observed as interconnected and highly distorted fringe patterns. These patterns diminish visibility and hence the effect becomes *saturated* at only 3 cm

---

[21] Defect geometry is random by construction to stimulate for real artwork cases but similar conclusions in regard to the deformative effect are drawn from geometrically model defects.

[22] The scale of the constructed defects used were in cm; smaller defects may overlap or constrain each other producing fringe patterns with a variety of fringe separation, but the deformative effect should follow the same slope as has been evidenced by real artworks and models.

distance (another choice of resolution would affect the scale and not the slope). Another variety of samples, including edge saddles in multilayered structures, generates fringe densities that vary in separation without affecting the dependence of deformative effect on distance. On the other hand, defects with increased separation are uniquely visualized, distinct, and easily located by comparatively minimized local surface influence. Thus the experimental observations highlight another important finding in regard to deterioration studies relating the condition of the work of art not only with the singular importance of an existing defect but with the defect tendency to increase, which supersedes the localized influence — actual size — of the defect. Since the presence of closely spaced defects will have a multiplicative influence on the surface, generating higher deformation than each one alone, deteriorated works of art are susceptible to further deterioration even from small and otherwise negligible defects. Antique and deteriorated art thus requires a careful approach in terms of structural restoration to prevent or delay future irreversibly induced deformations. Along these lines, knowledge of the entire interrelated presence of structural inconsistencies of defects must be taken into account when treating a work of art.

From visual study of fringe patterns the effect of the potential growth of each defect can be traced to indicate to the conservator possible *defect interconnections* and, of even more practical importance, the exact direction of growth, thus providing information for setting up accurate stabilization and conservation that could prevent the visualized tendency for future deterioration.

Finally, it should be noted that a conservation strategy may be radically affected by produced holographic interferograms, since experimental findings prove that relatively small defects may produce a greater effect on the surface, which may constitute a greater risk to the integrity of an object than larger ones, since the former are unstable and tend to grow, whereas the latter may prove to be more stable and affect less the overall deterioration process.

Nevertheless, thorough investigation with optical holographic interferometry provides a helpful tool for understanding of deterioration processes, defect growth and manifestation, and structural and mechanical instabilities that no other method can provide. Hence in this respect, optical holographic interferometry (and following its digital and electronic counterparts) is clearly distinct from complementary interferometry techniques such as shearography, which can dedicated for use in essential routine defect detection applications.

5.3.3.2 Direct Diagnosis: Defect Detection and Identification Procedures

Here we give examples of defect detection in artwork conservation.

The defect indicative fringe patterns can be extracted by

(1) *Intensity profiles*: Defects in artworks[23] are structural imperfections caused by mechanical fatigue due to repetitive dimensional processes or mechanical fracture caused by sudden deformation or accidents during handling. Each artwork depending on its technique of construction tends to develop specific

---

[23] Artworks and samples provided by cooperative partners, Laboratoire des Researche des Monument Historique, Benaki Museum of Athens, National Gallery of Athens, and others.

types of defects. Therefore prior to identifying a defect in a particular artwork one should know the possibilities for defect appearances. Interference fringe systems can be viewed as a coded representation of minute displacements (optical path changes) that a specific object can develop. For a particular fringe pattern (or characteristic) to appear, corresponding influences in artwork structure should be present. The underlying cause of fringe pattern generation to complete the diagnosis requires *a priori* knowledge about the artwork as has been already stated. Figure 5.15 shows an example where a circularlike fringe pattern characteristic, representing a detachment or void, may appear in many different works of art of various constructions, materials, or supports.

When the intensity distribution is extracted from a regular fringe pattern, e.g., as in the normal distribution of bright and dark fringes in the statuette after restoration of Figure 5.15, its sinusoidal nature is apparent. When the intensity profile is drawn across a defected area, this continuity is disrupted, and the fringe density (fringe/cm, for any one axis, or spatial frequency) and amplitude become the components of direct defect diagnosis.

A technical object with displacement in one direction, e.g., mechanical tilt, generates uniform fringes of mostly equivalent spacing and amplitude, as is shown in Figure 5.16a. On the other hand, artworks with multidirectional thermal load and with hidden defects generate chaotic fringe amplitude and abrupt fringe spacing changes, as in Figure 5.16b. Defects are also manifested through interrupted fringe continuity.

The regions of abrupt fringe changes, as in Figure 5.16c, correspond to defective areas. From a fringe intensity evaluation (Figure 5.16c,d), the exact defective areas and their extent can be located. Since the number of pixels per cm is known, the defect dimensions are correlated to the exact coordinates from which the intensity has been drawn and are extracted in spatial (x,y) coordinates. The number of fringes is indicative of abrupt change, attributed to the existence of defects.

(2) *Visual examination*: For the purpose of qualitatively analyzing an interferogram, a three-step defect identification procedure should be followed:

At first glance, fringe systems may look complicated and suggest abstract features difficult to put into given defect categories. After a more careful examination, fringe systems prove to share common characteristics repeatedly found even in the same interferogram. Works of art generate as many fringe systems as there are hidden defects, which usually are many, since they are continuously undergoing dimensional changes affecting structural integrity in general and defect generation and growth in particular.

1st Step: Whole-Body Fringe System
As has been stated in Section 5.3, the examination of the overall displacement distribution starts from a rigidly attached part of the artwork, here the edges. In the example shown in Figure 5.17, the canvas painting attached to two panel wood supports is covered with bright and dark fringes that formulate the whole-body fringe pattern.[24] The difference though in the spatial frequency and the resulting double whole-body fringe pattern (upper and lower parts) shows an absence of canvas in the lower part of the painting, as was later also confirmed by x-ray imaging. There

---

[24] Experiment carried out during an intership of the author at Aquila Univ. 1994, Dipartimento di Energetica with supervisor Prof. Domenica Paoletti, and the painting was borrowed from Aquila Museum.

Fresco with known defect: Structured detachment (concentric-like fringes)

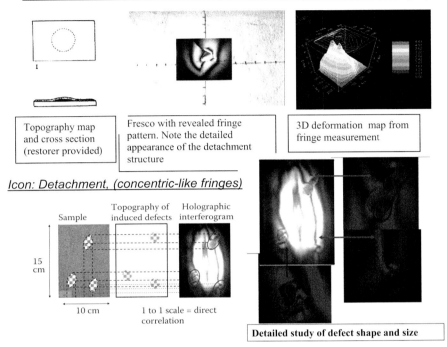

| Topography map and cross section (restorer provided) | Fresco with revealed fringe pattern. Note the detailed appearance of the detachment structure | 3D deformation map from fringe measurement |

Icon: Detachment, (concentric-like fringes)

Detailed study of defect shape and size

Real statuette with subsurface void (circular-like fringes)

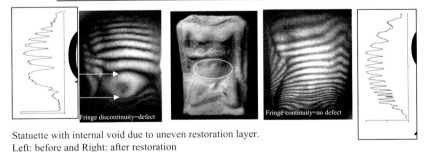

Statuette with internal void due to uneven restoration layer.
Left: before and Right: after restoration

FIGURE 5.15 Characteristic examples from defect detection, localization, isolation and sizing from known defected samples of typical construction materials as for frescos and icons and of real statuette. Note the consistency of circular fringe systems representing detachments/voids.

BOX 6: STEP WISE DIRECT DIAGNOSIS	
1st step	Identification of whole body fringe system
2nd step	Localization of fringe pattern irregularities
3rd step	Defect classification of local fringe pattern irregularities

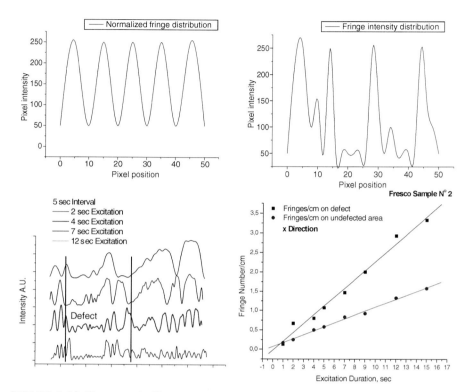

FIGURE 5.16 These graphs illustrate the uniform distribution of fringe intensity in a whole body fringe system (a) without defects and (b) with defects. Parts (c) and (d) show the fringe distribution extracted by intensity profiles drawn across interferogram (c) localize the defect in object coordinates, and (d) differentiation of fringe density among whole body and defect fringe density.

FIGURE 5.17 Painting exhibiting two whole-body fringe systems generated by panel separation along a central x-axis. The difference in the density of fringes visualizes the differential response of the upper in comparison to the lower part due to missing canvas at the lower region.

are significant implications for this information, since it determines the reaction and sensitivity of the artwork to the induced load for both parts, as can be seen by the dual spatial frequencies formation. A direct estimate from an average value of fringes from various viewing angles indicates the relative magnitude of the whole-body displacement. Alternatively, measurements of $\delta_{1...v} = 2\pi N_{1...v}$ for a displacement coefficient matrix are performed from three or four viewing directions from a zero-order fringe reference (stable point) as input to a series of linear algebraic equations.[25] The matrix is determined by geometry and laser wavelength [7].

The whole-body fringe patterns extracted from the interferogram show the unique reaction of the artwork at the moment of investigation. It may be used to visualize previously unknown structural problems or indicate further requirements for structural examination. The reaction can thus be used as a sensitivity indicator of structural responses for comparison at later times or seasonal examination under different conditions. Envionmental studies can also be performed on the basis of whole-body fringe pattern analysis.

2nd Step: Localization of Irregularities
The overall fringe pattern shows at a glance a number of localized irregularities that are due to various defect types. Fringe patterns are extracted from a holographic interferogram of a painting according to classification of common features. By zooming in with a CCD camera-lens system onto these localized fringe patterns as in Figure 5.18, these features can be extracted individually by the defect characteristic feature guide of Figure 5.19.

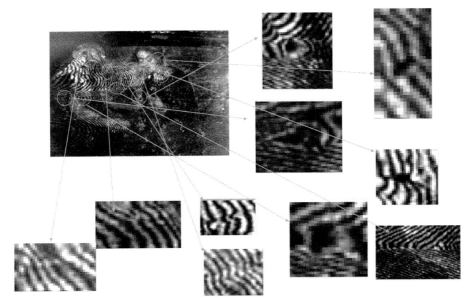

FIGURE 5.18 Independently of the amount of irregularities of an artwork is found consistency of common fringe features facilitating the classification of defect type.

---

[25] The mathematical elaboration is tedious, and automatic evaluation with software is used instead.

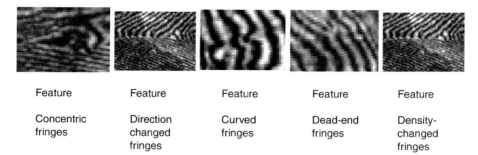

Feature	Feature	Feature	Feature	Feature
Concentric fringes	Direction changed fringes	Curved fringes	Dead-end fringes	Density-changed fringes

FIGURE 5.19 Characteristic patterns with similar features represent the same defect cause facilitating the classification of defect type. (Patterns isolated from real paintings.)

In the evaluation it is observed that, independently of the amount of irregularities in an original artwork, there is found a consistency of common features.

Therefore despite the seeming enormous range of fringe patterns, there will usually be distinguishing basic fringe characteristics that, independently of size or orientation, repeatedly appear in distinct locations in the interferogram. These features include (a) fringe concentricity, (b) fringe bending, (c) fringe cut, (d) fringe density change, and (e) fringe curvature. These repeated features have a variety of topological alterations, e.g., instead of circular fringes, elliptical ones are also common, but they describe a similar defect and correspond to the five defect categories summarized in Section 5.3.1.

3rd Step: Identification of Irregularities
From numerous experiments on actual highly complex artworks, it has been statistically proven that there are common features (defect categories) that dominate the visualization of discontinuous fringe patterns. Another commonly found is the defect associated with a localized lack of (or much higher density of) fringes, usually termed spatial frequency change or density changed fringes. This interferogram was recorded with a carrier frequency technique such that the fringe spatial frequency is constant over the entire area of each part. In this particular example (noted earlier in Figure 5.17), the canvas was destroyed in the lower part, in which higher spatial density is manifested but is of the same magnitude all over the same area. Thus one may use this example to support the protective role of canvas in minimizing dimensional changes of wood paintings as well as the capability of HINDT of visually exhibiting differences in bulk and support materials. In this section the main interest lies in the characteristic features that have been isolated; each corresponds to the same defect cause, and they can be summarized as shown in Figure 5.19: the shapes of these fringe systems are not absolute but each does characterize the features of various defect types.

The above procedure can be performed in any artwork with a variety of fringe patterns concluding in the three broad defect categories shown below in the classification table.[26] Thus it is employed as a rule in the diagnosis of defect

---

[26] The three broad defect categories are of the most interest and commonly found in artwork structural problems.

Defect Classification Table

CATEGORY A

- Detachments between multilayers
- Material voids in bulk
- Inclusions in surface contaminated layers
- Unevenly applied restoration layers
- Inclusions between layers

  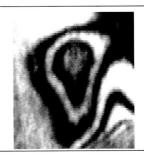

CATEGORY B

- Crack in the support
- Material separation

CATEGORY C

- Stress surface disruption
- Defect propagation

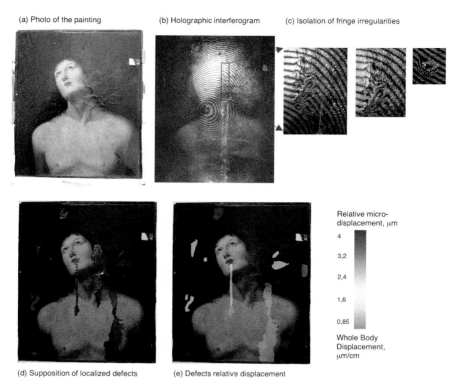

(a) Photo of the painting   (b) Holographic interferogram   (c) Isolation of fringe irregularities

(d) Supposition of localized defects   (e) Defects relative displacement

Relative micro-displacement, μm

4
3,2
2,4
1,6
0,85

Whole Body Displacement, μm/cm

FIGURE 5.20 (a–e) (See color insert following page 144.) Example of application on real painting on conservation direct structural diagnosis: Saint Sebastian (painting attributed to Rafael, National Gallery of Athens). Holographic interferograms provided a variety of small local discontinuitites that were isolated one by one and classified as type to defect according to the classification table. Calibration of whole-body fringes allows setting of priority risk indicator for structural restoration. Superposition of defects on painting with fringes shown on left, and on right the priority for restoration.

identification.[27] This broad categorization is possible according to the experimental findings, the conservators concerned, and the commonly presented structural defects in artworks.

Examples of application on direct diagnosis with qualitative structural mapping assessment, according to the described procedure, are shown in a painting of the martyrdorm of Saint Sebastian, attributed to the school of Rafael, in Figure 5.20. Locating and superposing the detected irregular schemes outputs a detailed map of extended or limited structural defects, as is shown in the next example with a painting (Figure 5.20a,e). The procedure may further lead an experienced conservator quickly and safely to make conclusions and suggest a priority list for

---

[27] Defects of each category can be represented by formation of the type of fringe patterns chosen by relevant interferometric results from realistic samples. Note: in artwork cases when a carrier frequency is not employed, the concentricity of whole-body fringe formation is mostly due to thermal effect and is not apparent with other means of excitation in applications of experimental mechanics.

restoration, by setting a priority risk assessment for structural restoration that would follow Figure 5.20e.

(3) *Fringe evaluation by computer processing:* Whereas qualitative analysis of interferograms is primarily based on the visual identification of characteristic irregularities and extraction of common features, there are several methods used for the automatic processing of interferograms with the aid of computer-based specialized software. Many commercial products are available nowadays either as computer programs or as combined hardware and software packages. Nevertheless, fringes generated from works of art are more complicated than the usual technical objects tested, so any product will most likely require specialized adaptation, an example of which is shown here.[28] The software used for fringe processing is based on skeleton lines extracted from *a priori* knowledge of exhibited interferogram features. This approach is similar to the one used in qualitative analysis previously described but is mostly dependent on operator *a priori* knowledge for interferogram formation as well as the technical skills in interferogram processing. The fringe analysis can also lead to a setting of priority risk assessment for structural restoration that would follow (Figure 5.21).

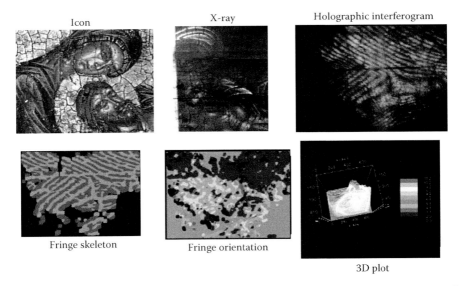

FIGURE 5.21 Computer processing of interferogram of Byzantine icon (Benaki Museum of Athens) correlated with x-ray imaging. Visible holes are seen in icon, x-ray and interferogram. An extended areas of the holes' influence on the surface is revealed only from the interferogram. By fringe numbering and processing the invisible deteriorating condition is quantified and visualized as shown in the 3D plot.

---

[28] In treating this issue it is important to refer to the work on structural mechanics and inverse engineering problems with significant input in this direction acknowledged to the Bremen Institute of Applied Beam Technology and specially to the group of Werner Juptner and Wolfgang Osten, which has explicitly studied the mechanics of defect indicative fringe formation delivering a rule-of-thumb for defect identification equally important for artworks as to other engineering applications.

A Byzantine icon fully documented with conventional conservation means was investigated by high-resolution holographic interferometry revealing an extended deterioration area in the bulk due to neighboring holes visible on the surface. The worm tunneling between the holes has decomposed the bulk material, and despite the lack of evidence of further deterioration due to the holes, holographic interferograms revealed a rather extended area of influence between the holes. The influence was revealed on the surface with characteristic curvature fringes between the holes indicating their influential extent and future propagation. Careful computer-aided fringe processing by extraction of skeleton lines indicates in the 3D the defect magnitude and the region for restoration.

A second example is the early El Greco painting shown in Figure 5.22. It belongs to the Benaki Museum of Athens and was investigated for restoration strategy prior to exhibition. It exhibited a very particular reaction to an induced excitation of its front surface and particularly in the region of its central theme. The scene of the Madonna with the baby Christ in her hands and the surrounding Magi generated totally different fringe patterns. The central scene shows a distinct fringe system due to cumulative effects of past repetitive restorations. The interferogram revealed a highly irregular fringe distribution with many small localized superficial detachments contributing to high mechanical instability. General defect propagation affecting the whole front surface is in such displacement that it resembles detachment from the support and is contacted only on the edges of the frame. After tedious elaboration (in FRINGE PROCESSOR[29]), the resultant plot shown in Figure 5.22 resembles displacement as of a membranelike surface.

Indeed, by simulating the experimental parameters with the *a priori* knowledge in regard to the painting condition, the simulated interferograms produced the same type of fringe characteristics in terms of fringe distribution as shown in Figure 5.23.

Loss of cohesion and further displacement of the membranelike surface may result in catastrophic failure and destruction of the priceless painting; the processed simulated results confirm the magnitude of the effect. Prevention of failure was set as priority for restoration strategy. Conservation actions were followed by interferometric inspection, and the results confirmed the stabilization of the front surface as shown in the photograph of the interferogram of the central theme and the 3D plot of the consequent fringe processing. The defect manifested in the 3D plot is due to a preexisting known crack in the wood support (Figure 5.24).

5.3.3.3 Preventive Applications

(1) Photomechanical Effects of Laser Cleaning

Shock wave effects due to laser ablation of selective surfaces were studied with the arrangement shown in Figure 5.25. A holographic interferometry camera was aligned with the UV laser cleaning workstation to record holograms during UV operation. With minor alterations, the setup serves for both reflecting and transmitting subjects.

---

[29] BIAS-FRINGE PROCESSOR™.

FIGURE 5.22 (See color insert.) The interferogram revealed highly irregular fringe distribution with extended superficial detachments and defect propagation affecting the whole front surface to react as loosely attached to frame membrane.

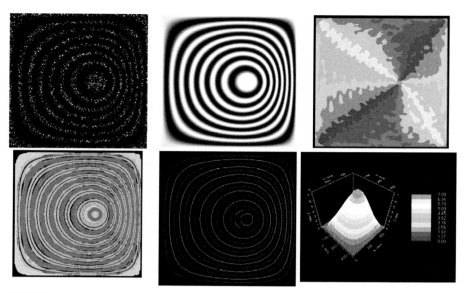

FIGURE 5.23 Simulated experimental parameters with *a priori* knowledge in regard to the painting condition; the interferograms produced the fringe particularity indicative of membrane like surface reaction.

The comparative procedure is based on recording the initial state of the object and taking the second recording after the ablation pulse. The procedure was in progress for as long as undesirable material was being removed from the surface. The whole-field property allows spatially resolving the monitored entire sample surface including that outside of the ablated area. Figure 5.26 shows the growth of the detachment layer area with prolonged application of laser pulses.

Methods of monitoring intervals after the ablation pulse or pulse accumulation can be uniquely developed according to the aim of the inspection. In the example shown in Figure 5.27, monitoring the effect of cumulative pulse ablation on the surface many hours after laser action is shown. Such interesting (and increasing) deterioration effects have been examined in terms of UV wavelength, energy per pulse, and pulse duration (nsec, fsec).[30]

The long-term effects can be studied with a sequential comparative method of investigation and assessment of interventive actions with probable long-term alterations[31] [68,89,90] can be made.

(2) Environmental Influences

By assessing the whole-body fringe pattern and employing alternate use of cw and pulsed lasers, simulated environmental conditions of interest can be studied. Since a few fringe pairs (representing multiples of a few microns) are sufficient

---

[30] Fluence, repetition rate, and wavelength have been studied in a series of experiments [68,83]

[31] To avoid overlapping of information, more information on photomechanical studies by HINDT can be found in Chapter 6.

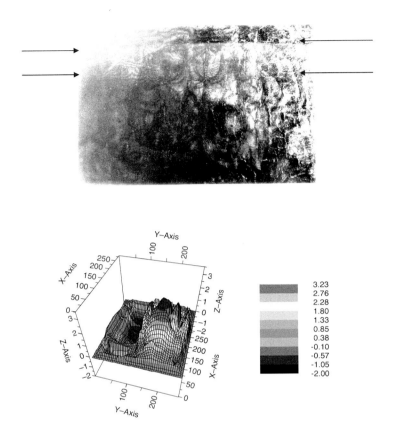

FIGURE 5.24 The central theme after restoration — after dry-out of consolidation treatment- and the updated 3D plot witnessing only the pre-existed known crack in the wood support (arrows).

to provide a safe and sensitive indicator of an artwork's response, such a test can be extrapolated to gain insight into the object's reaction to expected (larger) environmental changes, e.g., in the case of transportation due to exhibition in other countries. In the example shown in Figure 5.28, hygroscopic materials (of which moisture-sensitive museum objects normally consist) have been put into an air-tight chamber while relative humidity and temperature were stabilized. Upon removal of the salts used to reach the environmental stability, the inter-ferometric response proved to be analogous to the rate of change and the magnitude of the RH/T change. Thus the environmental influences can be directly studied by the artwork responses under concern resulting in a distinct assessment for condition settings [53,91–94].

5.3.4 SPECKLE PATTERN INTERFEROMETRY

The wide spread technique of Speckle Pattern Interferometry is closely related to holography principles and applications [11]. It is commonly termed Electronic Speckle Pattern Interferometry (ESPI) or Digital Speckle Pattern Interferometry

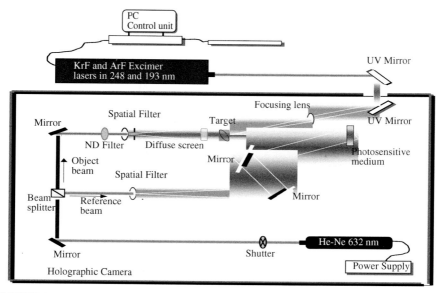

FIGURE 5.25 Schematic arrangement of holographic interferometry and laser cleaning workstation for study of shock wave effect.

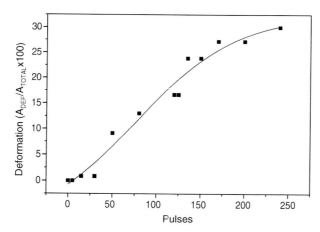

FIGURE 5.26 Growth of detachment of layer during ablation.

(DSPI) or TV-Holography. Although other means are used for recording and processing, sensitivity and in most of the cases experimental methodologies and most of restrictions, are common. Hence most of the procedures described in the chapter can be equally satisfyingly used or suitably adjusted for speckle interferometry. An example is given in Figure 5.29 where is shown an artwork sample investigated under same experimental conditions with Holographic Interferometry and Speckle Interferometry. The material separation across the diagonal of the sample is well observed at both images [95].

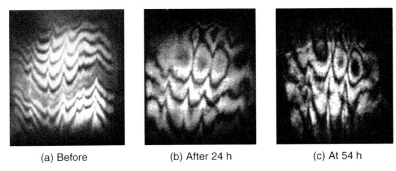

| (a) Before | (b) After 24 h | (c) At 54 h |

FIGURE 5.27 Photographs from samples irradiated for laser surface removal in long term.

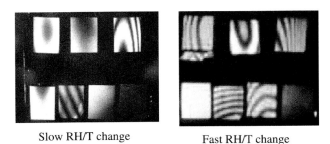

Slow RH/T change   Fast RH/T change

FIGURE 5.28 Whole-body interference employed with alternate use of CW and pulse laser to study environmental conditions of interest.

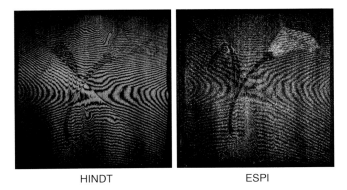

HINDT   ESPI

FIGURE 5.29 A decorative sample with material separation across its diagonal investigated simultaneously under the same thermal excitation with HINDT and DSPI.

It is a technique that has been employed in many studies of artworks with specific interest to separate the in-plane versus out-of-plane deformation of an artwork in real time and out of laboratory conditions [43, 47, 48, 95–96].

Nowadays ESPI is widely used satisfying ease of use and fast operation with high sensitivity. There are also quite few commercial systems with dedicated software that can be found in the market. The performance though of any system developed for industrial purposes cannot be warranted for application on works of art due to the complexity of latter. Thus any technique should be feasibly tested and adjusted in regards to the conservation problems before being considered suitable for works of art analysis [88, 102].

5.4 RELATED TECHNIQUES[32]

5.4.1 DIGITAL SPECKLE SHEAROGRAPHY (DSS)

Speckle shearography is a related coherent optical method for the nondestructive testing of technical components with respect to surface and subsurface flaws [97,98]. The surface of the object under test is imaged onto an electronic sensor (e.g., a CCD array), and the difference of the two intensity distributions I_1 (initial state) and I_2 (final state after loading), representing the two states to be compared, is computed. The special geometry of shearography is the generation of the reference wave produced by interfering the light that is scattered from the object with its sheared version at the sensor (Figure 5.30a) the principle of self-reference and consequently insensitivity against rigid body displacements is ensured. As for the shearing component, an optical wedge (a prism) in front of the sensor lens or a tilted mirror in one arm of a Michelson interferometer is mostly used. Detected are only differences of the displacements between the sheared image points. These differences can be approximated as strain components if the shear is sufficiently small. Figure 5.30b shows a typical shearogram of a small plate that is loaded in the middle.

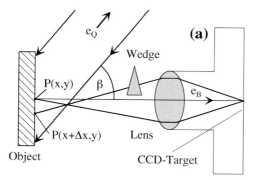
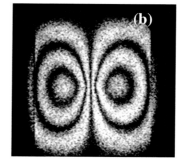

FIGURE 5.30 (a) Schematic setup for shearography (the shearing Δx of the object waves is made by a wedge, e_Q-illumination direction, e_B-observation direction); (b) shearogram of a centrally loaded plate.

---

[32] This section is part of the paper by V. Tornari, A. Bonarou, E. Esposito, W. Osten, M. Kalms, N. Smyrnakis, and S. Stasinopulos, "Laser based systems for the structural diagnostic of artworks: an application to XVII Century Byzantine icons," SPIE 2001, Munich Conference, June 18–22, 2001, vol. 4402. *Note from the author*: The above paper illustrates several examples of Byzantine icons with the combined application of the techniques briefly described in Chapter 5 (HINDT, DSS, LDV).

If we assume an image shear δx in the x-direction, the resulting intensity distribution $I(x,y)$ can be written:

$$I(x,y) = \left[I_1(x,y) - I_2(x,y) \right] = A(x,y) + B(x,y) \cdot \sin\left[\frac{\delta(x,y) - \delta(x + \Delta x, y)}{2} \right] \quad (5.4)$$

where $A(x,y)$ and $B(x,y)$ consider the additive and multiplicative disturbances (electronic noise, modulation V, background intensity I_0, speckle noise). Equation (5.4) shows that the phase term results from the difference of the phase differences of the nonsheared component $\delta(x,y)$ and the sheared component $\delta(x + \Delta x, y)$, which are caused by the load-induced surface displacements $\mathbf{d}(x,y)$. This phase difference can be approximated by

$$\frac{\delta(x,y) - \delta(x + \Delta x, y)}{2} \approx \frac{\partial \mathbf{d}(x,y)}{\partial x} \cdot \mathbf{S}(x,y) \cdot \Delta x \cdot \frac{\pi}{\lambda} \quad (5.5)$$

with the sensitivity vector $\mathbf{S}(x,y) = [\mathbf{e}_B(x,y) + \mathbf{e}_Q(x,y)]$ depending upon the setup geometry. The resulting interference pattern represents, in the first approximation, the derivation of the displacements in the direction of the image shear. This interpretation, however, is only valid for small shear δx. For rigid body displacements, $\mathbf{d}(x,y) = \text{const}$, and consequently $\partial \mathbf{d}(x,y)/\partial x = 0$. The sensitivity of this method can be controlled with the image shear.

The object is illuminated by a laser and observed by a CCD camera (sensor). A shearing element such as an optical wedge is placed in front of the lens. Consequently the camera sees both the usual image and a displaced (speckled) image of the object. Two states of the object are stored in the memory of the computer, the unloaded and the loaded one. Loading can be done thermally or mechanically. However, it must be ensured that only mild stress is induced in the object (e.g., a temperature increase by a few °C or a pressure difference of some hundred Pa) to guarantee the nondestructive testing principle. A subsequent digital processing in the computer is done using specialized software[33] [52]. An adequate loading of the object under test is of great importance for the success of the NDT experiment. The objective of loading is the generation of surface displacement gradients in the region of the subsurface flaws to be detected. The object response to an applied load depends on several factors: the material, the size and location of the defect, the stiffness of the structure, and the kind of load. Three types of loading have proven to be useful: (i) a thermal load caused by a variable illumination infrared lamp or by an intensive thermal flash, (ii) a mechanical load caused, for example, by a change of the pressure in the test environment, and (iii) a vibrational load caused by a mechanical shaker.

---

[33] The Fringe Processor™ was used in the presented paper.

5.4.2 LASER DOPPLER VIBROMETRY

Laser Doppler vibrometry is a nondestructive technique used successfully in the defult inspection of artworks[34] [99,102]. It is based on measurement of vibrations of an object surface to infer the presence of hidden defects, in particular, voids and delaminations. The standard laser Doppler vibrometer (LDV) is a noncontact velocity transducer working on the principle of measuring the Doppler frequency shift of laser light scattered from a moving object by means of an interferometer. The on-board electronics converts the Doppler signal to an analog voltage proportional to the instantaneous velocity of the object. The combination of an interferometer with two moving mirrors driven by galvanometric actuators makes it possible to direct the laser beam to the desired measurement points. Such an instrument, named the scanning laser Doppler vibrometer (SLDV), whose schematic is shown in Figure 5.31, can quickly perform a series of velocity measurements on a grid of points over the object surface under test, reaching top scan velocities of 100 point/sec.

These techniques are effectively used in structural dynamic testing, biological and clinical diagnostics, fluid–structure interaction, on-line monitoring of industrial plants, acoustics, and fault detection, to list only a few. Furthermore, the coupling of laser vibrometers and advanced scanning systems seems to open up new possibilities, e.g., in the field of measurements in tracking mode on moving objects. [100,101]

When a coherent radiation of frequency f, emitted by a source, illuminates an object moving at velocity u, the radiation observed from the source direction (backscatter configuration) undergoes a Doppler shift (f_D) proportional to the surface velocity. Such a frequency shift is given by

$$f_D = \frac{2u}{\lambda}\cos\theta \qquad (5.6)$$

with $u\cdot\cos\theta$ being the velocity component along the laser beam direction.

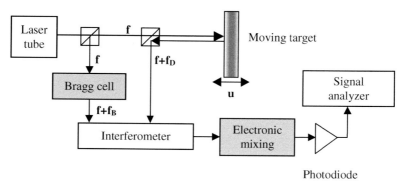

FIGURE 5.31 Schematic of an LDV.

---

[34] In SLDV for structural diagnosis, it is important to refer to the work of the Department of Mechanical Engineering of the University of Ancona and the group of Professors Enrico Primo Tomasiri and Enrico Esposito who systematically investigated the application of LDV for conservation applications.

By scanning the laser beam across an object, one can observe the frequency shift content of the velocity signal that will show significant differences in the presence of a hidden defect, e.g., a void, with respect to an integral object or to an integral area of the same defective object. Figure 5.32 reports three frequency spectra

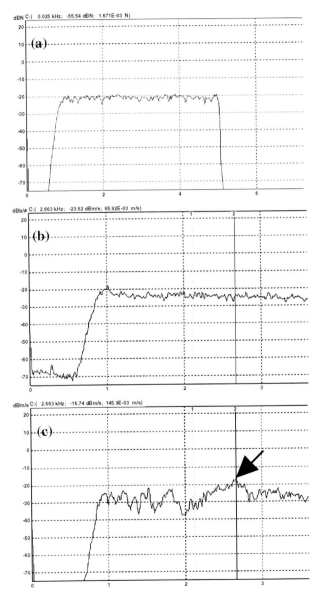

FIGURE 5.32 Vibration spectra of sample: (a) white noise signal input to amplifier; (b) laser aimed at a point inside an defected area; (c) laser aimed at a point inside defected area.

of a simple structure [96] presenting a void, that demonstrate how starting from a white noise exciting acoustic signal, Figure 5.32a, the void can be detected by the presence of resonance peaks (the black arrow in Figure 5.32c points at the most evident one in the defect area). On the other hand, the spectrum measured in the nondefective area of the same sample is flatter (more uniform), and no evident alteration of the input signal can be seen, Figure 5.32b.

5.5 COMPLEMENTARY INTEGRATION, LASERACT

The techniques shown in this chapter have been employed to generate a transportable structural diagnostic sensor based on their complementary advantages under the auspices of an EC funded fifth FWP.[35] The delivered compact prototype system is shown in Figure 5.33 during the first on-field investigation in Victory church, Valetta Malta. The system was developed and optimized according to the art conservation requirements, and relevant investigation methodology and diagnostic procedures were developed and embedded in user-friendly software. The whole investigation by means of the multitask system is driven by a pc interface through a fully automated procedure

FIGURE 5.33 (See color insert.) Laseract Multitask system for onfield investigation of wall painting.

[35] Laser Multitask Non-Destructive Technology in Conservation Diagnostic Procedures, EVK4-CT-2002-00096, Lasers in the Conservation of Artworks, LACONA VI, Wien 2005, on print.

so that the conservator uses only own knowledge about the artwork under inspection. Museum objects, sculptures, wood and wall paintings, a fresco, and a wall fortification were examined, and the proven feasibility allows for further optimization. The introduction of new practices in structural diagnosis of artworks are expected because they are properly developed to surpass the fragmented applicability of current tools and practices and flexible enough to suit to variety of investigation demands.

5.6 SUMMARY

A completely noninvasive, nondestructive laser-based structural diagnostic investigation as may be applied in art conservation has been summarized. The laser techniques involved are of the main laser-based systems implemented in artwork structural diagnostics and were all used with simulated and real artworks providing thus an investigation with common reference to the results.

The spatial resolution of each differs according to the technical characteristics of the technique, starting from ½ of λ (laser wavelength dependence) in holographic recording independently of distance from the target sustaining the use of the technique for revealing and tracing minute surface displacements and bulk defect localization, sizing, and growing deterioration on the surface. The high resolution in combination with flexible geometrical systems and adaptive methodologies constitute an advance in artwork diagnostic inspection with laser techniques. Of outstanding importance in the application of the techniques in the cultural heritage field is the tolerance in a variety of methods allowing a broad application in demanding conservation research, for which some examples were summarized.

For laser Doppler vibrometry, spatial resolution is about 1 mm, but this is a limit that could be easily lowered by the use of laser heads equipped with microscope lenses; a full field of 1×1 mm is realizable with a measurement grid of many hundreds of points. In this case, the measurement time needed to explore fully the artifacts such as Byzantine icons will greatly increase, but a mixed approach can be used, where a first scan, used to identify problematic areas, is followed by a detailed scan of these same smaller areas. Shearography is a robust and compact technique to allow the fastest possible but rather rough evaluation compared to holography or high-resolution vibrometry of monuments and large-scale structures. Some special features of each technique differ, such as the speed and ease of acquisition of shearography and the far longer distances implied in remote access with vibrometry. The former consists of compressed components securing a small portable system for whole-field acquisition. The latter requires extensive but transportable equipment. Holographic transportable digitized systems for field applications are nowadays available, and holography has great potential as an additional conservation laboratory–based technique for conservation research. For targeting to trace small defects and their influence following the guidelines described in this chapter, any conservator, provided that a small, lightweight optical table with minimum optical components and an inexpensive low-energy laser exists, can start experimenting toward a better understanding of structural effects. Similar optical configurations can also be implemented for remote whole-field recording with digital speckle pattern interferometry. Both methods maintain the highest

TABLE 5.4
Comparative Table of Optical Coherent Techniques Used in Art Conservation Structural Diagnosis

Features	Shearography	ESPI	Holography
Laser	cw	cw/pulsed	cw/pulsed
Laser power	400 mW	2.3 mJ	2 W
Wavelength, nm	532	532	532
Beams	1	2	2
Illumination angle	$<10°$	$<10°$	$<10°$
Obj.–sensor distance, cm	50–150	20–50	0-
F.O.V., cm × cm	$>(19 \times 15)$	$<(17 \times 13)$	
Sensor type	CCD	CCD	FILM
Sensor size, mm	8.8×6.6	10.2×8.3	$>(63 \times 63)$
Resolution, pixel × pixel	768×576	1392×1040	1600 lines/mm
Shear size, mm	1–10		
Ratio OB/RB		1	1–1/10
RB–OB Angle		$<2.7°$	$0°–180°$
Thermal excitation, max	$+10°C$	$+10°C$	$+10°C$

NOTE: This comparative table is relative, indicating the parameters of techniques employed for examination of a particular artwork; values can only be evaluated in a comparative and not an absolute way.

content of information with full field recording beams without any theoretical limit in application of distant or moving subjects.

Generally, the techniques presented here have demonstrated good characteristics as diagnostic tools for artwork inspection, although system and procedural improvements are still required. Even at the present state, these laser-based systems offer a series of important advantages over traditional techniques and practices. To sum up, (a) no sample removal, (b) no contact and remote measurements, (c) whole-field fast and repeatable procedures, (d) high and moderated sensitivity, and (e) on-field transportability.

It is expected that laser-based interferometric techniques will become even more competitive candidates for structural diagnostic in cultural heritage work, since with the increased interest in modern flexible devices for faster and safer procedures in the field, the technology advances in systems integration and miniaturization, with relevant development in neural knowledge synthesis algorithms and compact laser systems, will keep developing.

ACKNOWLEDGMENTS

The author wishes to greatly thank Ed Teppo, Austin Nevin, and Athanassios Aravantinos for their generous support in reading and commenting on the editing of the material included in the chapter.

REFERENCES

1. Hecht, E. and Zajac, A., *Optics*, Addison-Wesley, Reading, MA, 1974.
2. Stroke, G., *Coherent Optics and Holography*, Academic Press, London, 1969.
3. Erf, R.K., *Holographic Nondestructive Testing*, Academic Press, London, 1974.
4. Kock, W.E., *Engineering Applications of Lasers and Holography*, Plenum Press, New York, 1975.
5. Okoshi, T., *Three-Dimensional Imaging Techniques*, Academic Press, London, 1976.
6. Michelson, A., *Michelson Interferometry*, (http://www.colorado.edu/physics/phys540).
7. Vest, C.M., *Holographic Interferometry*, John Wiley, 1979.
8. Collier, R.J., Burckhardt, C.B., and Lin, L.H., *Optical Holography*, Academic Press, New York, 1971.
9. Hariharan, P., *Optical Holography*, Cambridge University Press, Cambridge, U.K., 1984.
10. Abramson, N., *The Making and Evaluation of Holograms*, Academic Press, London, 1981.
11. Jones, R. and Wykes, C., *Holographic and Speckle Interferometry*, 2nd ed., Cambridge University Press, 1989.
12. Saxby, G., *Practical Holography*, 2nd ed., Prentice-Hall International, Hemdt Hempstead (U.K.), 1994.
13. Heflinger, L.O., Wuerker, R.F., and Brooks, R.E., Holographic interferometry, *J. Appl. Phys.*, **37**, 642–649, 1966.
14. Amadesi, S., Gori, F., Grella, R., and Guattari, G., Holographic methods for painting diagnostics, *Appl. Optics*, **13**, 2009–2013, 1974.
15. Asmus, J.F., Guattari, G., Lazzarini, L., and Wuerker, R.F., Holography in the conservation of statuary, *Stud. Conserv.*, **18**, 1973.
16. Tanner, L.H., A study of fringe clarity in laser interferometry and holography, *J. Sci. Instr./J. Physics E/Series*, **2**, 517–522, 1968.
17. Robinson, D.W. and Williams, D.C., Automatic fringe analysis in double exposure and live fringe holographic interferometry, *SPIE Optics Eng. Meas.*, **599**, 134–140, 1985.
18. Osten, W. and Juptner, W., Digital Processing of Fringe Patterns, in *Handbook of Optical Metrology*, Rastogi, P.K., Ed., Artech House, Boston, 1997.
19. Gabor, D., Microscopy by reconstructed wavefronts, *Proc. R. Soc.*, **A197**, 454–487, 1949.
20. Gabor, D., Microscopy by reconstructed wavefronts, *Proc. R. Soc.*, **64**(2), 449–469, 1951.
21. Leith, E.N. and Upatnieks, J., Wavefront reconstruction with diffused illumination and three-dimensional objects, *J. Opt. Soc. Am.*, **54**, 1295–1301, 1964.
22. Powell, R.L. and Stetson, K.A., Interferometric analysis by wavefront reconstruction, *J. Opt. Soc. Am.*, **55**, 1593–1598, 1965.
23. Burch, J.M., The application of lasers in production engineering (1965 Viscount Nullfield Memorial paper), *Prod. Eng.*, **44**, 431–442, 1965.
24. Collier, R.J., Doherty, E.T., and Pennington, K.S., Application of moire techniques to holography, *Appl. Phys. Lett.*, **7**, 223–225, 1965.
25. Stetson, K.A. and Powell, R.L., Hologram interferometry, *J. Opt. Soc. Am.*, **55**, 1570A, 1965.
26. Haines, K.A. and Hildebrand, B.P., Surface deformation measurement using the wavefront reconstruction technique, *Appl. Opt.*, **5**, 595–602, 1966.

27. Brooks, R.E., Heflinger, L.O., and Wuerker, R.F., Interferometry with a holographically reconstructed comparison beam, *Appl. Phys. Lett.,* **7**, 248–249, 1965.

28. Thompson, B.J., Holographic particle sizing techniques, *J. Phys. E. Sci. Instrum.,* **7**, 781–788, 1974.

29. Zivi, S.M. and Humberstone, G.N., Chest motion visualised by holographic interferometry, *Med. Res. Eng.,* **9**, 5–7, 1970.

30. Sampson, R.C., Holographic interferometry applications in experimental mechanics, *Exp. Mech.,* **8**, 405–410, 1970.

31. Abramson, N. and Bjelkhagen, H., Industrial holographic measurements, *Appl. Opt.,* **12**, 2792–2796, 1973.

32. Wendendal, P.R. and Bjelkhagen, H.I., Dynamics of human teeth in function by means of double pulsed holography: an experimental investigation, *Appl. Opt.,* **13**, 2481–2485, 1974.

33. Wuerker, R.F. and Heflinger, L.O., Pulsed laser holography, *Soc. Photo-Optical Instr. Engineers,* **9,** 1970.

34. Amadesi, S., Altorio, A.D., and Paoletti, D., Sandwich holography for paintings diagnostics, *Appl. Optics,* **21**(11), 1889–1890, 1982.

35. Amadesi, S., Altorio, A.D., and Paoletti, D., Real and non real time holographic non-destructive testing (HNDT) for paintings diagnostics, in SPIE, *The Max Born Centenary Conference,* 1982.

36. Amadesi, S., D' Altorio, A., and Paoletti, D., Single-two hologram interferometry: a combined method for dynamic test on painted wooden statues, *J. Optics,* **14**, 243–146, 1983.

37. Paoletti, D., Schirripa, S.G., and D' Altorio, A., The state of the art of holographic nondestructive testing in works of art diagnostics, *Rev. Phys. Appl.,* **24**, 389–399, 1989.

38. Bertani, D., Cetica, M., and Molesini, G., Holographic tests on the Ghiberti panel: the life of Joseph, *Stud. Conserv.,* **27**, 61–64, 1982.

39. Takeda, M., Ina, H., and Kobayashi, S., Fourier transform method of fringe pattern analysis for computer-based topography and interferometry, *J. Opt. Soc. Am.,* **72**, 156–160, 1982.

40. Keller, J.B., Inverse problems, *Am. Math. Monthly,* **83**, 107–118, 1976.

41. Juptner, W., Kreis, T., Mieth, U., and Osten, W., Applications of neural networks and knowledge-based systems for automatic identification of fault indicating fringe patterns, *Proc. SPIE Interferometry,* **94**, 2342, 1994.

42. Osten, W., Juptner, W., and Mieth, U., Knowledge assisted evaluation of fringe patterns for automatic fault detection, *SPIE Interferometry VI: Applications,* **2004**, 256–268, 1993.

43. Boone, P.M., Use of close range objective speckles for displacement measurement, *Opt. Eng.,* **21**(3), 407–410, 1982.

44. Rastogi, P.K., Comparative holographic moire interferometry in real time, *Appl. Optics,* **23**, 924–927, 1984–1986.

45. Fuzessy, Z. and Gyimesi, F., Difference holographic interferometry, *Proc. SPIE Interferometry,* **398**, 240–243, 1983.

46. Fuzessy, Z. and Gyimesi, F., Difference holographic interferometry: technique for optical comparison, *Opt. Eng.,* **32**, 2548–2556, 1993.

47. Gulker, G., Hinsch, K., Holscher, C., Kramer, A., and Neunaber, H., *In-situ* application of electronic speckle pattern interferometry (ESPI) in the investigation of stone decay, *Proc. SPIE /Laser Interferometry: Quantitative Analysis of Interferograms,* **1162**, 156–167, 1990.

48. Boone, P.M. and Markov, V.B., Examination of museum objects by means of video holography, *Stud. Conserv.,* **40**, 103–109, 1995.

49. Gulker, G., Hinsch, K.D., and Kraft, A., Deformation monitoring on polychrome Chinese terracotta warriors using TV-holography. in LACONA IV, Paris 1998 .

50. Schirripa-Spagnolo, G., Ambrosini, P., and Paoletti, D., Image decorrelation for In-situ diagnosis of wooden artifacts, *Appl. Opt.*, **36** (32) 8358-8362, 1997.

51. Juptner, W., Non destructive testing with interferometry, *Phys. Res.* (Fringe 03, Academie Verlag), 315–324, 1993.

52. Osten, W., Elandaloussi, F., and Mieth, U. The bias fringe processor — a useful tool for the automatic processing of fringe patterns in optical metrology, in *Third International Workshop in Optical Metrology,* Series in Optical Metrology, Akademie Verlag Bremen, 1998 pp. 98–106.

53. Tornari, V. and Papadaki, K., Continuos wave and pulse holographic interferometry used for monitoring environmental effects on materials used in artefacts, *Proceedings of the First International Congress on Science and Technology for the Safeguard of Cultural Heritage in the Mediterranean Basin,* Catania, Siracusa, November 1995.

54. Tornari, V., Zafiropulos, V., Vainos, N.A., Fotakis, C., Stassinopolous, S., and Smyrnakis, N., Investigation on growth and deterioration of defects in Byzantine icons by means of holographic interferometry, in *Restauratorenblatter, Sonderband — LACONA II.,* Verlag Mayer, Vienna, 2000.

55. Castellini, P., Esposito, E., Paone N., and Tomasini, E.P., Non-invasive measurements of damage of frescoes, paintings and icon by laser scanning vibrometer: experimental results on artificial samples and real works of art, in *Proceedings of the Third International Conference on Vibration Measurements by Laser Techniques: Advances and Applications,* SPIE, Ancona, Italy, 1998.

56. Boone, P.M., Markov. B.V., Burykin, M.N., and Ovsyannikov, V.V., Coherent-optical localization and assessment of importance of damage and defects of cultural heritage, *Optics Lasers Eng.*, **24**(2–3), 215–229, 1996.

57. Dreesen, F. and von Bally, G., Color holography in a single layer for documentation and analysis of cultural heritage, in *Optics Within Life Sciences (OWLS IV): New Technologies in the Humanities,* Springer Verlag, Berlin, Heidelberg, New York, 4, 79–82, 1997.

58. Anglos, D., Solomidou, M., Zergioti, I., Zafiropulos, V., Papazoglou, T.G., and Fotakis, C., Laser induced fluorescence in artwork diagnostics. an application in pigment analysis, *Appl. Spectrosc.,* **50**, 1331–1334, 1996.

59. Snook, R., Laser techniques for chemical analysis, *Chem. Soc. Rev.,* **26**, 319–326, 1997.

60. Maravelaki, P.V., Zafiropulos, V., Kylikoglou, V., Kalaitzaki, M.P., and Fotakis, C., Laser induced breakdown spectroscopy as a diagnostic technique for the laser cleaning of marble, *Spectrochim. Acta B,* **52**, 41–53, 1997.

61. Xu, L., Bulatov, V., Gridin, V.V., and Schechter, I., Absolute analysis of particulate materials by laser-induced breakdown spectroscopy, *Anal. Chem.,* **69**, 2103–2108, 1997.

62. Georgiou, S., Zafiropulos, V., Tornari, V., and Fotakis, C., Mechanictic aspects of excimer laser restoration of painted artworks, *Laser Phy.,* **8**, 307–312, 1998.

63. Georgiou, S., Zafiropulos, V., Anglos, D., Balas, C., Tornari, V., and Fotakis, C., Excimer laser restoration of painted artworks: procedures, mechanisms and effects, *Appl. Surf. Sci.,* **5048**, 1998.

64. Cooper, M., *Laser Cleaning,* Butterworth-Heinemann, Oxford, 1998,

65. Salimbeni, R., Pini, R., and Siano, S., A variable pulse duration Q-Switched Nd:YAG laser system for conservation, in LACONA IV.

66. Kautek, W., Pentzien, S., Conradi, A., and Puchinger, L., Diagnostics of parchment laser cleaning in the UV, visible and infrared wavelength range: a systematic SEM, TEM, IR and GC study, in LACONA IV.

67. Hinsch, K.D., Frice-Begermann, T., Gulker, G., and Wolf, K., Speckle correlation for the analysis of random processes at rough surfaces, *Optics Lasers Eng.*, **33**, 87–105, 2000.

68. Bonarou, A., Tornari, V., Antonucci, L., Georgiou, S., and Fotakis, C., Holographic interferometry for the structural diagnostics of UV laser ablation of polymer substrates, *Appl. Phys. A*, **73**, 647–651, 2001.

69. Burgio, L. and Clark, R.J.H., *Spectrochim. Acta B*, **Part B 57**, 1491, 2001.

70. Fotakis, C. European facility for ultraviolet lasers, in ULF DG XII - ERBFMGECT 950021.

71. Marras, L., Fontana, R., Gambino, M.C.., Greco, M., Materazzi, M., Pamploni, E., Pezzati, L., and Poggi, P., Integration of imaging analysis and 3D laser relief of artworks: a powerful diagnostic tool, in *Springer Proc. Phys.*, 2003.

72. Castillejo, M., Martin, M., Oujja, M.. Santa maria, J., Silva, D., and Torres, R., Evaluation of the chemical and physical changes induced by KrF laser irradiation of tempera paints, *J. Cult. Heritage*, **4**, 257–263, 2003.

73. LASERART 2062, EC DG XII - SMT project, 1996–1999.

74. Tornari, V., Tsiranidou, E., Orphanos, Y., Farsari, M., Kalpoulos, C., Fotakis, C., Doulgeridis, M., and Aravantinos, A., On interference-generated defect indicative patterns for the validation of application on artworks structural diagnosis, *J. Cult. Heritage*, Dec. 2004.

75. Tornari, V., Zafiropulos, V., Vainos, N.A., Fotakis, C., Osten, W., Elandaloussi, F., and A holographic systematic approach to alleviate major dilemmas in museum operation, in EVA 98 Conference on Electronic Imaging and the Visual Arts, 13–16 November, Berlin, BIAS, 1998.

76. Mieth, U., Osten, W., and Juptner, W., Investigation on the appearance of materials faults in holographic interferograms, *Fringe*, 2001.

77. Tsiranidou, E., Orphanos, Y., Yingjie, Y., and Tornari, V., Detection dependencies of subsurface defect mapping in fringe-based analysis, in *International Society for Optical Engineering Proceedings*, AIVELA, SPIE International Society for Optical Engineering .

78. Tornari, V., Bonarou, A., Zafiropulos, V., Fotakis, C., Symrnakis, N., and Stassinopoulos, S., Structural evaluation of restoration processes with holographic diagnostic inspection, in LACONA IV, Paris, 10–14 September, *J. Cult. Heritage*, 2001.

79. Tornari, V., Tsiranidou, E., Orphanos, Y., and Kalpouzos, C., Dynamics of alteration in excitation-depended structural diagnostic techniques, in CLEO , 2005.

80. Schirripa Spagnolo, G. and Paoletti, D., Digital speckle correlation for on-line real-time measurement, *Optics Comm.*, 15 November, **132**(1–2), 24–28, 1996.

81. Hinsch, K.D., Gulkor, G., Hinrichs, H., and Joost, H., Artwork monitoring by digital image correlation, in LACONA V, *Springer Proc. Phys .*, 2003.

82. Tornari, V., Zafiropulos, V., Bonarou, A., Vainos, N.A., and Fotakis, C., Modern technology in artwork conservation: a laser based approach for process control and evaluation, *J. Optics Lasers Eng.*, **34**, 309–326, 2000.

83. Tornari, V., Antonucei, L., Bonarou, A., Georgiou, S., and Fotakis, C., Mechanical deformations in polymers by UV laser ablation, in Fourth International Conference on Vibration Measurements by Laser Techniques, 21–23 June, SPIE, 2000.

84. Athanassiou, A., Andreou, E., Bonarou, A., Tornari, V., Anglos, D., Georgiou, S., and Fotakis, C., Examination of chemical and structural modifications in the UV ablation of polymers, *Appl. Surf. Sci.*, 197–198, 757–763, 2002.

85. Tornari, V. and Bayer, S., Laser induced front and rear side ablation, in Fourth International Symposium on Laser Precision Microfabrication, LMT, LPM, Munich, 2003.

86. Scalise Esposito Lorenzo, E. and Tornari, V., Measurement of stress waves in polymers generated by UV laser ablation, *Optics Lasers Eng.*, **38**, 207–215, 2002.

87. Tornari, V., Zafiropulos, V., Fantidou, D., Vainos, N.A., and Fotakis, C., Discrimination of photomechanical effects after laser cleaning of artworks by means of holographic interferometry, in OWLS V: Biomedicine and Culture in the Era of Modern Optics and Lasers, 13–16 October, Heraklion, Crete, 208–212 1998.

88. Tornari, V., Falldorf, C., Esposito, E., Dabu, R., Bolas, K., Schipper, D., Stefanaggi, M., Bonnici, H., and Ursu, D., Laser multitask non destructive technology in conservation diagnostic procedures, in LASERACT 2003–2006.

89. Tornari, V., Fantidou, D., Zafiropulos, V., Vainos, N.A., and Fotakis, C., Photomechanical effects of laser cleaning: a long-term non-destructive holographic interferometric investigation on painted artworks, in *Holography Techniques and Applications,* SPIE, **3411**, 420–430, 1998.

90. Tornari, V., Bonarou, A., Zafiropulos, V., Antonucci, L., Georgiou, S., and Fotakis, C., Holographic interferometry sequential investigation of long-term photomechanical effects in the excimer laser restoration of artworks, in ROMOPTO Conference, Bucharest, SPIE, **4184**, 545–550, 2000.

91. Tornari, V., Bonarou, A., Zafiropulos, V., Fotakis, C., and Doulgeridis, M., Holographic applications in evaluation of defect and cleaning procedures, *J. Cult. Heritage* 1, **0**,S325–S329, 2000.

92. Tornari, V., Non invasive laser measurement for diagnosing the state of conservation of frescoes and wooden icons, in Fourth European Commission Conference for the Research on Protection, Conservation and Enhancement of Cultural Heritage, 22–24 November, Strasbourg, Session B, 74-80, 2000.

93. Tornari, V., Tsiranidou, E., and Orphanos, Y., Holographic interferometry in research of structural diagnosis, in ITECOM, EC Conference, 2003.

94. Osten, W., Active optical metrology — a definition by examples, in *Proc. SPIE.,* **3478**, 11-25, 1998.

95. Tsiraindou, E., Orphanos, Y., Kalpouzos, C., Tornari, V., and Stefanuggi, M., Time dependent defect inspection assessed by combination of laser sensing tools, submitted LACONA VI, September 2005.

96. Young, C., Quantitative measurement of inplane strain of canvas paintings using ESPI. *Proc. Appl. Opt. Div.*, Conference IDP Publishing, Brighton, 1998, 79–84.

97. Hung, Y.Y., Image-shearing camera for direct measurement of surface strains, *Appl. Opt.*, **18**, 1046-1105, 1979.

98. Hung, Y.Y., Shearography for nondestructive evaluation of composite struchures, *Opt. Lasers Eng.*, **24**, 161-182, 1996.

99. Castellini, P., Esposito, E., Legoux, V., Stefanaggi, M., and Tomasini, E.P., On field validation of non-invasive laser scanning vibrometer measurement of damaged frescoes: experiments on large walls artificially aged, *J. Cult. Heritage,* **09/00xyz**, S349–S356, 2000.

100. Castellini, P., Revel, G.M., and Tomasini, E.P., Laser Doppler vibrometry: a review of advances and applications, *The Shock and Vibration Digest*, **30** (6), 443-456, 1998.

101. Castellini, P. and Santolini, C., Vibration measurements on naval propeller rotating in water, *Proc. 2nd Intl. Conf. Vibration Measurements by Laser Techniques Advances and Applications*, Ancona Italy, SPIE, **2868**, 186-194, 1996.

102. Tornari, V., Bonarou, A., Esposito, E., Osten, W., Kalms, W., Smyrnakis, N., and Stasinopulos, S., Laser based systems for the structural diagnostic of artworks: and application to XVII century Byzantine icons, SPIE 2001, Munich Conference, June 18-22, 2001, **4402**.

6 Overview of Laser Processing and Restoration Methods

6.1 INTRODUCTION

Lasers are finding increasing use in the restoration of artworks, as their use offers a number of distinct advantages over conventional methods. In fact, the implementation of lasers in restoration closely corresponds to their use in a wide spectrum of highly successful applications, ranging from the etching of polymers in microelectronics [1–6], to the mass analysis of biopolymers in analytical chemistry [7], to photorefractive keratectomy and laser-based excision of tissue in medicine [8,9]. In all cases, laser irradiation is exploited in order to effect precise, accurate, and highly controlled material removal either for eliminating unwanted material or for shaping or structuring the substrate. Consequently, similar principles underlie all these applications. The extensive information that has been accumulated through this wide range of applications permits a general presentation of the underlying processes that may be helpful in the further development and study of laser restoration.

There are many demanding challenges in the field of artworks restoration [10,11], for example, (a) cleaning accumulated pollutants, photodegradation by-products, and dirt from painted and/or sculpted artworks in various media (e.g., stone, wood, metal, canvas), (b) removal of impurities and contaminants from textiles, paper, and parchment, and (c) removal of overpainting and varnish in order to uncover an original painting. Cleaning may be desirable for maintaining the aesthetics of artworks, but often it is absolutely imperative for prolonging their lifetimes.

Because of the high complexity of artworks, traditional methods have had in many cases only partial success in meeting these objectives. Furthermore, the success of these techniques depends critically on the expertise of the conservator. The traditional methods rely on the use of mechanical or chemical means [10,11]. Mechanical means (e.g., abrasive, jet spraying, or scalpel) are difficult to control accurately. Even with extensive experience, human limitations can lead to inadvertent material removal from the substrate and to changes of its texture. On the other hand, solvents may penetrate into the substrate and result in irreversible damage. Some chemicals (e.g., methyl ether ketone, methylene chloride, phenol) are quite aggressive, thereby introducing health hazards. Most importantly, owing to the ever-increasing environmental pollution effects, there is growing demand for techniques enabling efficient restoration of a high number of artworks with increasingly complex degradation characteristics.

In comparison with conventional techniques, laser processing offers several advantages such as precise control, site selectivity, versatility, lack of mechanical or chemical contact with the substrate, and a wide diversity in the types of substrates possible to process. Furthermore, automation through on-line monitoring and control can speed and improve the quality of the restoration process. The potential of laser processing has been most clearly demonstrated in the microelectronic industry where, owing to the ever-decreasing device size, there has been extensive work on the development of techniques for substrate structuring on submicron level with minimal collateral effects [12–15]. Even for the most experienced operator, removal of layers with submicron resolution by any other means is simply impossible. As a result, laser treatment techniques have emerged as the unquestionable tool for material removal in a number of industrial cases, for tissue excision in several medical applications, and, more recently, in the field of laser restoration [16–19]. Note also that the versatility of the interaction implies the potential for other applications besides restoration. For instance, the same phenomenon may be exploited for the micro-etching of holograms for authentication and security purposes [20].

In the following, the basic principles and phenomenological models underlying laser material processing are first presented (Sections 6.2 and 6.3). The methods that have been developed for dealing with the various cleaning problems are described in Section 6.4. Besides the apparent issues of efficiency and effectiveness, the most important one concerns the nature and extent of any deleterious effects that may be induced to the substrate. Thus major emphasis is placed on the issue of the processes and effects that are induced upon laser irradiation (Section 6.5) and their dependence on laser and material properties (Section 6.6).

6.2 LASER-INDUCED MATERIAL EJECTION

Laser restoration of artworks relies on the material removal that is effected upon irradiation of substrates with intense laser pulses (Chapter 2). Depending on laser parameters and material properties, the removed material thickness may range from tens of nanometers to many micrometers [1–9]. With increasing laser fluence, the amount of removed material generally increases. The dependence of material removal rate or etching depth per pulse on laser fluence is illustrated in Figure 6.1a, the so-called etching rate curve. The phenomenon giving rise to material ejection is called laser ablation (from the Latin word *ablatio* meaning material removal). The fluence at which the etching depth increases sharply (or the intercept of the extrapolation of the linear part of the curve with the abscissa) is usually considered to represent the ablation threshold fluence, F_{thr}.

Unfortunately, behind this deceptively simple phenomenological description lies a number of complex processes. The highly dynamic nature is for instance illustrated in time-resolved photography of the material ejection process (Figure 6.1b). As a result, there is much controversy and ambiguity about the proper definition of the phenomenon. In much of the literature, ablation is simply identified with material removal. However, for practitioners, ablation has generally been identified with indiscriminate and nonselective material removal that is effected upon irradiation at high fluences. In the case of ceramics and semiconductors, the term congruent

(a)   (b)

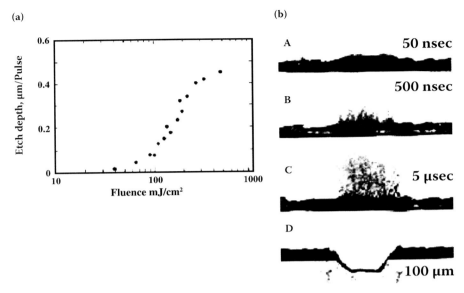

FIGURE 6.1 (a) Schematic of the etching depth as a function of the incident laser fluence. (b) Time-resolved photography of the material ejection process upon irradiation of a doped polymer (specifically for 2% wt biphenyl in *PMMA* at 248 nm). (The exact nature of the ejecta and the time scale may differ depending on employed laser fluence and material properties.) The final picture illustrates a cross section of a cleanly etched/laser ablated polymer surface. From Srinivasan and Braren, *Chem. Rev.*, **89**, 1303, 1989. With permission.

ablation is often used to denote the nonselective ejection of a material layer that is induced at high enough laser fluences.

It must be noted that the transition from the simple photochemistry at low fluences to the ablative regime is not abrupt. In most cases, a fluence range below the ablation threshold can be delineated in which laser irradiation results in morphological changes and even in some mass loss. In particular, for irradiation of polymers at weakly absorbed wavelengths (e.g., *PMMA* at 248 nm), swelling of the surface (of the order of μm) is usually observed. The actual ablation threshold is determined to be at much higher fluence. This swelling is usually ascribed to the formation of gaseous by-products accumulating within the bulk.

At least for simple molecular systems, there is now strong evidence that at these intermediate laser fluences, a thermal desorption/evaporation process operates [21–22]. In other words, in this fluence range, the ejection of compounds/species dispersed in a substrate correlates with their binding energy to the substrate (i.e., ejection signal proportional to $e^{E_{binding}/k_B T}$). In contrast, the ablative regime, as defined above, entails the volumetric/nonselective material ejection at higher fluences by other mechanisms (such as explosive boiling, stress-wave spallation, and photochemical processes). In this case, ejection intensities cannot be correlated with specific binding energies.

This differentiation is expected to be strictly observed in the processing of simple organic matrices composed of small oligomers (e.g., waxes). But even in the irradiation of a number of polymeric systems, a similar delineation of processes is indicated. Specifically, at low fluences, mass loss, as measured by sensitive quartz crystal balances, is observed (so-called Arrhenius tail in the low fluence portion of the etching rate vs. fluence curves) [3,23]. However, this is not accompanied by nonselective material removal. The mass loss to the polymer matrix is due to thermal desorption of fragments and by products that are formed upon laser irradiation and are weakly bound to the polymer matrix.

The effort in differentiating the operative mechanisms in the various fluence ranges may appear to be somewhat esoteric in nature. In fact, this is far from being the case. The operation of different material removal mechanisms can be exploited in dealing with different restoration problems (see Section 6.4).

6.3 ABLATION METRICS

6.3.1 ABLATION THRESHOLD FLUENCE, ETCHING CURVES, AND PHENOMENOLOGICAL MODELS

At least from the practical standpoint, the first aspect to consider concerns the efficiency and accuracy of material removal that can be attained with laser irradiation. This question is directly addressed by the examination of the etching curves (Figure 6.1a). The term ablation rate is often used to denote the material thickness removed per laser pulse. Various simple phenomenological models have been developed for describing the features of the etching curves. Two models have found most use in the literature, namely the so-called steady-state and blow-off models [1–9].

It can be assumed that material removal occurs for fixed absorbed energy density (per unit mass), once a threshold fluence value (F_{thr}) is exceeded. This assumption results in the steady-state (or stationary) model [3,9]. In this case, the etching depth δ scales linearly with F_{LASER}:

$$\delta = \frac{F_{LASER} - F_{thr}}{\rho E_{cr}} \text{ for } F_{LASER} \geq F_{thr} \qquad (6.1)$$

where E_{cr} represents the critical energy per unit mass (sometimes denoted as ablation enthalpy [9]) and ρ is the density. The formula presumes that the rate of energy deposition is balanced by the rate of energy removal due to material ejection (which accounts for the steady-state nature of the model). For this balance to be attained, material ejection must start early on during the pulse. Thus, strictly speaking, this model is applicable for microsecond or longer laser pulses.

On the other hand, for nanosecond pulses, the equilibrium implied by Equation (6.1) cannot be attained. In this case, it can be argued that material ejection is determined largely by the spatial distribution of the absorbed energy. The basic premise here is that for an incident fluence, all material within a depth exposed to a fluence above a threshold value (F_{thr}) is removed. This assumption results in the so-called blow-off model for nanosecond laser pulses [1]. Assuming Beer's law for the absorption process, the dependence of etching depth (δ) on incident fluence is now given by

$$F_{tr} = F_{LASER}e^{-\alpha_{eff}\delta} \geq F_{thr} \Rightarrow \delta = \frac{1}{\alpha_{eff}}\ln\left(\frac{F_{LASER}}{F_{thr}}\right) \text{ for } F_{LASER} \geq F_{thr} \qquad (6.2)$$

where F_{tr} is the fluence transmitted at depth δ, α_{eff} is the (effective) absorption coefficient (see Section 6.5.1), and F_{LASER} represents the incident laser fluence on the substrate, assuming no reflection. Thus the ejected material thickness increases gradually with increasing F_{LASER} (Figure 6.2); for this reason, it is also referred to as the layer-by-layer removal model. According to this model, what matters is the fluence absorbed at depth δ, whereas the energy absorbed in the ejected layers is essentially wasted (largely transformed into kinetic energy of the ejected material).

According to Equation (6.2), for high α_{eff}, sufficient energy is absorbed within a shallow depth to achieve efficient material removal, while penetration of light further into the bulk is much reduced, with a consequent limitation of any thermal and chemical effects there. As a result, material is removed with minimal morphological change or other side effects, and a highly smooth surface may be obtained. This is usually described as clean etching.

According to either Equation (6.1) or Equation (6.2), the basic parameter characterizing laser removal processes is the ablation threshold. The ablation threshold corresponds to the minimum fluence required to achieve nonselective ejection of a volume of material. Generally, the ablation threshold F_{thr} scales as E_{cr}/α_{eff} (the proportionality constant depending on the units employed for E_{cr}). Thus it depends strongly on the substrate absorptivity, while E_{cr} reflects the dependence on other substrate properties such as cohesive energy.

6.3.2 LIMITATIONS AND CAVEATS

Though convenient, the above functional dependences are largely idealizations based on specific assumptions and simplifications. As discussed in Section 6.5, various processes contribute to the material ejection process, and the exact shape of the

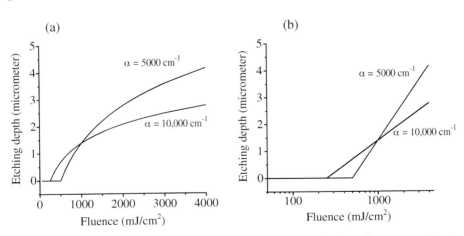

FIGURE 6.2 (a) Illustration of the dependence of the etching depth on F_{LASER} according to the blow-off model for two different absorption coefficients. For a high α, the superficial penetration of the laser radiation results in a small etch depth that increases slowly with F_{LASER}, whereas for low α, the threshold is at higher fluence; but once exceeded, the deeper penetration results in a much steeper slope of the etching vs. F_{LASER}. (b) The same data replotted in semilogarithmic format.

etching curves may differ according to material and laser parameters. Thus it is understandable that Equation (6.1) or Equation (6.2) may not have a general applicability or may fail to describe the complete etching curve. For instance, closer examination reveals that sections of the etching curves (Figure 6.3) have different dependences: below the threshold, an exponential dependence may be obeyed, whereas, close to the ablation threshold, δ may scale linearly with $(F_{LASER} - F_{thr})$ even for nsec laser pulses [3] (although the interpretation of this linear dependence may not be the one implied by Equation (6.1)). In addition, other processes such as absorption of incident radiation by the ejected plume may affect the shape of the etching curves. In all, etching curves may yield limited insight into the underlying physical processes.

It is important to note that the parameter values in Equation (6.1) and Equation (6.2) are usually chosen empirically, since in most cases they cannot be directly related with known material parameters. Thus plots of δ vs. $\ln(F_{LASER})$ yield straight lines, as expected from Equation (6.2), but in many cases, the slope deviates substantially from $1/\alpha$: where α is the small-signal absorption coefficient. The reasons for this deviation are discussed in Section 6.5.

Besides the above limitations, the measurement of the etching depth presents a number of subtleties. In particular, irradiation with successive laser pulses can result in different morphological changes and removal rates. As shown in Section 6.5.4, irradiation can result in chemical modifications to the substrate, with a consequent change in the absorption coefficient and therefore reduction of the fluence necessary for material ejection. Thus for F_{LASER} somewhat below the single-pulse ablation threshold, ablation may be induced after a certain number of pulses (incubation effect) [1–3]. Likewise, for ablation at fluences slightly above F_{thr}, the thickness

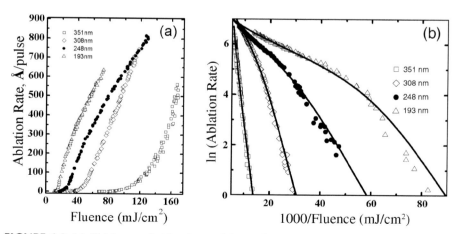

FIGURE 6.3 (a) Thickness of ablated material per single laser pulse as a function of laser fluence for irradiation of polyimide with ArF, KrF, XeCl, and XeF excimer lasers. (b) The same experimental points are presented in Arrhenius format. The solid lines represent the best approximation to the experimental data. Clearly different sections can be delineated with different slopes. From Bityurin et al., *Chem. Rev.*, **103**, 519, 2003. With Permission.

removed per pulse may vary with successive laser pulses, until a constant etching depth per pulse is attained. Thus in determining the ablation threshold, it is crucial to specify the number of pulses used; unfortunately in practice, this dependence is often disregarded.

Furthermore, different techniques yield different values for etching depth. Profilometry is very easy to use, but the method is prone to errors because the irradiated surface may be highly irregular (in fact, for weakly absorbing systems exhibiting swelling at moderate fluences, it may be very difficult to determine the fluence at which etching actually occurs). If the etching depth is very small, the application of the technique may require a multipulse protocol, in which case, the measurement may suffer from the limitations discussed in the previous paragraph. On the other hand, measurements of mass loss by quartz microbalance have been shown to be sensitive even to loss due to fragment desorption; but this does not necessarily correspond to ablation [3,23]. An accurate determination of ablation threshold can be obtained using piezoelectric transducers to measure the recoil pressures (Section 6.5.3). This method has been used effectively in ablation studies on polymers and tissues but little in the laser cleaning of artworks. Alternatively, a microphone system can be used to examine the acoustic transients in air caused by the material ejection [24].

Despite all these shortcomings, etching rate curves do provide a useful starting point.

6.3.3 Ablation Enthalpy and Efficiency

The term ablation enthalpy refers to the specific additional amount of laser energy absorbed to remove an additional mass, $\Delta E_{abs} / \Delta m$. In the steady-state model, enthalpy is a constant, whereas in the blow-off model it increases monotonically with increasing laser fluence.

The term ablation efficiency, $\eta_{abl} = \rho\delta / F_{LASER}$ (where ρ is the density of the material, and F_{LASER} is corrected for any losses due to reflection and/or scattering), is used to denote the total energy necessary to remove a given mass. For steady-state ablation, the efficiency is zero at threshold, increasing at higher fluences and reaching a plateau at high enough fluences [9]. In contrast, for the blow-off model, it reaches a maximum at moderate fluences, decreasing thereafter (Figure 6.4). This dependence can be understood by considering the premise underlying the derivation of Equation (6.2): for the blow-off model, the energy absorbed in the removed material is essentially wasted (i.e., it does not contribute to the material ejection process), becoming instead largely translational kinetic energy of the ejecta. At fluences just above the threshold, this amount is a relatively small percentage of the total absorbed energy, whereas at high fluences, its percentage increases. For practical applications, it would appear best to work at fluences close to the maximum efficiency.

In the early ablation literature, these concepts were used extensively. But, in practice, they give little physical insight, especially since the absorbed energy cannot be measured accurately (because of scattering of the laser beam by plume/plasma, etc.). Furthermore, there are other factors relating to the side effects of the laser processing that must also be taken into account. Thus, though these terms/parameters are still in use in practical applications, they should be used with caution.

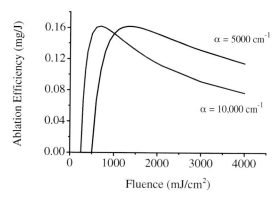

FIGURE 6.4 Illustration of the dependence of the ablation efficiency on F_{LASER} according to the blow-off model for two different absorption coefficients.

6.4 CLEANING METHODOLOGIES

6.4.1 Removal of Polymer Coatings

By far the most typical cleaning scheme concerns the removal of relatively thick, degraded, or unwanted coatings, e.g., of a polymeric coating from a substrate [25–31]. Of course, in the laser restoration of artworks, the substrates to be processed are much more complex, entailing stratified media of varying composition (with increasing depth from the surface), as indicated schematically in Figure 6.5a [17–19].

In these cases, material removal relies on the ablation phenomenon that is effected at high laser fluences. As follows from Equation (6.2), use of a wavelength that is relatively strongly absorbed by the polymer coating is required in order to achieve a good material removal rate. Paint coating removal has been attempted by a variety of laser systems including CO_2, Nd:YAG (1064 nm and its harmonics), and excimer and high-power diode lasers. However, the systematic comparison has been rather limited. Excimer lasers have been the most popular, if not the most useful, since commercial paints and polymeric coatings generally absorb well in UV. For example, varnish and its degradation by-products have $\alpha \approx 10^5$ cm^{-1} at 248 nm, thereby ensuring efficient and clean etching. From the etching rate curves ($\lambda = 248$ nm), it is seen that material removal can be controlled with a depth resolution of 0.1 to 1 µm per pulse. Given a laser repetition rate of 10 to 50 Hz, contaminated surface layers of 20 to 300 µm thickness can be removed in a reasonable amount of time by excimer laser. Illustrative examples for the restoration of painted artworks are provided in Chapter 8. Cleaning of canvas and other support material such as wood and silk has also been achieved.

Laser processing parameters (i.e., wavelength, fluence, pulse width, repetition rate, pulse overlap) are defined according to the criteria and the studies described in Section 6.5. However, because of the high variability in the nature of the materials employed in artworks, the optimal laser parameters are best defined through preliminary study of a representative object sample. Generally, the optimal fluences

(a)   **(b)**

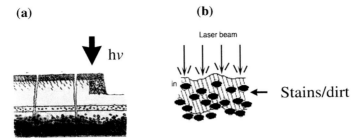

FIGURE 6.5 Schematic of (a) degraded coating removal from a painted artwork (i.e., case of removal of a coating) and of (b) isolated absorbing impurities/stains from a substrate.

for varnish removal are in the 200 to 600 mJ/cm² range, while the removal of overpainting requires about two to three times higher fluences.

The most important point, of course, is to achieve material removal with minimal, if any, side effects on the underlying material. In the case of varnish removal from painted artworks, it will be shown in Section 6.5.4.2 that this criterion can be met if a thin layer of varnish (usually 5 to 10 μm thick) is left intact on the pigmented medium. This layer acts as a blocking filter, preventing laser light from reaching the underlying pigments and binding medium.

6.4.2 SELECTIVE MATERIAL REMOVAL

In the previous examples, the etching depth is smaller than the film thickness that needs to be removed. Thus the concept of layer-by-layer ablation implied by Equation (6.2) suffices for the purposes of the application. If, however, the degraded material/dirt has a thickness comparable to or smaller than the etching depth or it consists of isolated particles on the surface of the substrate (Figure 6.5b), then a different approach must be used for effecting its removal without damage to the substrate. This objective may also be attained by laser irradiation, albeit only in certain cases.

The selective removal of impurities may be effected by exploiting the thermal evaporation process induced by irradiation at moderate fluences. To attain a satisfactory removal rate via evaporation, the fluence should be at least (assuming that the material removal depth is equal to the optical penetration depth α^{-1} [32]):

$$F_{\min} = \rho\alpha^{-1}\Delta H^{\mathrm{sub}}(1-R)^{-1} \tag{6.3}$$

where ΔH^{sub} represents the sublimation or evaporation enthalpy, ρ the density, and R the reflectivity. For proper cleaning, the threshold for substrate damage $(F_{\mathrm{sub,min}})$ should be much higher than the fluence required for contaminant removal, $F_{\mathrm{cont}}(F_{\mathrm{sub,min}} > F_{\mathrm{LASER}} > F_{\mathrm{cont,min}})$. Thus, assuming negligible heat conduction into the substrate, we must have

$$\frac{F_{\mathrm{cont,min}}}{F_{\mathrm{s,min}}} = \alpha_{\mathrm{sub}}\Delta H^{\mathrm{sub}}_{\mathrm{cont}}(1-R_{\mathrm{sub}})\rho_{\mathrm{cont}}/(\alpha_{\mathrm{cont}}\Delta H^{\mathrm{sub}}_{\mathrm{sub}}(1-R_{\mathrm{cont}})\rho_{\mathrm{sub}}) \ll 1 \tag{6.4}$$

(The subscript cont refers to contaminants and the subscript sub to the substrate.) Therefore, for this approach to be effective, contaminants should be highly absorbing and have low sublimation enthalpy, whereas the substrate should exhibit small or minimal absorptivity, a very high reflectivity, and very high ΔH^{sub}. The ideal situation would be for the contaminant to be highly absorbing and for the substrate to be highly reflecting with low absorptivity. In such a case, the laser beam is completely reflected from the substrate once the contaminant layer has been removed. Such a case is said to be self-limiting. Sometimes such ideal conditions are encountered in the removal of inorganics as described in Chapter 8 but, unfortunately, they are not common in the treatment of organic materials.

The efficiency of this selective impurity removal mechanism has been examined for parchments with carbonaceous contamination (graphite) [33–37]. For graphite, at $\lambda = 308$ nm, $R_{sub} = 0.35$, $\alpha_{cont} \approx 2 \times 10^5$ cm^{-1}, and $\Delta H^{sub}_{cont} \cong 6 \times 10^4$ J g^{-1}. The experimentally determined threshold fluence for the contamination removal was found to be $F_{min} \approx 0.4$ J/cm^2. On the other hand, for parchment, $\alpha_{sub} \leq 400$ cm^{-1}, so damage is induced only at much higher fluences (1.8 J/cm^2 for 100 pulses). Indeed, selective contaminant removal without any visible *morphological* modification to the parchment has been demonstrated by irradiation with 308 nm at 1.2 J/cm^2 [34]. However, since α_{sub} is not negligible, chemical alterations to the substrate are induced. In fact, the degree of the polymerization of cellulose is determined via chromatography to drop by 30% after a single shot at ≈ 1 J/cm^2 [33,34]. In contrast, direct photochemical modifications are not induced upon 532 nm and 1064 nm irradiation of paper, silk, or cotton of low absorptivity. For this reason, recent work has focused on the exploitation of these longer wavelengths [34–37]. However, natural cellulose materials such as lignin contain components that do absorb at these wavelengths, and as a result they suffer degradation and color changes (yellowing) even at 1064 nm.

Irradiation at substrate subthreshold fluences has also been used [38] for the removal of oxidized polymeric derivatives and waxes from metallic films. Though the etch rate at these fluences is very low (<1 nm/pulse), the approach constitutes a viable solution for removing very thin polymer films ($\approx$50 nm). In this case, again, absorption by the metallic substrate and/or heat conduction to it are the determining factors for the optimal laser processing.

6.4.3 Particle Removal Techniques

In the last decade, there has been extensive work on the development of laser-based methods for the removal of submicron-size particles from substrates, including polymers [39,40]. Particulates of this size adhere with relatively strong force [12,13], which scales with particle radius (r_p). On the other hand, in conventional techniques (e.g., in gas or liquid jet spray), the cleaning force is approximately proportional to the surface area of the particles [8]) and thus it decreases much faster with r_p than the adhesion force. Despite any advancements, conventional processing techniques become inefficient for the removal of small particles.

In contrast, laser irradiation has been shown to provide a good potential solution for addressing the above challenge. Laser-induced removal of the contaminants can be effected either via direct absorption of the laser light by the particles and/or substrate (dry laser cleaning) or via the prior application of a liquid film (steam laser cleaning). The particle removal process can be evaluated either by posttreatment optical or scanning electron microscopic examination of the sample or via light scattering by the remaining particles of an incident probing laser beam (since the intensity of the scattered light scales approximately linearly with the density of the scattering particles on the surface). The optical scattering method offers the additional advantages of direct on-line monitoring and of high temporal resolution [41].

In dry laser cleaning, when both the particles and the substrate are transparent to the irradiation wavelength, no cleaning takes place. In contrast, efficient cleaning can be effected upon irradiation at laser wavelengths that are absorbed either by the polymeric substrate or by particles. For instance, soot (< 500 nm in size) and debris (~ 5 μm in size) on polyimide (*PI*) have been removed by TEA-CO_2 laser irradiation [42]. SiO_2 particles of 400 nm and 800 nm in size, as well as polystyrene (PS) particles down to 110 nm in size (Figure 6.6), can be removed from *PI* using irradiation at 193 nm and 248 nm, which are absorbed by the polymer substrate [41,43–45].

Capitalizing on the previous studies, we can say that a number of mechanisms can contribute to particle removal:

- Direct ablation of particles.
- Particle expansion. This is effective for strongly absorbing particles on transparent substrates.
- Evaporation of particles. This is essentially the same as described by Equation (6.4). As explained in 6.4.2, such a mechanism can be important for very small particles with low vaporization enthalpy.
- Rapid thermal expansion of absorbing substrate. This is usually the most effective particle removal mechanism. Upon light absorption, the absorbing substrate expands (estimated surface acceleration of $\approx 10^8$ m/sec^2 [44]), thus resulting in particles gaining kinetic energy given by

$$E_{\text{KIN}} \approx m\upsilon^2 \propto \rho_{\text{p}} r_{\text{p}}^3 \left(\frac{F_{\text{LASER}}}{\tau_{\text{pulse}}} \right)^2 \qquad (6.5)$$

where the subscript p denotes particle. For nanosecond or shorter laser pulses, this energy can well exceed the adhesion energy and result in the particle removal [42]. According to this equation, the cleaning threshold is predicted to increase with decreasing particle size, in good agreement with the experimental results (Figure 6.6).

The cleaning efficiency is defined as $(1 - n/n_0)$, where n_0 and n represent, respectively, the particle density on the surface before and after irradiation. This depends strongly on the incident fluence (Figure 6.6). In many cases, high cleaning efficiencies can be attained at fluence levels well below the substrate damage threshold. For instance, for

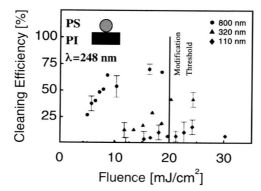

FIGURE 6.6 Single-pulse cleaning efficiency of polystyrene particles from *PI* ($\lambda = 248$ nm, $\tau_{pulse} = 30$ nsec, and average particle density of $\approx 2 \times 10^5$ particles/cm²) as a function of the incident laser fluence. The different symbols represent particles of different diameters. (From Fourrier et al., *Appl. Phys. A*, **72**, 1–6, 2001. With Permission.)

SiO₂ particles and for polymeric particles on *PMMA* or *PS*, single pulse efficiencies of up to 69 to 80% are attained at 248 or 193 nm, respectively [41,43–45]. However, the single pulse efficiencies decrease rapidly with decreasing particle size and are only about 10% for 110 nm particles [43]. The dependence of cleaning efficiency on the number of laser pulses applied, $\eta_C(N_{pulse})$, can be approximated by

$$\eta_c(N_{pulse}) \approx 1 - (1 - \eta_c)^{N_{pulse}} \qquad (6.6)$$

where η_c represents the cleaning efficiency for a single pulse. Thus cleaning efficiencies can be increased from 10–20% up to 80–90% by applying 10 to 20 pulses, until finally there is little benefit in applying more pulses. In practice, deviations from this simple law are usually observed, due to the different cleaning threshold fluences for particles of different sizes, to surface roughness, and to other factors.

Understandably, the range of fluences between the cleaning threshold and the surface modification of the polymer substrate (the fluence margin) is often quite narrow. Thus considerably more caution must be exercised than in the case of layer-by-layer material removal. Furthermore, a most important effect of the procedure has turned out to be the field enhancement under the particles [45], which may lead to the local ablation of the substrate at fluences much lower than in the absence of particles. Besides these issues, for polymeric substrates and particles, additional processes such as photochemical modifications become important, but these have not yet been addressed in the literature.

6.4.4 Liquid-Assisted Material Removal

Material removal is much enhanced when the substrate is irradiated when submerged in a liquid. Similarly, there is a high enhancement in the efficiency of particle removal from surfaces [46–49]. Such effects may turn out to be useful for treating substrates submerged within aqueous environments. Typically, a liquid film, a few tenths to several microns thick, consisting of water mixed with 10 to 20% alcohol (for improving liquid spreading on the substrate), is often applied to the substrate. In industrial applications (e.g., in microelectronics), usually a laser wavelength is employed that is strongly absorbed by the substrate but nearly transparent to the liquid. (In other cases, however, it is thought preferable to employ a (thin) layer of liquid that strongly absorbs the incident radiation, so that the effect on the substrate is minimized.) For sufficient deposited energy, the heat conduction from the absorbing substrate results in the vaporization of adjacent liquid (more precisely, in its explosive boiling; see Section 6.5.2.3). The high amplitude pressure wave that is generated by the rapidly growing bubbles results in a significant enhancement of the etching efficiency. In the case of particle contaminants on a surface, this pressure wave results in large enough particle acceleration (10^8 to 10^9 m/sec^2) to effect their detachment and subsequent ejection by the induced high-speed jet stream. A wide range of particulates including metallic, semiconductor, and polymeric can be removed efficiently in this way [3].

Because the forces exerted are about ten times higher than those developed by the thermal expansion of the substrate, steam laser cleaning efficiencies are much higher than those in dry cleaning. This permits the use of lower laser fluences, which is particularly important for heat-sensitive substrates or for particles that can melt. Furthermore, the cleaning efficiency is largely independent of the particle size [49], thereby permitting efficient cleaning even of particle collections with a wide size distribution.

For these reasons, steam cleaning has been considered for use on polymer substrates. For instance, debris removal from polymer surfaces has been demonstrated by using KrF laser radiation with transparent isopropanol films. Organic coatings can also be removed from a variety of substrates, including painted artworks. In particular, it has been suggested [50] that we pretreat a substrate with the application of a thin layer of water and/or alcohol. Such OH-rich solvents absorb strongly at 2.94 μm, so irradiation at this wavelength (by Er:YAG laser) will be preferentially absorbed by the solvent with reduction of the heat diffusion to the substrate. In fact, addition of a solvent may not be necessary if the medium tends to absorb water (humidity). Additional advantages accrue from the facts that 2.94 μm is eye-safe and the employed laser system is portable, inexpensive, and easy to use. However, it should be noted that the use of liquid on the highly sensitive painted surfaces presents always the danger of solvent penetration to the substrate, resulting in deleterious side effects. Furthermore, because of the high pressures developed in the bubbles, the potential for collateral damage is much higher than that for dry laser cleaning (this impact decreasing with longer pulse width). This is not a major concern in industrial applications, but it can be a crucial factor in the cleaning of highly fragile artworks.

6.5 ASSESSMENT OF THE LASER-INDUCED PROCESSES AND EFFECTS

Besides the issue of efficiency, the most important consideration in the selection of any cleaning technique concerns the nature and extent of side effects that may be induced to the substrate. It is customary to delineate the induced processes and effects into thermal, photochemical, and photomechanical. This division provides a convenient basis for the discussion of the mechanisms and effects of UV ablation of polymers. However, this delineation is rather formalistic, and as will become clear in the next sections, the three phenomena are closely interrelated.

6.5.1 PHOTOEXCITATION PROCESSES

It is common to use the tabulated small-signal absorption coefficients in Equation (6.2) in order to estimate etching depths. Unfortunately, the etching depths estimated in this way deviate significantly from the experimental results. One of the reasons for this is that absorptivity during ablation may differ significantly from the small-signal one.

Such deviations in the absorption coefficient are easily understood in the framework of the discussion in Chapter 2. First, the strong thermal and chemical transients that are induced during laser ablation can alter significantly the optical properties of the substrate. Several illustrative examples have been encountered in the irradiation of inorganics as well as of organics [51]. Alternatively, changes in the electronic excitation step, i.e., saturation or multiphoton processes, may dominate. Petit et al. [52] have presented detailed modeling of the influence of multiphoton excitation and chromophore saturation processes on the efficiency of material removal. Masuhara et al. [53] have demonstrated that cyclic multiphoton processes are particularly pronounced in the UV ablation of doped polymers.

Concerning applications, the important implication is that the optical penetration depth, and thus the extent of thermal effects and chemical modifications, may differ significantly from those expected on the basis of the small-signal absorption coefficient of the substrate. Unfortunately, there is no specific model to predict the relative importance of such processes for different substrates. In fact, even for the same system, dynamic changes of the absorption coefficient may differ significantly according to the wavelength used. Thus the extent of the contribution must be evaluated experimentally in each case.

At higher fluences, two additional factors become important in affecting incident light propagation. The formation of gaseous bubbles within the substrate can result in a significant scattering of laser light as it propagates within the substrate. Second and more importantly, scattering and absorption by the plume itself becomes significant and limits the light intensity reaching the substrate (Section 6.7) [54]. These factors are responsible for the plateau observed at high laser fluences in the etching rate curves (Figure 6.1).

6.5.2 THERMAL EFFECTS

6.5.2.1 Photothermal Mechanism

As described in Chapter 2, following absorption, a good part of the absorbed energy — at least for irradiation with typical nanosecond pulses — will be converted into heat. The extent of the subsequent heat diffusion is conveniently described in terms of *thermal diffusion length,* $l_{th} = 2(D_{th}t)^{1/2}$, or in terms of *thermal diffusion time*

$$t_{th} = \frac{1}{D_{th} \cdot \alpha_{eff}^2} \qquad (6.7)$$

Here D_{th} is the thermal diffusivity, and α is the absorption coefficient. Consider now that the material at the attained temperatures thermally decomposes [23,55–60]. Thermal decomposition (typically a unimolecular reaction) usually follows an Arrhenius equation with a rate constant $k(T) = Ae^{-E_{act}/R_G T}$, where A is the preexponential factor, E_{act} is the activation energy, and R_G is the universal gas constant. For organic material, typically, A ~ 10^7 to 10^{10} sec^{-1} and E_{act} ~ 50 to 200 kJ/mole. For a specific thermal transient, the number of bonds thermally decomposed up to time t can be expressed as $N_D(t) = \int_0^t A\exp[-E_{act}/kT(t')]dt'$, where $T(t')$ represents the termporal evolution of the temperature on the surface. If the decomposition results in small enough fragments or oligomers that have a small binding energy to the matrix, these can desorb in the gas phase. By doing so they remove energy, thereby lowering the substrate temperature. Figure 6.7b shows the temperature evolution with depth in a polymeric substrate following irradiation at three different fluences. For nanosecond or shorter laser pulses at high enough fluences, decomposition and material removal occurs fast enough that heat diffusion to the substrate is minimal. On the other hand, for microsecond laser pulses, heat diffusion and the consequent thermal degradation in the substrate are extensive enough. This simple factor is one of the crucial ones for the success of processing labile substrates with short laser pulses. This can be more precisely expressed in terms of $t_{th} = 1/\alpha_{eff}^2 D_{th}$ being much longer than the laser pulse duration. If this condition is satisfied, the process is said to proceed under thermal confinement.

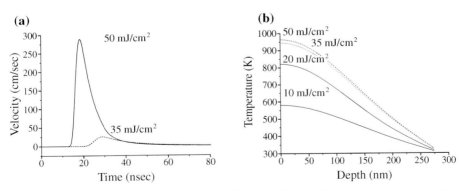

FIGURE 6.7 (a) Velocity of material removal for four different fluences as calculated by the bulk photothermal model [58,23]. At fluences < 35 mJ/cm², material ejection can be considered to be nil (the calculation is performed for $a = 3.2 \cdot 10^5$ cm^{-1}, $\lambda = 248$ nm, and the rest of parameters are for polyimide). (b) Estimated temperature profiles in the substrate (as a function of depth) at 100 nsec upon laser irradiation at the indicated fluences.

The above description provides the basis for the so-called photothermal mechanism of ablation. According to this mechanism, ablation is exclusively due to the thermal decomposition and desorption of material. In this case, the minimum energy (in J/cm^3) required for material ejection is given by $\rho c_P \Delta T + \Delta H_{transf} + \Delta H_{diff}$, where the first term represents the energy required for heating a mass to its decomposition temperature, ΔH_{transf} represents the energy required for polymer decomposition and desorption of products to the gas phase, and ΔH_{diff} is the energy lost by heat diffusion to the sublayers. Accordingly, $F_{thr} = (\rho c_P \Delta T + \Delta H_{transf} + \Delta H_{diff})/a$. Thus a high thermal diffusivity, D_{th}, results in high heat losses to the sublayers with a consequent increase in the ablation threshold. Detailed analytical description of the model is given in [3,23,25]. Briefly, the enthalpy form of the heat equation is usually employed because of its convenience in dealing with phase changes including melting [3,23]. Assuming materials removal and heat diffusion only along one axis (along the z-direction i.e., one dimensional), this becomes

$$\frac{\partial H}{\partial t} - \upsilon_{int} \frac{\partial H}{\partial z} = \frac{\partial}{\partial z}\left(\xi \frac{\partial T}{\partial z}\right) - \frac{\partial(Ie^{-\alpha z})}{\partial z}, \qquad (6.8)$$

where $H(T)$ represents the total enthalpy of the substrate (at temperature T.) Solution of this equation requires the specification of appropriate boundaries concerning, e.g., the energy loss at the surface due to material removal. A satisfactory solution has been obtained only for ablation with long pulses, in which case a steady-state condition is attained, as in Equation (6.1), where the rate of material removal during the laser pulse is constant. However, this condition is hardly reached in the irradiation with nsec pulses. The approximations necessary for solving the equation for the nonstationary case have been discussed amply in the literature [23]. In this case, the rate at which material is removed is specified either by the temperature attained at the surface or by the condition that the surface concentration of broken bonds reaches a specific level. The former condition leads to the so-called surface photothermal model, whereas the latter one leads to the volumetric photothermal model. The applicability of each particular model depends on the substrate absorptivity, with the surface model being more appropriate for substrates of a very high absoprtivity (e.g., metals), whereas the volume model is better for substrates of moderate absorptivities (e.g., polymers). Despite any approximations and shortcomings, this volume photothermal model has been the most successful in explaining various experimental aspects of polymer ablation. For further information on these formulations, the reader is referred to [23].

6.5.2.2   Illustrations and Implications

At least for high enough fluences (at which ablation can be considered to commence early during the laser pulse and to proceed at a constant interface velocity υ_{int}), minimization of thermal heat diffusion and thus of thermal effects can be attained if D_{th}/υ_{int} (i.e., the relative rate of heat diffusion to the sublayers vs. the rate of material removal) is small compared to the (effective) optical penetration depth. This supposes knowledge of v_{int} that depends on the particularities of the material. The extent of thermal decomposition can be more approximately estimated just by the comparison of α_{eff} to D_{th}. Since D_{th} for polymers and amorphous organic materials is typically $\approx 10^{-3}$ cm^2/sec, the thermal diffusion length for nsec pulses is estimated to be only ≈ 100 to 500 nm. Since the typical optical penetration depth of polymers in the UV is ~1 to 10 microns, the extent of bulk disruption is, to a first-order approximation, specified by the substrate absorptivity at the irradiation wavelength. For instance, in the case of painted artworks, optical microscopic examination demonstrates clean etching for ablation of varnish at 248 nm, a strongly absorbed wavelength ($\alpha \sim 10^5$ cm^{-1}). In contrast, irradiation at 308 nm results in pronounced collateral damage and morphological change. This can be ascribed to the lower absorption coefficient of varnish at 308 nm than at 248 nm (at 308 nm, $\alpha \sim 10^2$ cm^{-1}) [19], resulting in absorption and thermal degradation (as well as photochemical modifications) in the bulk. However, in other cases, thermal conduction clearly limits the efficiency of laser removal techniques. For instance, in the irradiation of contaminated parchment (Section 6.4.2 and Section 6.4.3.), the separation between absorbing impurities and substrate is comparable to the estimated thermal diffusion length. Thus in the IR irradiation of parchment (transparent at this wavelength) dirtied with absorbing impurities, water loss and ether cross-linking (resulting in an increase of cellulose degree of polymerization) are observed [33,34], as a result of the heat conducted from the absorbing impurities.

Irradiation may result in melting of the substrate even at fluences well below the ablation threshold. This may have deleterious effects on the appearance and integrity of the irradiated substrate. First, for multicomponent systems with constituents of different binding energies to the matrix, the enhanced diffusion of species within the molten zone can result in significant segregation effects or changes in the chemical composition through the preferential desorption of weakly bound species (Section 6.3).

Second, upon ablation, the back pressure exerted by the ejected material (Section 6.7) forces the liquid toward the edges of the irradiated area. Upon cooling, this material resolidifies with formation of irregular edges. Such morphological features are used extensively as a diagnostic tool of the interaction mechanism. Thus the lack of a melted zone around the irradiated area is often considered to be a direct indication of a nonthermal mechanism of ablation (photochemical implied). This, however, is not necessarily correct. As indicated by Figure 6.7, the depth of the molten material and duration of melting decrease much with absorption coefficient. Consequently, segregation effects may not occur on such fast time scales. Furthermore, the flow velocity of material is very much affected by the melt depth. For a thin layer of melt (i.e., as in the irradiation strongly absorbed wavelength), the presence of the nearby solid interface severely limits flow velocity. Thus even a simple thermal mechanism can result in clean processing with perfect sharp edges [61,62].

The influence of heat diffusion on the evolution and extent of chemical processes induced to the substrate will be discussed in Section 6.5.4.

6.5.2.3 Introduction to Explosive Boiling

A concept that has turned out most important in ablation studies is that of explosive boiling, describing the "boiling process" upon ultrafast heating of melts/liquids [9].

For a simple compound, the processes can be understood in terms of the usual (P,T) thermodynamic diagrams. For slow heating, the system follows the binodal, and the vaporization transition is well described by the Clausius–Clapeyron equation $P \approx P_0 \exp[-(\Delta H_{vap} / R_G)(1/T - 1/T_0)]$, where P represents the pressure at temperature T and P_0 the reference pressure at reference temperature T_0 (usually T_0 is the boiling point at $P_0 = 1$ atm).

However, the liquid–gas transformation requires the formation of gaseous bubbles. Their formation is energetically costly, as energy is required in order to form the necessary interface. The required energy is specified by the surface tension σ of the compound. The overall work (W) necessary for the formation of a bubble of radius r is $W = 4\pi r^3 / 3(\mu_v - \mu_l) + 4\pi r^2 \sigma$, where the first term, expressed in terms of the chemical potentials of the compound in the gas and liquid phases, gives the driving energy, whereas the second term represents the required energy (work) for interface formation. The rate of bubble nucleation scales as $e^{-W/k_B T}$. The ratio of the two terms $4\pi r^2 \sigma / [4\pi r^3 / 3(\mu_v - \mu_l)]$ scales as $(1/r)$. So for very small r, the surface tension term is much larger than the driving force. As a result, W has a large positive value and formation of nuclei is a rather slow process (microseconds to milliseconds). The implication is that for nanosecond laser pulses, the system can be heated to temperatures much higher than its boiling point, before bubble growth occurs. With increasing T, the driving force eventually becomes sufficiently high to overcome the surface tension limitation, i.e., the nuclei formation rate becomes competitive with the heating rate. This results in abrupt liquid–gas phase transformation and accounts for the explosive character of laser ablation.

These concepts have turned out to be most useful in considering laser ablation of metals, semiconductors, organic liquids, steam laser cleaning, etc. [46–49]. However, the extension of the concept to more complex systems, such as polymers, is not yet well-defined. The reason for this is simple: with increasing molecular complexity/size, thermal decomposition sets in at lower temperatures than the temperature for liquid–gas phase equilibrium. Thus strictly speaking, it is not possible to define the degree of overheating in relationship to a reference phase transformation temperature. Nevertheless, the term is sometimes indiscriminately used to describe the ablation of complex substrates. The reader is cautioned that this is not fully validated scientifically.

6.5.3 PHOTOMECHANICAL PROCESSES AND EFFECTS

6.5.3.1 Photomechanical Mechanism

Laser irradiation can result in the development within the sample of stress waves with amplitudes of several hundred bars [63–67]. These stress waves may be generated in different ways.

As described in Chapter 2, one source of stress is the thermoelastic wave generation, which results in a bipolar wave. The faster the heating, the higher the magnitude of the generated thermoelastic stress in the medium, with the ultimate efficiency attained for heating time much faster than the time required for stress to propagate through the irradiated depth, i.e., for $\tau_{pulse} < 1/\alpha_{eff}c_s$ (so-called stress confinement regime), where c_s is the sound velocity in the substrate.

Another source of stress waves derives from expansion of any gases produced by thermal or photochemical decomposition within the substrate [64]. This factor, for instance, has been invoked to account for the transient stresses of about 0.1 MPa detected in the UV irradiation of polyimide below the ablation threshold [66]. In the case of doped-*PMMA* irradiation with 150 psec pulses at 1064 nm, the thermoelastic mechanism and the expansion of decomposition by-products contribute about equally to the generated pressure at the ablation threshold [65].

If the pressure wave amplitude exceeds the substrate tensile strength (defined as the minimum tension pressure required for material fracture), then it can result in ejection of material essentially via fracture (Figure 6.8a). This constitutes the so-called photomechanical mechanism for material ejection. Since fracture can occur without the overheating of the material implied by the thermal mechanism, it offers the possibility for "cold" ablation. As a result, it has attracted much attention, in particular in medical applications. Indeed, the operation of a photomechanical-based ablation can be significant in the nanosecond laser irradiation of liquids, as well as in soft tissues (largely because of their low tensile strength). It can also be important in the irradiation of thin films, for example, of polymers on a substrate with nanosecond or shorter pulses. The wave produced becomes tensile upon reflection at the substrate–polymeric film interface and induces ejection of the material essentially as a whole. Although stratified structures characterized by weakly adhered layers are commonly encountered in artworks, specific examples exploiting this mechanism for the laser restoration of artworks have not been reported. On the other hand, for (thick) polymeric substrates (i.e., in the absence of interfaces), for typical UV

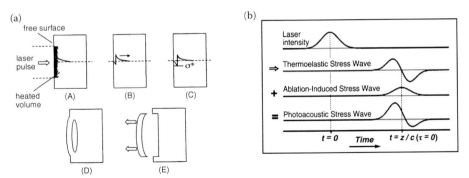

FIGURE 6.8 (a) Schematic illustrating the photomechanical mechanism. The stress wave signal of material removal developed by thermoelastic mechanism (A–B); at some depth the tensile strength σ^* is exceeded and the material fractures; (D, E) detachment and ejection of material from the front surface. (Reprinted with permission from Ref. 64.) (b) Schematic illustrating the stress wave signal developed by thermoelastic mechanism and the pressure wave generated by the back momentum due to the ejected material above the ablation threshold. The third curve represents the overall pressure transient detected above the threshold.

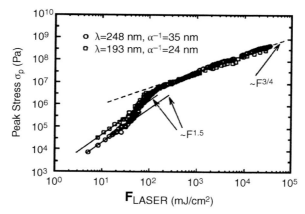

FIGURE 6.9 Peak pressure generated in polyimide (measured via piezoelectric attached to the substrate) vs. F_{LASER} for excimer irradiation at the indicated wavelengths. The lower abrupt change in slope is due to the onset of ablation whereas the change at higher fluence is due to the influence of plasma expansion. (From Zweig, Venugopalan, and Deutch, *J. Appl. Phys.*, **74**, 4181, 1993. With permission.)

nanosecond pulses, the generated stress waves turn out to be rather weak for being exclusively responsible for material ejection; but such mechanisms may play an important role for shorter laser pulses.

For irradiation above the ablation threshold, a third source of stress wave relates to the back momentum (recoil) exerted by the ejected material (independently of the mechanism responsible for its ejection) to the substrate. This results in a compressive wave propagating through the substrate. The peak stress amplitude (and its scaling with incident fluence) depends on the time scale of material removal, as well as on the nature of the process. The peak pressure can be estimated by assuming the ejected material to be an ideal gas expanding against the air and the substrate, resulting in the expression [57,66]

$$P_{abl} \propto \left[\frac{(F_{LASER} - F_0)}{\tau_{pulse}} \right]^{\frac{2}{3}} \tag{6.9}$$

where $F_0 \approx F_{thr}$ represents the energy that remains as heat in the substrate. The formula essentially relies on the assumption that any energy in excess of F_{thr} becomes kinetic energy of the ejected material. The predicted scaling has been demonstrated in the UV ablation of polyimide and of cornea. At even higher fluences, where plasma formation becomes significant, the pressure relates to the plasma expansion, in which case P_{max} scales as $F_{LASER}^{3/4}$ (Figure 6.9).

6.5.3.2   Illustrations and Implications

The measurement of laser-induced stress waves relies on the use of piezoelectric transducers or optical techniques. Generally, peak pressure amplitudes range from a few MPa at the ablation threshold up to several hundred MPa at high laser fluences [65,66]. Typical results from such measurements are illustrated in Figure 6.10. At fluences below the threshold, a bipolar thermoelastic wave is observed, whereas above the threshold, the positive amplitude dominates, owing to the compressive wave generated by the back momentum of the ejected material. Thus the temporal onset of material ejection can be directly specified from the evolution of the recorded pressure wave. In practice, it is easier to record the time-integrated pressure $(P_{\text{abl,total}} = \int_0^\infty P(t)dt)$. The onset of the recoil pressure due to the ejected material provides an unambiguous specification of the ablation threshold [63]. In all, the measurement of the stress waves provides an important diagnostic tool of the laser-induced material ejection.

Besides their diagnostic significance, laser-induced stress generation may be of importance for the success of laser processing. The high amplitude waves propagate through the substrate and may induce structural modifications in areas away from the ablation spot (in sharp contrast to the photochemical effects, which are confined to the irradiated area). This can be an important consideration in the processing of mechanically fragile substrates [9].

For examining this issue, double exposure holographic interferometry can be employed [68]. Holographic techniques afford the advantage of storing the full field image of the sample before and after irradiation, thereby permitting spatially resolved characterization of the induced effects over a large surface area. In contrast, piezoelectric techniques, though capable of high sensitivity and temporal resolution, yield information only about limited regions of the substrates. Two exposures are made on the same hologram, one before and one after irradiation of the sample, with a given number of ablating laser pulses. Any change in the optical path length of the

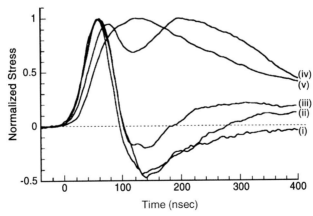

FIGURE 6.10 Stress transients resulting from TEA-CO_2 laser (τ_{pulse} = 30 nsec) irradiation of porcine dermis for radiant exposures below the ablation threshold (i), at the threshold (ii), and above the threshold (iii–v). Irradiation below the ablation threshold produces bipolar thermoelastic stress transients. Irradiation closely above the threshold results in a compressive pulse produced by the ablative recoil that is delayed relatively to the laser pulse. The time delay decreases with increasing laser fluence. (Adapted from Paltauf and Dyer, *Chem. Rev.*, **103**, 487, 2003. Copyright 1996 Biophysical Society.)

probing beam results in interference between the two successive holograms, $U(x,y) \propto U_0(x,y)\{1 + \cos\{(2\pi/\lambda)\Delta\phi(x,y)\}$. The phase term $\Delta\phi$ relates to induced refractive index changes. If $\Delta\phi = \int [n_R(x,y,z) - n_{R_0}]dz = N_{fringe}\lambda$, interference fringes are generated that can be related with the laser-induced modifications (Figure 6.11).

The application of this technique [69] has shown that upon irradiation of model (polymeric films cast on substrates) and of realistic samples, morphological changes can be induced at distances up to a few centimeters from the irradiated spot. Two types of deformations are observed, namely localized delaminations between the various substrate layers, and cracks (crazing). Generally, the extent of formation of these defects increases with an increasing number of laser pulses (Figure 6.12a). An initiation phase ($\approx$5 to 10 pulses) may be observed, in which defects are not detectable. This induction period must represent the laser-induced formation of defect nuclei in initially defect-free substrates. Following defect nucleation, there is a gradual increase in the size of initiated defects, as well as the appearance of new ones.

Given the large distance from the irradiated spot (Figure 6.12b) and the low laser repetition rate (1 Hz), thermal effects can be discounted, and the formation of these defects must be ascribed to the influence of propagating stress waves. Through the multiple reflections of the laser-induced waves back and forth between the various substrate layers, a complex wave pattern propagates in the system. During propagation, the acoustic waveforms are attenuated by sound absorption, diffraction, and plausibly nonlinear acoustic effects. Losses due to absorption are described by $P(z) = P_{abl}\exp(-a_{ac}z)$, where z represents the distance from the ablation spot, a_{ac} is the acoustic wave attenuation coefficient, and P_{abl} is the pressure exerted by the removal process. Generally, the attenuation coefficient for polymers scales as v_{ac}^2 (where v_{ac} is the acoustical wave frequency), ranging typically from 0.05 to 0.2 dB/m for longitudinal waves (for v_{ac} from $\approx$6 MHz to $\approx$30 MHz) and from 0.14 to 0.65 dB/m for shear waves [70]. (In acoustics, it is conventional to express intensity in relative units; decibels dB = $10 \log_{10} (I/I_{ref})$ where I is the wave intensity and I_{ref}

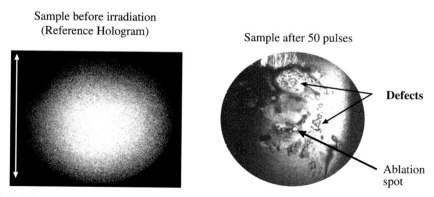

Sample before irradiation
(Reference Hologram)

Sample after 50 pulses

Defects

Ablation
spot

FIGURE 6.11 Illustration of interferograms recorded in the UV ablation of *PMMA* polymeric films cast on suprasil (250 pulses, $\lambda = 193$ nm, $F_{LASER} \approx 1$ J/cm^2). The interferogram clearly demonstrates defect formation at distances far away from the irradiation spot. For the purpose of enhancing defect formation, this particular film has not been annealed. (Reprinted from Bonarou et al., *Appl. Phys. A*, **74**, 647, 1993.)

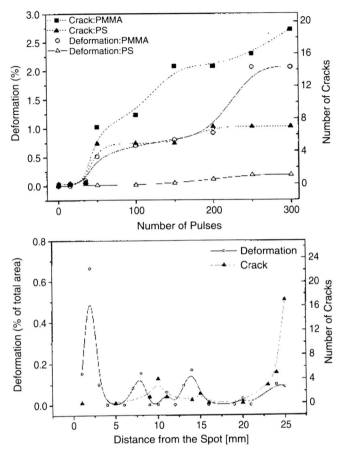

FIGURE 6.12 (Top) Growth of defects in the irradiation of *PMMA* and *PS* films on suprasil substrate at 193 nm as a function of successive laser pulses ($F_{LASER} \approx 1$ J/cm^2). (Bottom) Spatial distribution of the defects following UV ablation of nonannealed *PMMA* films at 193 nm ($F_{LASER} \approx 1$ J/cm^2, 200 pulses). The lines serve only as a guide to the eye. The ordinate values represent the average distance of the center of the defects, independently of their size, from the irradiation spot. The delaminated area is expressed as a percentage of the total sample area, while the cracks are quantified by their number. (From Bonarou et al., *Appl. Phys. A*, **73**, 647, 2003. With permission.)

is a reference wave intensity. Because sound intensity is proportional to the square of pressure so dB $= 10 \log_{10} (P/P_{ref})^2$. The attenuation factor of the sound intensity, expressed in dB/m, is equal to 8.686 a_{ac}.) Thus 1/e loss of intensity should occur over 5 to several centimeters for low-frequency longitudinal waves and somewhat smaller for the shear waves. Coating adhesion strength to substrates varies greatly [74]. In the case of artworks, the variability is even higher, depending on the history, environmental influence, and so forth. Thus the pressure developed in the ablation may carry enough energy to induce defects. The exact nature of induced deformations depends, of course, on the particular substrate structural properties. For instance, in the cleaning of realistic Byzantine icon models (i.e., samples composed of wooden panel support, sized ground, pigmented layers, and varnish), delaminations develop at the interfaces between the weakly adhered layers [19].

Thus the use of multipulse irradiation protocols may be detrimental for the structural integrity of especially mechanically fragile substrates. It is noted, however, that the previous studies employed rather high fluences (≥ 1.5 J/cm^2) that in practice are avoided. Furthermore, after sufficient time, the laser-induced deformations may gradually relax. This issue of reversibility of laser-induced defects, however, has not yet been examined.

6.5.4 CHEMICAL PROCESSES AND EFFECTS

The most crucial question in laser processing, especially of organic substrates, concerns the nature and extent of chemical modifications induced to the substrate and the influence of these changes on substrate integrity.

6.5.4.1 Photochemical Mechanism

The issue of chemical processes in ablation of molecular substrates is also closely related to the question of mechanisms involved, since photochemical processes have been suggested to contribute to, or even dominate, the material ejection in irradiation of photolabile systems [1,4]. According to this model, the formation of a large number of photofragments with high translational energies and the formation of gaseous photoproducts that exert a high pressure may result in material ejection. This possibility was advanced already in the first ablation studies [1] in order to account for the clean etching observed with UV laser pulses as compared with IR pulses. According to this model, because UV photon energy is largely "consumed" in bond dissociations, heat generation and diffusion are minimal. This argument seemed to provide a particularly simple and novel way to account for the clean etching achieved upon UV ablation. Thus this model became quite popular in the field and was an overly used explanation, especially in applications.

However, even for simple, well-defined molecular systems, it has turned out very difficult to assess the contribution of such a mechanism. In fact, the issue of photochemical vs. thermal mechanisms has been the most hotly debated one in the field of ablation. Overall, there is a growing consensus that for common organic and polymeric systems, a thermal mechanism is dominant even for ablation in UV (for nanosecond laser pulses). For instance, in a number of studies on polymers doped with highly photolabile compounds decomposing with a near unity yield, ablation has been found to be governed by the amount of absorbed energy. Only for irradiation at 193 nm, there are features that have not been reconciled with the thermal mechanism. For a few specifically designed polymers, which upon photolysis produce a very high number of gaseous by-products, there is some evidence in support of a photochemical mechanism [75].

However, recent theoretical and experimental work [9,74] suggests that although the extreme view of the exclusive contribution of a photochemical mechanism is unlikely, still chemical processes like bond decompositions may result in the disruption of the substrate structure, thereby facilitating material ejection. Furthermore, besides the pressure exerted by any gaseous by-products, heat released by exothermic reactions effectively contributes to material ejection. Thus the ablation threshold is estimated to be at lower fluence than it would be in the absence of such reactions.

Given all these controversies even for relatively simple polymers, it is very difficult to be sure that a photochemical mechanism is applicable in the irradiation of the chemically complex materials encountered in artworks [17–19].

6.5.4.2 Factors Affecting Chemical Effects in the Substrate

In practice, what really matters is the extent and nature of chemical modifications induced to the substrate upon laser-induced material removal. Given the very high laser irradiances used, pronounced chemical modifications would be expected to the substrates. However, as already underlined, UV ablation entails a number of processes occurring on a very fast time scale. As a result of the combined influence of these factors, chemical processes may differ qualitatively and quantitatively from the predictions of conventional photochemistry.

If the absorbing chromophores in the substrate undergo photofragmentation or other photoexcited reaction with a quantum efficiency η, then upon irradiation, the number density of dissociated/photomodified molecules scales with depth z (from the surface):

$$N_D(z) = \eta \frac{\sigma N F_{LASER} e^{-\alpha z}}{hv} \qquad (6.10)$$

where σ is the absorption cross section, N is the number density of chromophores, and hv is the photon energy. However, the removal of a layer via the etching process implies that the number of products remaining in the substrate is correspondingly lowered. As described by Equation (6.2) for the blow-off model, for $F_{LASER} > F_{thr}$ the material left behind is subject to the same F_{thr}, so that the amount of product in the remaining substrate should be constant. In other words, at these fluences, any by-products formed in addition to those at lower fluences are removed by the etching process (Figure 6.13).

The previous description is, however, correct only to a first approximation, because (1) the temperature changes (Section 6.5.2) may greatly affect the evolution of the chemical processes, and (2) the high laser intensities may cause excitation of molecules to higher electronic states that may react differently from the one assessed upon one-photon excitation. Thus differences can be observed between the ablative regime and the very low fluence regime concerning (a) the nature and (b) the efficiency of products formed.

The interplay of the various factors affecting chemical processes is best illustrated in studies using simple photosensitive organic compounds (dopants) dispersed within polymer films [76]. These doped systems constitute a good, even if idealized, model of the paint layer in artworks, which essentially consists of chromophores (pigments) dissolved or dispersed within an organic medium.

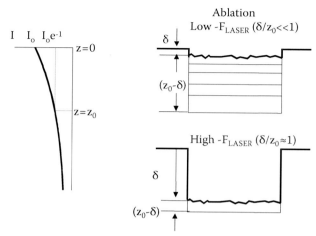

FIGURE 6.13 Schematic illustrating the different extent of product accumulation in the substrate for irradiation at fluences close to the threshold and well above it. d represents the etching depth.

For some dopants, no particular difference in their decomposition is found when irradiating above vs. below the ablation threshold [77–79]. This is the case, for instance, in the UV ablation of *PMMA* films (using 60 psec pulses at 248 nm or 266 nm) [78] doped with 5-diazo Meldrum's acid. Upon excitation at 266 nm with nanosecond pulses of 1,1,3,5-tetraphenylacetone (TPA) dopant in *PMMA*, TPA was found to decompose into two diphenylmethyl radicals and CO. The diphenylmethyl radical concentration grows linearly with fluence, whereas at higher fluences, excited radicals are also observed, formed via a two-photon excitation process [79].

The dopants employed in studies [77–79] dissociate to stable compounds (CO or N_2) with a decomposition quantum yield close to unity. Thus it is understandable that no particular change in their photolysis occurs as the fluence is raised above the threshold. On the other hand, many other chromophores, such as those encountered in artworks, dissociate into reactive radicals. Since reactivity of such radicals is strongly temperature dependent, it may be much affected by the heat diffusion effects discussed in Section 6.5.2.

Understandably, the nature and extent of the induced chemical modifications depend crucially on material and laser parameters and especially on substrate absorptivity. These dependences are discussed in detail in the next sections.

6.6 ASSESSMENT OF THE INFLUENCE OF PARAMETERS

6.6.1 INFLUENCE OF MATERIAL PROPERTIES

Even if the initial extreme suggestions for the operation of drastically different mechanisms depending on the nature of the absorbing chromophores have been abandoned, the physical and chemical characteristics of the substrate affect various aspects of the laser-induced material ejection processes.

The first pertains to the influence of the molecular weight (*MW*) of the polymer, though very little has been reported on this. Yet this parameter can be crucial in such diverse laser applications as medicine and artwork restoration since it varies considerably for the substrates involved. Initial work indicates that the ablation threshold as well as the etching efficiency may decrease with higher polymer molecular weight [76,83]. This decrease can be ascribed to the fact that the minimum energy required for material ejection depends on the cohesive energy of the system. For simple molecular systems of large cohesive energy, a much higher temperature is required for explosive boiling to occur. For polymeric systems, with increasing molecular weight, a much larger number of bonds must be broken for the chains to decompose into small units and oligomers as necessary for desorption and ejection in the gas phase.

The consequences of the dependence of the etching process on molecular weight for the laser restoration of painted art works have been explored very little. Yet some inferences can be drawn. First, the ablation threshold and material removal efficiency may vary with the degree of polymerization of varnish. In practice, this result matches the empirical finding of conservators that "harder" materials require higher fluence. Second, because of the difference in etching efficiencies, a higher amount of by-products remains in the substrate in the ablation of polymers of higher molecular weight. Third, since the degree of polymerization usually decreases from the superficial (oxidized) layers to the lower, nonoxidized ones, etching efficiency will increase as the cleaning continues into the depth. Therefore it is imperative that laser fluence be appropriately adjusted during the procedure.

In comparison with the molecular weight, the influence of polymer chemical composition on the ablation process has been extensively examined. The aim of most of these studies has been the mechanistic understanding of the phenomenon [6,72–82]. A wide range of different techniques have been applied to characterize the products formed upon ablation, and a wide range of different organics and polymers have been examined. Although no general model exists to enable *a priori* specification of the chemical effects that may be induced, there is enough information collected for various materials to provide a good starting point. The chemical processes and effects in the ablation of polymers have recently been presented in detail [6]. Studies on doped polymer systems have been specifically reviewed [76]. Here we summarize only some key results that may be relevant to materials that may be encountered in painted artworks, etc.

For most common polymers, decomposition appears to be dominated by depolymerization reactions where the polymer chain fragments into its constituent monomers and oligomers. Figure 6.14 illustrates suggested decomposition pathways for two typical polymers, *PI* and *PMMA*. It must be noted that the extent of depolymerization upon laser ablation may be quite different from that expected from conventional thermogravimetric measurements. As shown in Section 6.5.2, the heating rates attained with nanosecond pulses are so high that typical depolymerization processes may not compete (having much lower rates). This result has been nicely illustrated by a study on the IR ablation of *PMMA* with 100 psec pulses showing that less than 10% of the polymer has decomposed at the onset of ablation [65].

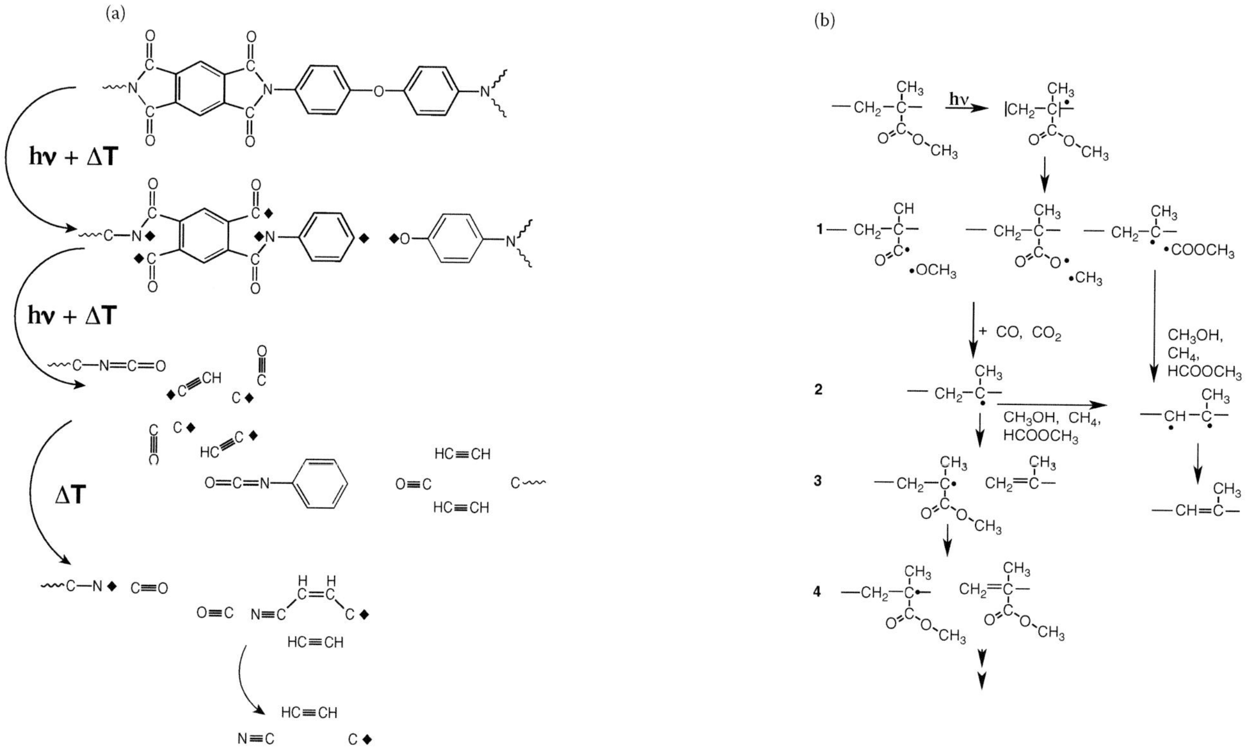

FIGURE 6.14 (a) Suggested scheme for the decomposition of *PI*. ◆ indicates a broken bond. (b) Suggested scheme for the laser-induced (308 nm) decomposition of *PMMA*. Steps 1 and 2 show the photolysis of the ester side chain, the typical small products detected in mass spectrometry measurements, and the double bonds that are created during incubation. Step 4 shows the photochemical and thermal activated reaction to release *MMA*. (From Lippert and Dickinson, *Chem. Rev.*, **103**, 453, 2003. With permission.)

Depending on the particular units present, radicals remaining in the substrate may react with the ambient O_2 to form hydroxyl or peroxyl units. As a result, the wettability of the polymer may be significantly affected. However, such effects are significant for polymers that are initially hydrophobic and are not usually of concern for varnishes or other materials employed in artworks. On the other hand, polymers containing extensive aromatic groups are often carbonized at fluences close to the ablation threshold [6]. Besides the detrimental effect for the substrate integrity and appearance, carbonization is difficult to deal with because carbon has a higher ablation threshold than the polymeric substrate. Fortunately, most materials encountered in artworks do not present the extensive aromatic conjugation that leads to significant carbonization.

6.6.2 WAVELENGTH DEPENDENCE AND IMPORTANCE OF ABSORPTIVITY

Wavelength is the most important parameter that needs specification in implementations of UV ablation [1–5]. It follows from Equation (6.2) that irradiation at a strongly absorbed wavelength must be selected to ensure efficient etching and good surface morphology. These are usually the only two criteria used for optimal wavelength selection. However, note that for very high α_{eff}, the etching depth may be very small. This is often the case for irradiation at 193 nm, where absorptivity of organics is generally very high. Thus in combination with the fact that ArF lasers are characterized by very low output energy, processing at 193 nm can be very time-consuming and is often not practical.

In addition, as described above (Section 6.5.4.1), a photochemical mechanism is often invoked for irradiation at wavelengths absorbed mainly by photodissociable units. Such a mechanism is considered to result in a reduced thermal load to the substrate with a consequent improvement in the morphology of the processed surface.

The rest of this section focuses on the influence on the induced chemical modifications since it is of prime concern to conservators.

To illustrate the influence of wavelength on the induced chemical modifications, we examine the dopant-deriving by-products formed in the irradiation of polymers doped with the photolabile iodonaphthalene or iodophenanthrene [83–90] (Chapter 2). To this end, Figure 6.15a illustrates the intensity of ArH (Ar = Nap or Phen) product formed in the substrate upon irradiation at two excimer wavelengths (Figure 6.15). Clearly, at the two wavelengths, the dependence is qualitatively the same but quantitatively very different. With increasing α_{eff} the onset of the nonlinearity in product yield is observed at lower fluences α. This can be ascribed to the higher temperature attained with increasing absorptivity. However, the most important feature to note in this figure is that the amount of ArH by-product that remains in the substrate is greatly reduced following ablation at strongly absorbed wavelengths. At the simplest level, this dependence can be understood in that the absorption coefficient determines the relative ratio of etching depth vs. optical penetration depth and thus the depth over which products remain in the substrate.

A more fundamental understanding of the factors involved derives from the examination of the kinetics of product formation at the different wavelengths (Figure 6.16). At weakly absorbed wavelengths such as 248 nm for *PMMA* and

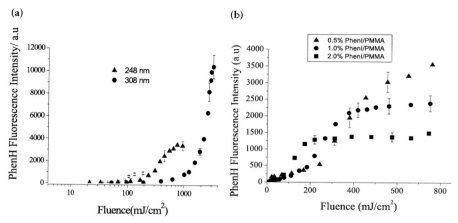

FIGURE 6.15 (a) F_{LASER}-dependence of aryl (PhenH) by-product emission in the UV irradiation of doped *PMMA* at the indicated excimer wavelengths (dopant: PhenI at 0.5% wt) with a single pump pulse. (b) F_{LASER}-dependence of the aryl (PhenH) product remaining in the substrate following irradiation (each time with a single pulse) of *PMMA* films doped with PhenI at 248 nm for three different PhenI concentrations. At this wavelength, *PMMA* absorbs very weakly and thus absorption is almost exclusively due to the PhenI dopant. Note that although the concentration of the photolabile chromophore increases, the product remaining in the substrate at fluences above the threshold is much reduced. (From Bounos et al., *J. Appl. Phys.*, **98**, 084317-1, 2005. With permission.)

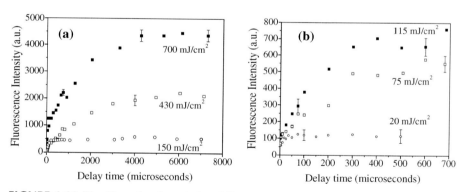

FIGURE 6.16 PhenH product laser-induced fluorescence intensity as a function of the delay time between the pump and the probe beams in irradiation of PhenI/*PMMA* films (0.5% wt) at the laser fluences (representing one well below the threshold, a fluence close to the corresponding threshold, and a fluence well above the threshold) at (a) irradiation at 248 nm and (b) irradiation at 193 nm.

308 nm for *PS*, the rate of ArH formation is much higher. Furthermore, formation of ArH products continues for well over ≈1 msec, whereas in contrast, for irradiation at strongly absorbed wavelengths (e.g., 193 nm), product formation is quenched within some 100s μsec. These differences in product formation kinetics account largely for the different F_{LASER} dependences observed in Figure 6.15 for the product amount remaining in the substrate.

   This difference in product formation kinetics can be understood within the frame-work of discussion on thermal effects (Section 6.5.2). Reactions of radicals within polymers usually follow a pseudo-unimolecular Arrhenius equation, i.e., $[Product] = [R](1 - e^{-k(T)t})$, where $[R]$ represents the concentration of the radicals produced by the photolysis and $k(T)$ is the reaction rate constant $(k(T) = A e^{-E_{act}/R_G T(t)})$. The rates for such reactions, thus, are determined by the temperatures attained upon laser irradiation and by the heat diffusion time $t_{th} \approx 1/(a_{eff}^2 D_{th})$. Indeed, the observed difference in the time scale of ArH formation in Figure 6.16 corresponds well to the different thermal diffusion times ($\approx 10^{-3}$ to 10^{-2} sec at weakly absorbed wavelengths and $\approx 10^{-5}$ to 10^{-4} sec for the strongly absorbed wavelengths). In other words, for irradiation at a strongly absorbed wavelength, the fast drop of the temperature in the remaining substrate results in the fast quenching of any activated reactions. Detailed simulations and modeling of the observed kinetics can be found in [90], where it is shown that the effective reaction depth can be approximated as

$$I_{rxn} = \sqrt{\frac{1}{\alpha_{eff}^2} \ln\left[\frac{(\alpha_{eff} F_{LASER}/\rho c_P)[\ln A - \ln(\alpha_{eff}^2 D_{th})]}{(E_{act}/R)}\right]} \tag{6.11}$$

For typical values, this reaction depth turns out to be ≈ 7 to 15 μm for weakly absorbing systems and ≈ 1 μm for strongly absorbing ones. Therefore, for the same absorbed energy density $\alpha_{eff} F_{LASER}$, by-product formation decreases with increasing α_{eff} (assuming all other factors are the same).

   A simple extension of the previous analysis also accounts for the dependence of the extent of formation of recombination (i.e., via diffusion) by-products on substrate absorptivity. To this end, Figure 6.17 illustrates the intensity of Nap_2 by-product formed in the irradiation of NapX/*PMMA* (X = Br, I) at the excimer laser wavelengths 248 nm and 308 nm. Most importantly, formation of the bi-aryl species is much reduced at 248 nm, at which α_{eff} is much higher than at 308 nm [7]. With

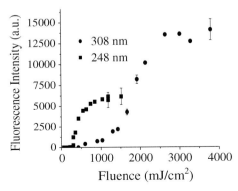

FIGURE 6.17 Nap_2 by-product laser-induced fluorescence recorded after the ablation of napI-doped *PMMA* samples (1.2% wt, $\alpha_{eff\ 248nm} = 2000$ cm^{-1}, $\alpha_{eff\ 308nm} = 700$ cm^{-1}) at the indicated excimer laser wavelengths. There is hardly any bi-aryl species formation upon ablation at 193 nm. (From Bounos et al., *J. Appl. Phys.*, **98**, 084317-1, 2005. With permission.)

increasing absorptivity, the depth of melt is smaller and the duration of polymer melting shorter (Section 6.5.2). As a result, radical diffusion is much reduced, with a consequent reduction or minimization of recombination by-product formation.

The previous discussion clearly underlines the importance of high substrate absorptivity for the success of laser processing schemes. *It is generally assumed that the necessity for irradiation at strongly absorbed wavelengths is due to the need for attaining efficient etching and good morphology of the treated area. However, as shown above, with a high substrate absorptivity, the extent of induced chemical modifications and by-product formation in the substrate is highly reduced, and so a high degree of photochemical protection is afforded to the remaining material.* This protection justifies the use of lasers for processing of even highly thermally sensitive and photolabile substrates.

Based on the previous results, it is estimated — see Equation (6.11) — that for varnishes with a high absorptivity ($\alpha \approx 10^5$ cm^{-1}) at 248 nm, chemical modifications due to ablation should be localized to within ~1 μm of the etched depth. Thus if during material removal a thin layer of varnish is left unprocessed, then the influence on the pigmented medium will be negligible. This has, indeed, been demonstrated in studies on realistic models [17–19,91,92] as described in subsequent chapters.

There is another factor that may be involved that has not been adequately addressed in the literature. Equation (6.2) suggests that the extent of photofragmentation and by-product formation remaining in the substrate is constant with laser fluence for $F_{LASER} \geq F_{thr}$. However, this neglects the possibility that fragments and by-products that are formed within the thermally affected zone below the ejected material and that are weakly bound to the matrix may diffuse to the surface and thermally desorb (postablation desorption, Section 6.5.2.2). This possibility has been directly examined in the KrF laser ablation of varnishes and in a number of studies on polymers [23]. This effect may be an important factor for further contribution to the success of laser processing schemes, since small (usually the most reactive) species formed even below the etched depth may be efficiently removed. The importance of this contribution should depend on substrate absorptivity. With increasing substrate absorptivity, formation of by-products is confined closer to the surface and thus the probability of desorption of weakly-bound species increases.

6.6.3 DEPENDENCE ON NUMBER OF LASER PULSES

For reasons of simplicity, the discussion of thermal effects in Section 6.5.2.2 was focused on the effects of a single laser pulse. When multiple pulse protocols are used, it is important to ensure that the repetition rate is low enough to avoid progressive accumulation of residual heat that will result in a larger thermal damage zone (see also Section 2.3.1.1).

In contrast to thermal effects, photochemical and photomechanical effects cannot be resolved so easily, since they may accumulate independently of laser repetition rate. The dependence of mechanical effects on multiple protocols has already been discussed in Section 6.5.3.2. Here we concentrate on the chemical effects.

At low and moderate laser fluences, thermal desorption mechanism may be operative (Section 6.5.2.2) so that by-products that are weakly bound to the matrix may be able to diffuse through the molten zone and desorb. However, removal of by-products that are tightly bound to the matrix remains inefficient. Thus, with successive laser pulses, these species accumulate in the substrate and may react further to form additional by-products. For irradiation close to the threshold, their accumulation in the substrate can result in a pronounced dependence of the material removal process on successive laser pulses. For some systems, the accumulated chemical modifications to the superficial substrate layers result in the reduction of the etching efficiency, often to the point of insignificant ablation. Subsequent irradiation at higher fluences removes the modified layer with restoration of the ejection signal [94]. On the other hand, for weakly absorbing systems (such as *PMMA* at 248 nm), the irradiation may result in highly conjugated species that absorb more strongly than the initial compound. In this case, etching efficiency increases with successive laser pulses. In semicrystalline polymers, volume changes may be observed and have been ascribed to amorphization of the crystalline domains and the accumulation of trapped gases.

In contrast, for irradiation at fluences above the ablation threshold, accumulation of by-products is much reduced, owing to the efficient removal of even the larger, tightly bound species. In this case, equilibrium is established between new species produced and species removed. The degree of accumulation will depend on the etching depth vs. optical penetration depth. For a high absorptivity, accumulation occurs only to a small depth, and modifications to the bulk are minimal.

These trends have been well documented in studies on doped polymers [85,86]. A similar trend is also observed for chromophores being part of the polymer chain. For instance, in the irradiation of polystyrene at 248 nm with successive pulses at low fluence, the *PS* fluorescence peak at ~320 nm decreases gradually to a very low level, with the parallel growth of a broad emission band at ≈440 nm (Figure 6.18). These spectral changes can be ascribed to the fact that *PS* undergoes fragmentation to benzyl and/or phenyl radicals (thereby accounting for the decrease of the emission at ≈320 nm) with the parallel formation of polyene structures (highly conjugated

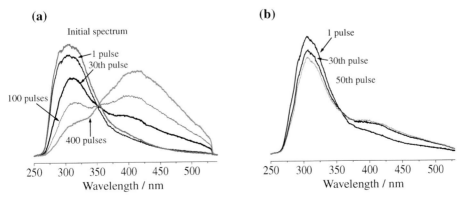

FIGURE 6.18 Evolution of the laser-induced fluorescence probe spectra of polystyrene recorded following irradiation at 248 nm (a) at ≈150 mJ/cm², a fluence below the ablation threshold, and (b) at ≈1000 mJ/cm², i.e., a fluence above the ablation threshold. Accumulation of degradation by-products with successive laser pulses is significant at low laser fluences.

products deriving from the polystyrene chains), indicated by the increasing emission at 440 nm. In contrast, in the irradiation above the threshold, the ratio of degradation/PS emission never turns to the extreme ratio observed at low fluences, evidently because of the efficient removal degraded species [87].

In all, multipulse irradiation protocols are often detrimental to the chemical integrity of substrates. Of course, in practice, the number of required pulses is specified by the etching efficiency as compared to the amount of material that must be removed [93]. Thus a compromise may have to be drawn between efficient material removal and minimization of chemical effects.

6.6.4 FEMTOSECOND ABLATION

Ablation with femtosecond (fsec) pulses has attracted significant attention, because of several advantages that it provides for material processing [95–98]. However, the effects of the femtosecond pulses on the irradiated substrate are only now starting to be examined. Furthermore, the mechanisms underlying femtosecond ablation are still obscure. For this reason, no effort is made here to present a systematic presentation of the effects of femtosecond irradiation; rather, an overall perspective of likely possibilities is noted. Ablation with femtosecond pulses is reviewed by Krüger and Kautek [96].

First, the heat-affected zone is minimal, so several of the side effects described for longer (nanosecond) pulses are reduced or minimized. Second, because material ejection occurs well after the pulse, there is no plasma shielding, so maximum coupling of energy into the substrate is attained. Third, the efficient energy use (i.e., negligible loss due to thermal diffusion) enables processing at much lower fluences than with nanosecond pulses. Most importantly, the high intensities attained with femtosecond pulses result in highly efficient multiphoton processes. This enables processing of substrates that are transparent or weakly absorbing at the irradiation wavelength (e.g., the threshold for ablation of *PMMA* with ~500 fsec pulses at 248 nm is ~5 times lower than that for nanosecond pulses) [95]. Furthermore, the efficient operation of multiphoton processes suggests greatly reduced optical penetration depth and thus a highly reduced depth of by-product formation and accumulation. This should enable successful processing of a wide range of artworks, even in the near absence of a protecting varnish layer. Generally, the quality of structuring/ processing with femtosecond pulses far surpasses that attained with nanosecond ablation.

Indeed, in the irradiation of doped systems with 500 fsec pulses at 248 nm, dopant photolysis and by-product formation are found to be very limited above the ablation threshold. Most importantly, recombination or other by-products is not observed even after extensive irradiation [83]. Likely because of the limited thermal dissipation in ablation with femtosecond pulses, the mobility of any radicals is highly restricted. It is also tempting to argue that with the shorter pulses, material ejection occurs too fast for formation of by-products to compete. Though these suggestions are reasonable, they are not sufficient to account for the extremely clean chemistry that is observed in irradiation at high femtosecond fluences.

In conclusion, the chemical effects in the UV ablation with femtosecond laser pulses are, qualitatively and quantitatively, highly defined and limited. Thus, besides the well-acknowledged advantages described above, femtosecond ablation also affords a high degree of control over the induced chemical modifications. Thus processing with femtosecond lasers is expected to be highly important in the treatment of molecular substrates. This potential has been recently demonstrated in the processing of biological substrates [98]. It is important to note that because of these specific features, femtosecond laser technology has already enabled new material processing schemes that have not been possible with nanosecond laser technology, e.g., formation of submicron structures via 2-photon-induced polymerization [95–98]. Accordingly, we can expect that besides the evident extension of laser techniques to material removal, such as varnish, other restoration schemes will be possible by femtosecond processing.

Of course, at present, femtosecond laser technology remains too expensive and its availability limited to specific laboratories.

6.7 ABLATION PLUME/PLASMA DYNAMICS

Although the conservator's interest lies mainly in the substrate condition after cleaning, the dynamics of material ejection directly affects the process. First, for irradiation with nanosecond pulses at fluences moderately above the threshold, material ejection is already significant during the laser pulse. Thus scattering and absorption of the incident radiation by ejected material reduces the amount of light that reaches the substrate (Figure 6.19). In fact, at fluences where plasma formation occurs, shielding becomes so extensive that etching depth is severely limited. Plasma formation is initiated by the formation of free electrons either via multiphoton ionization or through thermionic ionization of the ejected material. Subsequently, these free electrons seed an ionization avalanche, creating many more electrons and ions in the process.

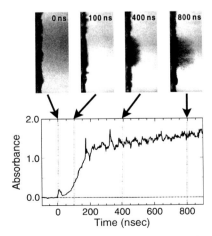

FIGURE 6.19 Time-resolved photography and transient absorption measurements of a probing beam to illustrate how the ejection of material affects the transient absorption. The substrate is *PMMA* doped with biphenyl, and ablation is effected with an excimer laser at 248 nm. (From H. Fukumura et al., *J. Phys. Chem.*, **99**, 750, 1995. With permission.)

Besides seriously limiting the fluence that reaches the surface, the formation of intense plasma can have deleterious effects on the substrate by causing further heating of the substrate and/or by the emission of UV light that may be absorbed by the chromophores of the substrate. Second, plasma formation and material ejection in the plume (as described in Section 6.5.3) may have serious consequences for the mechanical integrity of the substrate. In particular, the back momentum exerted by the expanding plume may cause pronounced secondary ejection of material. (The term plume is used here to denote the expanding cloud of material above the irradiated surface.) Note that although this is also a mechanical effect, it should be differentiated from the photomechanical mechanism discussed in Section 6.5.3. This effect is often exploited in applications for enhancing cleaning efficiency. Under these conditions, however, control of side effects is much reduced, and the resolution (accuracy) of material removal is much degraded.

6.8 EXPERIMENTAL TECHNIQUES: POSSIBILITIES AND LIMITATIONS

In the field of laser restoration, studies have been largely limited to simple measurements such as etching depth as a function of fluence. However, as technology advances and methods are improved, the need for more detailed examination by more elaborate approaches/techniques of the processes becomes important. Furthermore, these techniques can be used as diagnostics for monitoring the success of laser removal processes. For these reasons, a brief description of some plausible methods (besides the ones described in the previous chapters) is presented here.

6.8.1 CHARACTERIZATION OF EJECTA

Mass spectrometry. This method has been used extensively in ablation studies [6]. The technique permits identification of the ejected species, their kinetic energies, and, with more demanding design, their angular distribution. This information can be most useful in elucidating the underlying processes and also as an analytical tool. However, the nature and distribution of the gas phase ejecta are affected by secondary absorption and reaction processes in the plume. For conservators, instead, the main interest lies in knowing the chemical modifications effected to the substrate, rather than the nature of the ejected species. The most important limitation of its use, however, in the case of artworks, is that its implementation requires the placement of the substrate in a vacuum system.

The advent of atmospheric mass spectrometry obviates this requirement (albeit with reduction of the information extracted). Thus atmospheric mass spectroscopy can be expected to develop into a powerful analytical method in several fields, including characterization of artworks.

Laser-induced breakdown spectroscopy (LIBS). Analytical information, mainly about atomic constituents, can be obtained directly through the analysis of the plasma emission spectrum produced during ablation. The spectrally and temporally resolved detection of atom or small fragment emissions yields information about the elemental composition of the ablated volume. The independence of the LIBS process from charging effects (and thus from substrate conductivity), in combination with high sensitivity for pigment constituents, makes LIBS a powerful tool for monitoring laser restoration of painted artworks.

6.8.2 EXAMINATION OF EJECTION PROCESS

Photoscattering by ejected material. This method is one of the simplest approaches to characterize the onset of material ejection. It requires only a laser beam traversing close to and parallel to the laser irradiated area and a simple detector (e.g., a photodiode) to detect the deflection (due to transient changes in the refractive index of the atmosphere above the laser-heated surface) and scattering of the beam caused by the ejected plume, and one can easily establish the onset of material ejection. Unfortunately, implementation of this technique in ambient conditions can be problematic. Some care is needed due to the multitude of signals. However, even under moderate vacuum, the signal is much simplified. In fact, even under ambient conditions, a careful experimenter can use this method to extract detailed information.

There are also much more elaborate methods, relying on flash photography, for visualizing the details and time evolution of the ejection process. The use of these techniques has been particularly powerful in elucidating the dynamics of ablation of a wide range of materials.

6.8.3 TEMPERATURE MEASUREMENT

This has already been discussed in the general framework of the temperature changes effected upon pulsed laser irradiation in Chapter 2, where common caveats and limitations of such measurements were also considered.

6.8.4 EXAMINATION OF IRRADIATED SURFACE

Chromatography. By far, this method is the most common technique in restoration methods of artworks. Yet it can be quite erroneous in the quantitative assessment of the chemical modifications induced upon ablation. This technique requires extraction of at least milligrams of material, which translates into several microns of material depth. However, laser-induced changes are usually localized to within few microns at most.

There are a number of spectroscopic techniques that can be used for the characterization of irradiated areas, as well as for on-line monitoring of the laser–material interaction. For instance, **laser-induced fluorescence** [99], **Raman** [100], and **FTIR** spectroscopy can be used for the chemical characterization of the irradiated area. The potential of these techniques as diagnostic tools for artworks is described in other chapters. However, the reader should be forewarned that the extraction of *quantitative* information by these spectroscopic techniques may be limited, because the highly modified substrate morphology that is often created upon ablation can affect the propagation of the probing beam and thus the intensity of the induced spectroscopic signals.

Structural information about the painting can be derived from **broadband reflectography** [101]. In particular, the UV absorptivity of varnish increases strongly upon degradation and oxidation. Thus reflectography provides a convenient tool for examining the laser cleaning process by identifying "fresh" material of higher reflectivity that is exposed as dirt and debris are removed.

Piezoelectric measurement. As described in Section 6.5.3, this method constitutes a highly powerful tool. Alternatively, **holographic** methods are powerful for the examination of photomechanical effects over extended areas of the irradiated objects and can be used for the on-line monitoring of UV laser processing schemes. Furthermore, the versatility of the technique allows its adaptation to the different detection requirements posed by the diverse types of processed objects. For instance, the technique can be used directly in a reflection mode for the characterization of effects in nontransparent substrates. Real-time monitoring or pulsed probe lasers can be used for the study of the temporal evolution of the observed effects.

6.9 EXPERIMENTAL SETUPS

There are well-established experimental setups that have been developed for industrial applications of lasers and adapted for the purposes of laser cleaning schemes. The workstations generally consist of a substrate mounting stage, suitable optics for laser beam delivery and focusing, and diagnostic modules for the on-line monitoring and control of the cleaning process. The mounting stage can be a computer-controlled x-y-z mechanical translator. Alternatively, an optical fiber or motorized optical arms can be used for the accurate scanning of the beam over the substrate. On the other hand, experimental setups for steam laser cleaning are somewhat more involved. A laser cleaning setup designed for industrial use is described in detail elsewhere [41]. Cleaning rates on silicon wafers up to 200 cm^2/min are reported for this system.

Often it suffices to define an operational range of laser parameters for a successful treatment. This assumes invariability in the nature of substrate and material to be removed. Generally, however, this is not the case in artworks, and as a result laser processing parameters may have to be appropriately adjusted during the procedure. Therefore it is important to implement techniques for the *in situ* on-line control of the procedure, thereby safeguarding against damage. Monitoring can be achieved by a variety of optical and spectroscopic techniques described in the previous section.

6.10 CONCLUSIONS

This chapter has overviewed laser schemes for the cleaning of organic substrates and the basic principles underlying these methods. Laser irradiation has demonstrated a highly effective and efficient method of unwanted material removal. Certainly, experimental parameters must be carefully optimized in order to achieve proper cleaning. Furthermore, on-line monitoring is essential for safeguarding against damage. With these constraints, laser cleaning can be a highly accurate and versatile method, surpassing by far the degree of selectivity and control afforded by traditional cleaning methods.

Particular emphasis is placed on the side effects of the procedures. Certainly, the interaction of intense laser pulses with substrates is highly complex. In fact, in many respects, conventional photochemical concepts would suggest that the laser procedures would be particularly damaging for organic substrates. Yet it has been shown, at least empirically, that these limitations or deleterious influences can be avoided to a large extent. Studies on model systems provide insight into the factors involved in these interactions and can be most useful in establishing criteria for the systematic development of laser processing cleaning schemes.

Abbreviations and Symbols

A	Preexponential (rate) factor
α	Absorption coefficient
α_{eff}	Effective absorption coefficient
a_{ac}	Acoustic wave damping coefficient
β	Thermal expansion coefficient at constant temperature
C	Heat capacity
c_P	Specific heat capacity at constant pressure
c_V	Specific heat capacity at constant volume
c_S	Speed of sound
δ	Etching depth per pulse
$E_{binding}$	Binding energy to the substrate
D_{th}	Thermal diffusivity
D_{sp}	Species diffusion constant $\approx$
E_{act}	Activation energy
E_{cr}	Critical energy density for ablation
E_{KIN}	Kinetic energy
F_{LASER}	Laser fluence
F_{thr}	Threshold fluence for material removal
F_{tr}	Fluence transmitted
ΔH_{vap}	Evaporation enthalpy
ΔH^{sub}	Sublimation enthalpy
η_{abl}	Ablation efficiency
η_c	Laser-induced removal efficiency of particles
θ	Ratio τ_{pulse} / τ_{ac}
H	Enthalpy
$h\nu$	Photon energy
l	Laser intensity
J	Rate of bubble formation
$k(T)$	Reaction rate constant
κ_B	Boltzmann constant
κ_T	Isothermal compressibility
l_{th}	Thermal diffusion length
λ	Wavelength
M	Molecular mass
MW	Molecular weight

μ	Chemical potential
N	Number density of chromophores
N_{fringe}	Number of interference fringes
N_{pulse}	Number of laser pulses
η	Photodissociation quantum yield
n_R	Refractive index
n	Density of particles on surface after irradiation
n_0	Density of particles on surface before irradiation
ν_{ac}	Acoustic wave frequency
ξ	Thermal conductivity
P	Pressure
P_{abl}	Pressure exerted by the removal proccess
Pl	Polyimide
$PMMA$	Polymethylmethacrylate
PS	Polystyrene
R	Reflectivity
R_G	Universal gas constant
ρ	Density
r_P	Particle radius
σ	Absorption cross section
σ_P	Particle absorption coefficient
σ_{tens}	Surface tension coefficient
T	Temperature
t	Time
t_{th}	Thermal diffusion time
τ_{pulse}	Laser pulse duration
τ_{ac}	Time for an acoustic wave to traverse the irradiated volume
υ	Expansion velocity
υ_{ac}	Acoustic wave velocity
υ_{int}	Interface velocity
U	Wave amplitude in the hologram plane
W	Work
ϕ	Phase of the optical wave

REFERENCES

1. Srinivasan, R . and Braren, B., Ultraviolet laser ablation of organic polymers, *Chem. Rev.,* **89**, 1303–1316, 1989.
2. Miller, J.C., History, scope and the future of laser ablation, in Miller, J.C. (Ed.), *Laser Ablation Principles and Applications,* Springer Ser Muter, Springer-Verlag, Berlin, 1994.
3. Bauerle, D., *Laser Processing and Chemistry,* 3rd ed., Springer-Verlag, Berlin, 2000,
4. Dyer, P., Excimer laser polymer ablation: twenty years on, *Appl. Phys. A Mater.,* **77**, 167–173, 2003.
5. Georgiou, S. and Hillenkamp, F., Introduction: laser ablation of molecular substrates, *Chem. Rev.,* **103**, 317–320, 2003.
6. Lippert, T. and Dickinson, T.J., Chemical and spectroscopic aspects of polymer ablation: special features and novel directions, *Chem. Rev.,* **103**, 453–486, 2003.
7. Dreisewerd, K., The desorption process in MALDI, *Chem. Rev.,* **103**, 395–426, 2003.
8. Laude, L.D., *Excimer Lasers,* Kluwer Academic, Dordrecht, 1994.
9. Vogel, A. and Venugopalan, V., Mechanisms of pulsed laser ablation of biological tissues, *Chem. Rev.,* **103**, 577–644, 2003.
10. Reed, R., *Ancient Skins, Parchments and Leathers,* Seminar Press, London, 1970.
11. Wolbers, R., Sterman, R.N., and Stavroudis, C., *Notes for Workshop on New Methods in the Cleaning of Paintings,* Getty Conservation Institute, Marina del Rey, CA, 1990.
12. Luk'yanchuk, B.S., *Laser Cleaning*, World Scientific, Singapore, 2002.
13. Kern, W., *Handbook of Semiconductor Wafer Cleaning,* Noyes, New York, 1993.
14. Watkins, K.G., Larson, J.H., Emmony, D.C., and Steen, W.M., *Laser Processing: Surface Treatment and Film Deposition,* Kluwer Academic, 1995.
15. Georgiou, S., Laser cleaning methodologies of polymer substrates, *Adv. Polym. Sci.,* **168**, 1–49, 2004.
16. Asmus, J.F., More light for art conservation, *IEEE Circuits Dev. Mag.,* **2**, 6, 1986.
17. Fotakis, C., Anglos, D., Balas, C., Georgiou, S., Vainos, N.A., Zergioti, I., and Zafiropulos, V., Lasers and optics for manufacturing, *Opt. Soc. Am.,* **9**, 99–104, 1997.
18. Zafiropulos, V. and Fotakis, C., in *Laser Cleaning in Conservation: An Introduction,* Cooper, M., Ed., Butterworth Heinemann, Oxford, 1997, chap 6.
19. Georgiou, S., Zafiropulos, V., Anglos, D., Balas, C., Tornari, V., and Fotakis, C., Excimer laser restoration of painted artworks: procedures, mechanisms and effects, *Appl. Surf. Sci.,* **127–129**, 738–745, 1997.
20. Vainos, N.A., Mailis, S., Pissadakis, S., Boutsikaris, L., Parmiter, P.J.M., Dainty, P., and Hall, T.J., Excimer laser use for microetching computer-generated holographic structures, *Appl. Optics,* **35**, 6299–6303, 1996.
21. Zhigilei, L.V., Leveugle, E., Garrison, B.J., Yingling, Y.G., and Zeifman, M.I., Computer simulations of laser ablation of molecular substrates, *Chem. Rev.,* **103**, 321–348, 2003.
22. Yingling, Y.G., Zhigilei, K.V., Garrison, B.J., Koubenakis, A., Labrakis, J., and Georgiou, S., Laser ablation of bicomponent systems: a probe of molecular ejection mechanisms, *Appl. Phys. Lett.,* **78**, 1631–1633, 2001.
23. Bityurin, N., Luk'yanchuk, B.S., Hong, M.H., and Chong, T.C., Models for laser ablation of polymers, *Chem. Rev.,* **103**, 519–552, 2003.
24. Kruger, J., Niino, H., and Yabe, A., Investigation of excimer laser ablation threshold of polymers using a microphone, *Appl. Surf. Sci.,* **197**, 800–804, 2002.

25. Brannon, J.H., Tam, A.C., and Kurth, R.H., Pulsed laser stripping of polyurethane-coated wires: a comparison of KrF and CO2 lasers, *J. Appl. Phys.,* **70**, 3881–3886, 1991.

26. Manz, C. and Zafiropulos, V., Laser ablation strips thin layers of paint, *Opto & Laser Europe (OLE),* **45**, 27–30, 1997.

27. Daurelio, G., Chita, G., and Cinquepalmi, M., Laser surface cleaning, de-rusting, de-painting and de-oxidizing, *Appl. Phys. A,* **69**, S543–S546, 1999.

28. Galantucci, L.M., Gravina, A., Chita, G., and Cinequepalmi, M., An experimental study of paint-stripping using an excimer laser, *Polym. Polym. Compos.,* **5**, 87–94, 1997.

29. Liu, K. and Garmire E., Paint removal using lasers, *Appl. Optics,* **34**, 4409–4415, 1995.

30. Snelling, H.V., Walton, C.D., and Whitehead, D.J., Polymer jacket stripping of optical fibers by laser irradiation, *Appl. Phys. A,* **79**, 937–940, 2004.

31. Barnier, F., Dyer, P.E., Monk, P., Snelling, H.V., and Rourke, H., Fibre optic jacket removal by pulsed laser ablation, *J. Phys. D — Appl. Phys.,* **33**, 757–759, 2000.

32. Kautek, W., Pentzein, S., Rudolph, P., Kruger, J., and Konig, E., Laser interaction with coated collagen and cellulose fibre composites: fundamentals of laser cleaning of ancient parchment manuscripts and paper, *Appl. Surf. Sci.,* **127–129**, 746–754, 1998.

33. Kolar, J., Strlic, M., Müller-Hess, D., Gruber, A., Troschke, K., Pentzien, S., and Kautek, W., Near-UV and visible pulsed laser interaction with paper, *J. Cult. Heritage,* **1**, S221–S224, 2000.

34. Kolar, J. and Strlic, M., IR pulsed laser light interaction with soiled cellulose and paper, *Appl. Phys. A,* **75**, 673–676, 2002.

35. Kennedy, C.J., Vest, M., Cooper, M., and Wess, T.J., Laser cleaning of parchment: structural, thermal and biochemical studies into the effect of wavelength and fluence, *Appl. Surf. Sci.,* **227**, 151–163, 2004.

36. Kautek, W., Pentzien, S., Conradi, A., Leichtfried, D., and Puchinger, L., Diagnostics of parchment laser cleaning in the near-ultraviolet and near-infrared wavelength range: a systematic scanning electron microscopy study, *J. Cult. Heritage,* **4**(suppl. 1), 179–184, 2003.

37. Sportun, S., Cooper, M., Stewart, A., Vest, M., Larsen, R., and Poulsen, D.V., An investigation into the effect of wavelength in the laser cleaning of parchment, *J. Cult. Heritage,* **1**(suppl. 1), S225–S232, 2000.

38. Lu, Y.F., Lee, Y.P., and Zhou, M.S., Laser cleaning of etch-induced polymers from via holes, *J. Appl. Phys.,* **83**, 1677–1684, 1998.

39. Zapka, W., Ziemlich, W., and Tam, A.C., Efficient pulsed laser removal of 0.2 μm sized particles from a solid surface, *Appl. Phys. Lett.,* **58**, 2217–2219, 1991.

40. Imen, K., Lee, S.J., and Allen, S.D., Laser-assisted micron scale particle removal, *Appl. Phys. Lett.,* **58**, 203–205, 1991.

41. Chaoui, N., Solis, J., Alfonso, C.N., Fourrier, T., Muehlberger, T., Schrems, G., Mossbacher, M., Bauerle, D., Bertsch, M., and Leiderer, P., A high-sensitivity in situ optical diagnostic technique for laser cleaning of transparent substrates, *Appl. Phys. A,* **76**, 767–771, 2003.

42. Coupland, K., Herman, P., and Gu, B., Laser cleaning of ablation debris from CO$_2$-laser-etched vias in polyimide, *Appl. Surf. Sci.,* **127–129**, 731–737, 1998.

43. Fourrier, T., Schrems, G., Muelberger, T., Heitz, J., Arnold, N., Bauerle, D., Mosbacher, M., Boneberg, J., and Leiderer, P., Laser cleaning of polymer surfaces, *Appl. Phys. A,* **72**, 1–6, 2001.

44. Halfpenny, D.R. and Kane, D.M., A quantitative analysis of single pulse ultraviolet dry laser cleaning, *J. Appl. Phys.,* **86**, 6641–6646, 1999.

45. Mosbacher, M., Munzer, H.J., Zimmerman, J., Solis, J., Boneberg, J., and Leiderer, P., Optical field enhancement effects in laser-assisted particle removal, *Appl. Phys. A*, **72**, 41–44, 2001.

46. Dongsik, K., Park, H.K., and Grigoropoulos, C.P., Interferometric probing of rapid vaporization at a solid-liquid interface induced by pulsed-laser, *Int. J. Heat Mass Trans.*, **44**, 3843–3853, 2001.

47. Yavas, O.J., Leiderer, P., Park, H.K., Grigoropoulos, C.P., Poon, C.C., and Tam, A.C., Enhanced acoustic cavitation following laser-induced bubble formation: long-term memory effect, *Phys. Rev. Lett.*, **72**, 2021–2024, 1994.

48. Lee, Y.P., Lu, Y.F., Chan, D.S.H., Low, T.S., and Zhou, M.S., Steam laser cleaning of plasma-etch-induced polymers from via holes, *Jpn. J. Appl. Phys.*, **37**, 2524–2529, 1998.

49. Tam, A.C., Park, H.K., and Grigoropoulos, C.P., Laser cleaning of surface contaminants, *Appl. Surf. Sci.*, **127/129**, 721–725, 1998.

50. de Cruz, A., Wolbarsht, M.L., and Hauger, S.A., Laser removal of contaminants from painted artworks, *J. Cult. Heritage*, **1**, S173–S180, 2000.

51. Walsh, J.T., Flotte, T.J., and Deutsch, T.F., Er:YAG laser ablation of tissue: effect of pulse duration and tissue type of thermal damage, *Lasers Surg. Med. A*, **9**, 314–326, 1989.

52. Pettit, G.H., Ediger, M.N., Hahn, D.W., Brinson, B.E., and Sauerbrey, R., Transmission of polyimide during pulsed ultraviolet-laser irradiation, *Appl. Phys. A*, **58**, 573–579, 1994.

53. Fujiwara, H., Fukumura, H., and Masuhara, H., Laser ablation of pyrene-doped poly(methyl methacrylate) film: dynamics of pyrene transient species by spectroscopic measurements, *J. Phys. Chem.*, **99**, 11844–11853, 1995, and references therein.

54. Lazare, S. and Granier, V., Ultraviolet-laser photoablation of polymers — a review and recent results, *Laser Chem.*, **10(1)**, 25–40, 1989.

55. Schmid, H., Ihlemann, J., and Luther, K.E.A., Modeling of velocity and surface temperature of the moving interface during laser ablation of polyimide and poly(methyl methacrylate), *Appl. Surf. Sci.*, **139**, 102–106, 1999.

56. Schmidt, H., Ihlemann, J., and Wolff-Rottke, B., Ultraviolet laser ablation of polymers: spot size, pulse duration, and plume attenuation effects explained, *J. Appl. Phys.*, **83**, 5458–5468, 1998.

57. Venugopalan, V., Nishioka, N.S., and Mikic, B.B., The thermodynamic response of soft biological tissues to pulsed ultraviolet laser irradiation, *Biophys. J.*, **69**, 1259–1271, 1995.

58. Bityurin, N., Arnold, N., Luk'yanchuk, B., and Bauerle, D., Bulk model of laser ablation of polymers, *Appl. Surf. Sci.*, **127–129**, 164–170, 1998.

59. Lee, I.Y.S., Wen, X.N., Tolbert, W.A., Dlott, D.D., Doxtader, M., and Arnold, D.R., Direct measurement of polymer temperature during laser ablation using a molecular thermometer, *J. Appl. Phys.*, **72**, 2440–2448, 1992.

60. Kuper, S., Brannon, J., and Brannon, K., Threshold behavior in polyimide photoablation — single-shot rate measurements and surface-temperature modeling, *Appl. Phys. A*, **56**, 43–50, 1993.

61. Weisbuch, F., Tokarev, V.N., Lazare, S., and Debarre, D., Ablation with a single micropatterned KrF laser pulse: quantitative evidence of transient liquid microflow driven by the plume pressure gradient at the surface of polyesters, *Appl. Phys. A — Mat. Sci.*, **76**, 613–620, 2003.

62. Tokarev, V.N., Lazare, S., and Belin, C., Viscous flow and ablation pressure phenomena in nanosecond UV laser irradiation of polymers, *Appl. Phys. A — Mat. Sci.*, **79**, 717–720, 2004.

63. Gusev, V.E. and Karabutov, A.A., *Laser Optoacoustics,* American Institute of Physics, New York, 1993.

64. Paltauf, G. and Dyer, P.E., Photomechanical processes and effects in ablation, *Chem. Rev.,* **103**, 487–518, 2003.

65. Hare, D.E., Franken, J., and Dlott, D.D., Coherent Raman measurements of polymer thin-film pressure and temperature during picosecond laser ablation, *J. Appl. Phys.,* **77**, 5950–5960, 1995.

66. Zweig, A.D., Venugopalan, V., and Deutch, T.F., Stress generated in polyimide by excimer-laser irradiation, *J. Appl. Phys.,* **74**, 4181–4189, 1993.

67. Siano, S., Pini, R., and Salimbeni, R., Variable energy blast modeling of the stress generation associated with laser ablation, *Appl. Phys. Lett.,* **74**, 1233–1235, 1999.

68. Vest, C.H., *Holographic Interferometry,* Academic Press, New York, 1971.

69. Bonarou, A., Antonucci, L., Tornari, V., Georgiou, S., and Fotakis, C.. Holographic interferometry for the structural diagnostics of UV laser ablation of polymer substrates, *Appl. Phys. A,* **73**, 647–651, 2001.

70. Callister, W.D., *Materials Science and Engineering: An Introduction,* John Wiley, New York, 1997.

71. Wu, S., *Polymer Interface and Adhesion,* Marcel Dekker, New York, 1982.

72. Fukumura, H., Mibuka, N., Eura, S., and Masuhara, H., Mass spectrometric studies on laser ablation of polystyrene sensitized with anthracene, *J. Phys. Chem.,* **97**, 13761–13766, 1993, and references therein.

73. Lippert, T., Nakamura, T., Niino, H., and Yabe, A., Laser induced chemical and physical modifications of polymer films: dependence on the irradiation wavelength, *Appl. Surf. Sci.,* **109–110**, 227–231, 1997.

74. Yingling, Y.G., Zhigilei, L.V., and Garrison, B.J., The role of the photochemical fragmentation in laser ablation: a molecular dynamics study, *J. Photochem. Photobiol. A,* **145**, 173–181, 2001.

75. Lippert, T., Laser applications of polymers, in *Polymers and Light, Adv. Polym. Sci.,* **168**, 51–246, 2004.

76. Lippert, T., Yabe, A., and Wokaun, A., Laser ablation of doped polymer systems, *Adv. Mater.,* **9**, 105–119, 1997.

77. Fujiwara, H., Nakajima, Y., Fukumura, H., and Masuhara, H., Laser ablation dynamics of a poly(methyl methacrylate) film doped with 5-diazo meldrum's acid, *J. Phys. Chem.,* **99**, 11481–11488, 1995, and references therein.

78. Lippert, T. and Stoutland, P.O., Laser–material interactions probed with picosecond infrared spectroscopy, *Appl. Surf. Sci.,* **109–110**, 43–47, 1997.

79. Arnold, B.R. and Scaiano, J.C., Laser ablation of doped polymers: transient phenomena as the ablation threshold is approached, *Macromolecules,* **25**, 1582–1587, 1992.

80. Larciprete, R. and Stuke, M., Direct observation of excimer laser photoablation products from polymers by picosecond-UV-laser mass spectroscopy, *Appl. Phys. B,* **42**, 181–184, 1987.

81. Webb, R.L., Langford, S.C., Dickinson, J.T., and Lippert, T.K., Sensitization of *PMMA* to laser ablation at 308 nm, *Appl. Surf. Sci.,* **127–129**, 815–820, 1998.

82. Hahn, C., Lippert, T., and Wokaun, A., Comparison of the ablation behavior of polymer films in the IR and UV with nanosecond and picosecond pulses, *J. Phys. Chem. B,* **103**, 1287–1294, 1999.

83. Rebollar, E., Bounos, G., Oujja, M., Domingo, C., Georgiou, S., and Castillejo, M., *Appl. Surf. Sci.,* **248**, 254 –258, 2005.

84. Lassithiotaki, M., Athanasiou, A., Anglos, D., Georgiou, S., and Fotakis, C., Photochemical effects in the UV laser ablation of polymers: implications for laser restoration of painted artworks, *Appl. Phys. A,* **69**, 363–367, 1999.

85. Athanassiou, A., Andreou, E., Anglos, D., Georgiou, S., and Fotakis, C., UV laser ablation of halonaphthalene-doped *PMMA*: chemical modifications above versus below the ablation threshold, *Appl. Phys. A,* **86**, S285–S289, 1999.

86. Athanassiou, A., Lassithiotaki, M., Anglos, D., Georgiou, S., and Fotakis, C., A comparative study of the photochemical modifications effected in the UV laser ablation of doped polymer substrates, *Appl. Surf. Sci.,* **154–155**, 89–94, 2000.

87. Andreou, E., Athanassiou, A., Fragouli, D., Anglos, D., Georgiou, S., Laser and material parameter dependence of the chemical modifications in the UV laser processing of model polymeric solids, *Laser Chem.,* **20**, 1–21, 2002.

88. Athanassiou, A., Andreou, E., Bonarou, A., Tornari, V., Anglos, D., Georgiou, S., and Fotakis, C., Examination of chemical and structural modifications in the UV ablation of polymers, *Appl. Surf. Sci.,* **197–198**, 757–763, 2002.

89. Athanassiou, A., Andreou, E., Fragouli, D., Anglos, D., Georgiou, S., and Fotakis, C., A comparative examination of photoproducts formed in the 248 and 193 nm ablation of doped *PMMA*, *J. Photochem. Photobiol. A,* **145**, 229–236, 2001.

90. Bounos, G., Athanassiou, A., Anglos, D., Georgiou, S., and Fotakis, C., Product formation in the laser irradiation of doped poly(methyl methacrylate) at 248 nm: implications for chemical effects in UV ablation, *J. Phys. Chem. B,* **108(22)**, 7052–7060, 2004.

91. Castillejo, M., Martin, M., Oujja, M., Silva, D., Torres, R., Manousakis, A., Zafiropulos, V., Van de Brink, O.F., Heeren, R.M.A., Teule, R., Silva, A., and Gouveia, H., Analytical study of the chemical and physical changes induced by KrF laser cleaning of tempera paints, *Anal. Chem.,* **74(18)**, 4662–4671, 2002.

92. Pouli, P. and Emmony, D.C., The effect of Nd:YAG laser radiation on medieval pigments, *J. Cult. Heritage,* **1**(1), S181–S188, 1 August, 2000.

93. Krajnovich, D.J., Near-threshold photoablation characteristics of polyimide and poly(ethylene terephthalate), *J. Appl. Phys.,* **82**, 427–435, 1997.

94. Krajnovich, D.J. and Vazquez, J.E., Formation of "intrinsic" surface defects during 248 nm photoablation of polyimide, *J. Appl. Phys.,* **73**, 3001–3008, 1993.

95. Preuss, S., Spath, M., Zhang, Y., and Stuke, M., Time resolved dynamics of subpicosecond laser ablation, *Appl. Phys. Lett.,* **62**, 3049–3051, 1993.

96. Baudach, S., Kruger, J., and Kautek,W., Femtosecond laser processing of soft materials, *Rev. Laser Eng.,* **29**, 705–709, 2001.

97. Vogel, A., Noack, J., Huettmann, G., and Paltauf, G., Femtosecond-laser-produced low-density plasmas in transparent biological media: a tool for the creation of chemical, thermal, and thermomechanical effects below the optical breakdown threshold, *Proc. SPIE,* **4633**, 23–37, 2002.

98. König, K., Riemann, I., and Fritsche, W., Nanodissection of human chromosomes with near-infrared femtosecond laser pulses, *Opt. Lett.,* **26**, 819–821, 2001.

99. Anglos, D., Solomidou, M., Zergioti, I., Zafiropulos, V., Papazoglou, T.G., and Fotakis, C., Laser-induced fluorescence in artwork diagnostics: an application in pigment analysis, *Appl. Spectrosc.,* **50**, 1331–1334, 1996.

100. Castillejo, M., Martin, M., Silva, D., Stratoudaki, T., Anglos, D., Burgio, L., and Clark, R.J.H., Analysis of pigments in polychromes by use of laser induced breakdown spectroscopy and Raman microscopy, *J. Mol Struct.,* **550–551**, 191–198, 2000.

101. Balas, C., An imaging colourimeter for noncontact tissue color mapping, *IEEE Trans. Biomed. Eng.,* **44**, 468–474, 1997.

7 Laser Cleaning of Polymerized Substrates: Removal of Surface Resin from Paintings

7.1 INTRODUCTION

In previous chapters it was demonstrated that laser-based techniques play an important role in the preservation of cultural heritage in terms of analysis and diagnostics. The laser has also been used for the removal of black encrustations from marble sculptures and stonework [1–7] as well as aged (yellowed) varnish from paintings [8–14]. The successful implementation of these techniques relies on the close collaboration of conservators with laser scientists for defining the process end point. Current trends consider the need for scientific approaches that may supplement or even replace some traditional empirical methods being used in art conservation. Extensive literature on laser-based conservation techniques exists from LACONA Conference proceedings [15–18].

Cleaning of artworks and antiquities by lasers provides the opportunity for the selective removal of undesired surface material. In principle, it is possible to leave the original delicate (underlying) surface or substrate entirely unaffected. In some cases, on-line monitoring and *in situ* control of laser cleaning, using the LIBS technique described in Chapter 3, have proven valuable [4–6,19–21].

In this book the presentation of laser cleaning applications has been based on the nature of the material to be removed. Here we consider complex polymeric substrates where mainly UV laser pulses of nanosecond or picosecond duration are necessary. Such materials include oxidized resins, with or without inorganic inclusions, that have been used as protective coatings on top of precious painted substrates. More complicated situations may be encountered when the original surface layer has been covered with a number of overpaint layers (e.g., see [12,22]). In such cases, the organic binder can be an oil medium (e.g., linseed oil) that has been polymerized with time. As a second class of materials presented in the next chapter, we consider the laser cleaning applications concerning inorganic encrustation on marble or other stone surfaces. On these largely inorganic surfaces the laser ablation mechanisms are different, with the main characteristic being the self-limiting character of the ablation. Finally, in Chapter 9 the reader may find laser cleaning applications on paper, parchment and metal.

This chapter mainly includes laser cleaning of polymerized resins. Laser cleaning of modern paintings (e.g., removal of soot from an acrylic paint) has been intentionally left out, since it refers to a particular class of substrates. It is divided into eight more sections, starting with some general issues related to the optimization of UV laser cleaning parameters. Next a review of recent studies on ablation efficiency, light transmission, and chemical alterations is presented. Then test case studies of laser cleaning applications on painted surfaces are given. Following, the reader may find some representative experimental results on laser ablation of paint layers from composite materials. Finally, and before the conclusions, the results of the application of the Er:YAG laser in the removal of surface layers are summarized.

7.2 OPTIMIZATION OF LASER PARAMETERS: GENERAL CONSIDERATIONS

The ablation of polymers by short UV laser pulses has been studied extensively owing to the plethora of applications [23–25]. A comprehensive review of the subject can be found in the books of Bäuerle [26] and Luk'yanchuk [27].

Let us review first some important findings in the laser ablation of polymers, which resemble the laser ablation of aged resins. In the UV laser ablation of polymers, the ablation by-products typically consist of small fragments (atoms, small molecules, and their ions) and medium-sized fragments, such as monomers, as well as larger molecular fragments [28–30]. These fragments are created in the breaking of covalent bonds by direct photodissociation or thermal decomposition. Photochemical and photothermal ablation could be considered limiting cases of the photophysical mechanism, where the removal of mainly electronically excited species from the surface is taken into account [31–34].

The photochemical modifications induced by UV laser ablation of doped PMMA films were investigated by Georgiou et al. [9] and Athanassiou et al. [35], by simulating the process in natural resins. However, we should mention that polymerized materials found in paintings are rather complex. For example, aged polymerized resins may consist of hundreds of different organic molecules irregularly cross-linked in a compact 3D matrix. There are also different salts in the form of clusters as well as crystals or particulates embedded into the polymeric network. Recent studies [11,36,37] have revealed the presence of a gradient in polymerization and oxidation through the film thickness adding complexity to the ablation process (see insert in Section 7.6).

The basic guidelines for choosing the proper exposure parameters for *safe* laser ablation of unwanted surface material without affecting the underlying substrate will be presented here. First we make the assumption that a laser wavelength for the ablation of these materials lies in the UV up to the maximum acceptable laser wavelength $\lambda_L = 248$ nm or, in some cases, 266 nm. At these UV wavelengths, the optical penetration depth equals $l_\alpha = \alpha^{-1} \ll d_p < d_{tot}$, where α is the linear absorption coefficient, d_p is the thickness of the polymerized material to be removed, and d_{tot} is the total thickness of resin. A typical painting cross section is shown in Figure 7.1.

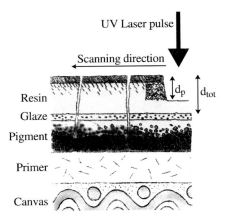

FIGURE 7.1 Representation of a typical cross section of a painting.

The above condition generally ensures the safety of the substrate (pigmented medium) that is under the remaining resin. For highly polymerized natural resins, l_α is of the order of 0.1 μm for $\lambda_L = 193$ nm and 1 μm for $\lambda_L = 248$ nm, while d_{tot} is usually 10 to 100 μm. For the pulse duration, we assume use of readily available nanosecond lasers—without ignoring the ultrashort UV pulse lasers, which are not yet available for mass applications. Another point that must be clearly noted here is the usual choice of the KrF laser over the ArF excimer laser. Although the 193 nm ArF laser offers a better ablation result than the 248 nm KrF because of its smaller l_α value, the overall efficiency of the ArF excimer laser is an order of magnitude less. This is due to both the commercially available energy levels and the much lower ablation rates. Therefore from the application point of view, we concentrate on $\lambda_L = 248$ nm with the proviso that, for interventions requiring a higher resolution, the better choice may be a shorter wavelength.

Apart from the laser wavelength and pulse duration, the most important parameter is the laser energy density or fluence. It is of major interest to quantify the transmission of laser light through the resin layer as a function of incident fluence. Aged varnish usually consists of unidentified complex polymerized material with random inorganic inclusions. An experimental method has been developed for finding the optimal fluence for each particular application. Three different assessment approaches have been considered based on ablation efficiency, light transmission, and chemical alterations. Each approach is addressed separately here.

7.3 ABLATION EFFICIENCY STUDIES

Common challenges to be addressed in painting conservation have to do with (i) varnish layer darkening due to aging associated with polymer phase formation, oxidation, cross-linking, and photochemical degradation, (ii) accumulation of various surface pollutants that may be further polymerized and cross-linked to the resin, and (iii) overpaints on the original painting that must be removed. Removal of these various surface materials, most often found in combination, must be accomplished in such a way that the integrity of the original work is guaranteed.

Laser cleaning is based on the accurate removal of a well-defined surface layer under controlled conditions. In the case of a homogeneous layer, in other words, where we have constant physicochemical properties in that outermost portion of the aged varnish to be removed, the ablation rate is constant for a given fluence level. Therefore, proper control of the number of pulses results in a material removal of uniform thickness (Figure 7.1). It has been found that for depths between d_p and d_{tot} the ablation rate changes, resembling that of fresh (unaged) resin (see insert on pages 251–252).

Figure. 7.2a. shows characteristic ablation rate curves obtained from model and real samples using the KrF excimer laser (emitting at 248 nm, pulse duration of 25 nsec). The four representative polymerized materials presented here are two types

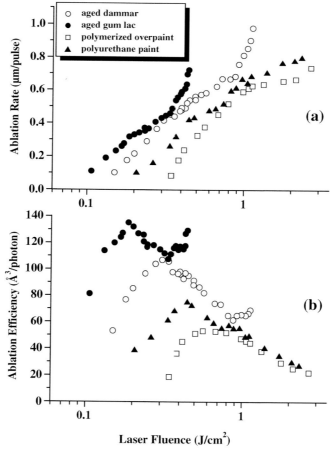

FIGURE 7.2 Ablation rate (a) and ablation efficiency data (b) for four different complex polymerized materials using KrF excimer laser. The error is less than 10%.

of artificially aged varnish (dammar and gum lac), a multilayer polymerized linseed oil-based overpaint, and a black polyurethane paint film. Here a note must be put on the accelerated aging process that was chosen, namely the UV including xenon arc radiation, which has been proven to obtain the most extreme degradation consequences in natural resin varnishes [38,39]. A problem encountered is that accelerated light aging of natural varnishes does not result in discoloration [40,41], which is one of the main characteristics of aged varnish. This effect is attributed to bleaching of the chromophores that are responsible for discoloration during the intense irradiation [40–42]. Accelerated aging under xenon arc radiation, excluding UV wavelengths and radiation under fluorescent tubes, generates a lower degree of cross-linking than that of natural aged varnishes [43,44]. Aging using xenon arc radiation with UV wavelengths to simulate outdoor exposure has shown that natural varnishes generate a considerable degree of polymerization [38,44].

The artificially aged resin films presented here were made by applying the film on a quartz plate via spin-coating and then exposing the prepared sample to cw UV radiation (centered at 360 nm) and elevated temperature [11]. This exposure was stopped when the film absorbance stabilized. The ablation rate vs. fluence is then measured and, in some cases, the dependence on wavelength and pulse duration is also determined [45]. The data plotted in Figure 7.2a shows the mean thickness [45] of various materials removed per pulse as a function of fluence (in units of J/cm^2). It can be seen that the depth resolution attained by the KrF excimer laser can be as good as 0.1 μm per pulse, which is not possible using mechanical or chemical techniques. Figure 7.2a also demonstrates the precise control possible in varnish removal. The ablation rate can be adjusted by choosing the fluence and number of pulses per site. The abscissa has a logarithmic scale to demonstrate the linear response of ablation rate to the logarithm of fluence [26,28], except at very low fluence values (because of the so-called Arrhenius tail, which is not presented in the data set). For every polymerized material this linearity is always strictly localized within a relatively narrow fluence range.

The ablation efficiency curves in Figure 7.2b provide the optimum fluence for ablating each material. Here the word optimum refers to the fluence at which the ablation becomes most efficient in terms of ablated volume per incident photon, and it should not be confused with the fluence corresponding to the highest depth step resolution (i.e., just above the ablation threshold). In the next section it is shown that, at optimum fluence, the transmission is least. The ablation efficiency is a quite significant parameter in cases where large surfaces or thick varnish layers are encountered, and therefore the goal is to remove safely the maximum possible volume of material per incident photon. The data presented in Figure 7.2b indicates, for instance, that the optimum fluence for ablation of artificially aged dammar and gum lac are 0.3 and 0.2 J/cm^2, respectively, while it is about 0.6 J/cm^2 for the oil-based polymerized overpaint. Another important observation in Figure 7.2b is the similar experimental trends. As the fluence increases, the ablation efficiency rises until it reaches a maximum. This maximum corresponds to the point where saturation is reached in the bond breakage within the outermost zone near the surface and where the plasma shielding starts to occur. This point may be evident by the onset of plume generation at fluence values at, or just above, the optimal fluence. At this fluence, the excited chromophores [46] (light absorbing sites within the 3D macromolecular lattice of the polymerized phase) have reached a maximum density. For even higher fluence levels, the efficiency drops. Especially in the case of the artificially aged varnishes, as we further increase the fluence above a certain level (0.9 and 0.35 J/cm^2 for dammar and gum lac, respectively), the ablation efficiency starts to increase for a second time. Such high fluence levels are then sufficient for additional material removal efficiency through vaporization. At this fluence range photomechanical effects, thermoelastic stresses, and shock wave formation may become important factors in laser ablation. Their direct contribution to laser ablation becomes significant when the ablated polymerized material includes inorganic particulates (e.g., overpaint) or when there is delamination that can lead to ejection of flakes. In such cases, a large percentage of the material leaves the surface as small particles or flakes. When removing uncontaminated robust aged varnish, though, the shock wave produced seems to have a negligible effect, especially when working near the optimum laser fluence. Of course, this situation applies to the contribution of photomechanical effects in the removal process itself. The possible long-term consequence of the shock wave is a different issue that is currently under investigation (e.g., see [47] and Chapter 5).

Another phenomenon that should be mentioned is so-called incubation. The word is frequently used in laser ablation, and it describes the effect of the first few pulses on the surface. The incubation generally results in different ablation rates for the first few pulses interacting with a surface. In general, incubation phenomena may be very important for fluence levels near the ablation threshold [26]. Although a detailed analysis is essential to study the effects occurring in this low fluence regime, it turns out that in the applications under consideration the optimal fluence is well above the ablation threshold, and incubation is found to be negligible.

Here it is worthwhile to present some results obtained using ultrashort laser pulses. Figure 7.3 shows the ablation rate and efficiency data of artificially aged dammar resin (as in Figure 7.2) using pulses of 500 fsec and 25 nsec duration at $\lambda_L = 248$ nm. Surprisingly, the maximum ablation efficiency for both pulse durations is reached at about the same fluence (~0.3 J/cm^2). For this particular resin and for the encountered low degree of polymer formation (five to seven monomers), the ablation rate depends more on fluence than on peak power density (see the considerable overlap of data in Figure 7.3). One of the many different reasons for such a behavior may be the long relaxation times of electronically excited bonds of polymer complexes in this material.

Finally, for a highly polymerized naturally aged varnish the situation is expected to be quite different. Figure 7.4 shows results for three different pulse durations (500 fsec, 5 psec, and 25 nsec) at $\lambda_L = 248$ nm on a highly polymerized varnish, which has undergone natural aging for more than 60 years. Here the corresponding relaxation times must be of the order of picoseconds. Such short relaxation times have been measured in polymers [26,27]. As a result, it is expected that the ablation mechanisms are clearly not thermal when a picosecond or subpicosecond UV pulse interacts with a highly polymerized, naturally aged resin. This may be seen in the SEM images of Figure 7.5 of the same sample tested for Figure 7.4.

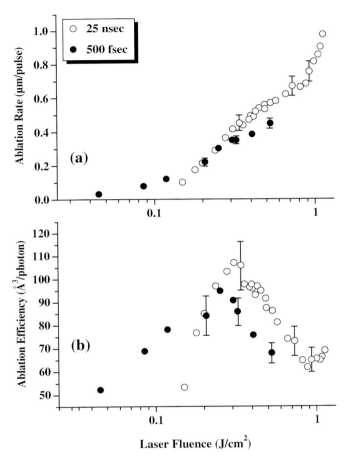

FIGURE 7.3 Ablation rate (a) and ablation efficiency data (b) of artificially aged dammar resin using laser pulses of 25 nsec and 500 fsec duration (λ_L = 248 nm).

Figure 7.5a shows the SEM image of the (original) reference surface, while Figure 7.5b through Figure 7.5d show the resin after removal of ~1 micron film thickness with pulses having durations of 500 fsec, 5 psec, and 25 nsec at λ_L = 248 nm. The laser ablation was carried out under scanning mode operation with overlapping laser spots (Section 7.6). The white particles in the images are salt clusters containing mainly Ca and Na, which are usually present in the natural resins. The surface after ablation using nanosecond pulses has a very smooth texture as if it had undergone melting (Figure 7.5d). Additionally, the inorganic particles have also undergone a melting–resolidification transformation, resulting in the sharp peaks seen. On the other hand, the resin surface and the inorganic particle morphology are more similar to the reference after ablation by subpicosecond or picosecond pulses (Figure 7.5b and Figure 7.5c, respectively). This is the main advantage of ultrashort UV laser pulses, in addition to the higher depth step resolution than with the nanosecond pulses. In contrast, the nanosecond UV laser has proved to be much

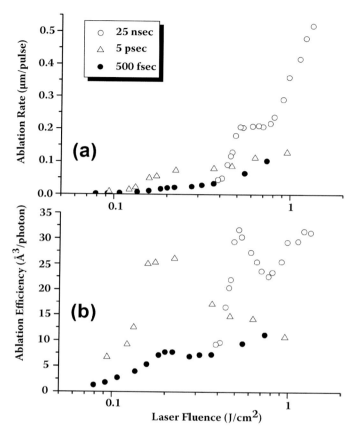

FIGURE 7.4 Ablation rate (a) and ablation efficiency data (b) of a naturally aged unknown resin using laser pulses of 25 nsec, 5 psec and 500 fsec duration (λ_L = 248 nm). The error is less than 10%.

more efficient, and in addition it is readily available. Thus it is the better choice for regular laser interventions, where the ~1 µm order of magnitude resolution in depth step is adequate. The "melting" texture of the surface is not of concern because it only extends to ~1 µm below the surface ($\sim l'_\alpha$).

7.4 LIGHT TRANSMISSION STUDIES

During the laser ablation of polymerized resins, some of the light may be transmitted and reach the underlying substrate. Therefore quantification of the optical penetration depth is important. The goal here is to measure the transmitted energy and find the optimal exposure parameters to minimize the transmitted light, which could disrupt the substrate.

The method presented here permits the measurement of the transmission of laser light through model resin samples on quartz. A similar study on polymers, e.g., polyimide, has been reported by Pettit et al. [48]. During laser ablation, the energy

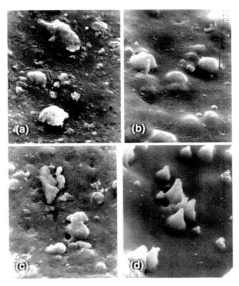

FIGURE 7.5 SEM pictures (×3500, 40° angle of observation) of reference surface (a) and after the removal of 1 μm film using laser pulses of 500 fsec, (b), 5 psec, (c), and 25 nsec (d).

passing through the sample is measured for each successive laser shot and as the resin film becomes thinner. These data show the energy of transmitted UV light as a function of depth.

If the intensity of the beam that irradiates the film surface is I_0, the intensity of the transmitted light is I_{trans}, and the intensity of the absorbed and scattered light is I_1, then the measured absorbance, A, of the *remaining* (after each laser pulse) resin film is given by

$$A = \log\left[\frac{I_0 - I_1}{I_{trans}}\right] = \alpha\, d \qquad (7.1)$$

where α is the effective linear absorption coefficient and d is the film thickness remaining after every laser pulse. Here for simplicity we assume that α is constant over the film thickness, although this is not quite true in aged resins [11,36,37]. Equation (7.1) results in

$$\log\left[\frac{I_{trans}}{I_0}\right] = \log\left[1 - \frac{I_1}{I_0}\right] - \alpha\, d \qquad (7.2)$$

According to Equation (7.2), the results may be plotted as $\log(I_{trans}/I_0)$ versus d. Such a graph is presented in Figure 7.6 for an artificially aged mastic film prepared on a quartz plate. For this test, the laser beam is appropriately shaped and directed onto the sample. A sensitive energy meter measures the transmitted energy for each consecutive pulse. For each fluence level used, a series of pulses was fired until the entire mastic film was removed. The d values are calculated using the mean ablation rate data and the total film

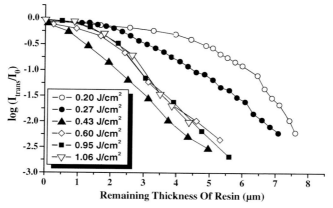

FIGURE 7.6 Light transmission measurements of aged mastic during laser ablation ($\lambda_L = 248$ nm, 25 nsec pulse duration), presented as $\log(I_{trans}/I_0)$ versus d, for six different laser fluence values. All points represent consequent laser pulses, except at 0.20 J/cm². The corresponding errors for $\log(I_{trans}/I_0)$ and d are less than 5% and 10%, respectively.

thickness. The light reflected by the surface is taken into account by normalizing the I_{trans} measurements, comparing them with the transmission after complete removal of the film. The slope of the graph represents the effective absorption coefficient of the film at the test wavelength. This graph provides the following information:

1. The total energy fluence of transmitted light. This can be calculated if initial film thickness, fluence, ablation rate, and number of pulses fired are known. A knowledge of the total transmitted energy density is crucial when the substrate is sensitive to the laser light.
2. The optimal laser fluence can be determined, where optimal in this case is defined in terms of least transmitted fluence at a certain d value. Although optimal is defined differently here from in Section 7.3, it turns out that the two definitions give the same value as shown below.

Here it is interesting to compare the transmission data with the ablation efficiency data obtained from the same samples. Figure 7.7a and Figure 7.7b show the results of ablation rate and ablation efficiency studies, respectively, of the same aged mastic sample as in Figure 7.6. The six different fluence values used to obtain the transmission data presented in Figure 7.6 are marked in Figure 7.7b for comparison. At the low fluence of 0.2 J/cm² (just above the ablation threshold), the transmission of laser light is rather high. For slightly higher fluence levels we observe a sharp drop in transmission, reaching a minimum at 0.43 J/cm², a fluence that corresponds to the highest ablation efficiency. For higher fluence levels the transmission curves tend to coincide, possibly owing to plasma absorption and shielding. Figure 7.7c shows the film transmission as a function of fluence when the remaining mastic resin is only 3 μm thick. A comparison between Figure 7.7b and Figure 7.7c shows that *the optimum fluence defined in terms of transmission coincides with the optimum fluence defined in terms of ablation efficiency*. This has been found to be a general

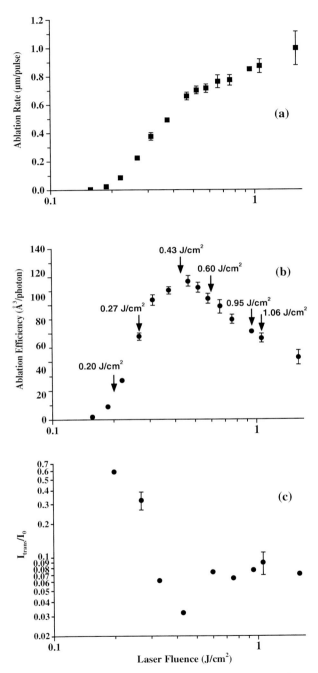

FIGURE 7.7 Ablation rate (a) and ablation efficiency data (b) of artificially aged mastic resin (same as in Figure 7.6) using KrF excimer laser. Plot (c) shows the transmission as a function of fluence for a hypothetical remaining mastic film of 3 μm thickness. (Data extracted from Figure 7.6.)

rule for a series of different polymerized resins [36]. The results are generally in accordance with the three-level chromophore model of Pettit et al. [48].

The implications of these observations are essential for the safe use of pulsed UV lasers in the removal of topmost (superficial) resin from paintings or other delicate substrates. It is also important to note that in such applications the approach cannot be empirical and certainly cannot be the same as in the removal of inorganic encrustation. In the latter case (see Chapter 8), the end user starts from low fluence and progressively increases the level until a cleaning threshold is reached. In the case of polymerized resins, however, such an approach could endanger photosensitive substrates. The next section deals exactly with this important issue.

7.5 CHEMICAL ALTERATION OF SUBSTRATE

Many painting pigments consist of inorganic salts mixed in a binding medium, which can be linseed oil or egg-yolk tempera. A resin layer (varnish) is usually applied on top of the paint layers for visual effects and to protect against oxidation, dust, and dirt. It has been found [49] that resins also protect pigments from the ambient UV light that induces free radical formation and subsequent oxidation, in both the resin and the binding medium. At the same time, most salts are sensitive to direct laser irradiation undergoing chemical change or phase transformation [50–52]. As pointed out throughout this chapter, only the outermost portion of the aged resin must be removed from the paint surface. During laser ablation, if not done properly, the transmitted light (through the resin) may cause long-term deterioration of the remaining original painting layers, e.g., glaze and/or pigment. To address the matter, laser ablation was combined with gas chromatography coupled to mass spectroscopy (GC-MS) analysis. The goal was to detect possible oxidation by-products induced by laser radiation [10].

The process of degradation of the linseed oil medium under oxidation conditions involves simultaneous polymerization and depolymerization reactions. In general, unsaturated fatty carbon chains undergo various oxidation reactions resulting in unstable peroxides. These finally result in conjugated by-products, of up to nine carbon atoms. At the same time, 3D cross-linking occurs as a basic consequence of the presence of radicals. Oxidative cleavage (bond breakage accompanied by oxidation) and production of low-molecular-weight volatile by-products (as the end result of the possible oxidation processes) are routinely investigated by gas chromatography [53]. After transesterification of all possible glyceryl esters with hydrochloric methanol, a mixture of lower molecular weight methyl esters of all present fatty acids is produced and analyzed. Such a gas chromatogram, for example, for an aged linseed oil medium, reveals that (a) long unsaturated fatty chains (e.g., oleate) are decreased owing to oxidation; (b) saturated fatty chains are basically unchanged; and (c) a number of oxidation by-products such as dimethyl esters of azelaic (nonadioic) acid are formed. The quantities of esterified fatty acid residues such as methyl palmitate (C16:0), stearate (C18:0), oleate (C18:1), as well as monomethyl and dimethyl azelate (C9) have been previously used as measures of various oxidative routes in oils [53]. Here the two numbers correspond to the number of C atoms and the number of double bonds, respectively. Oleate is slowly oxidized (decreases in concentration), while azelate is produced as an oxidation by-product, thus enabling an evaluation of the aging process. Therefore, low C18:1/C18:0 or C18:1/C16:0 and high C9/C16:0 or C9/C18:0 ratios suggest a strongly oxidized medium.

Based on the above, model samples of cinnabar pigment in linseed oil covered with dammar varnish (~20 μm thickness) were made and artificially aged using a UV lamp and elevated temperature. Then for each sample a predetermined thickness of varnish was removed by choosing the level of fluence (KrF excimer) based on the ablation curves of dammar (see Figure 7.2). Note that the same dammar samples were used to obtain the ablation rate data and the GC-MS data. Finally, the ensemble (pigment, oil medium, remaining varnish) of each sample was analyzed by GC-MS [10].

In general, the results show no increase of oxidation by-products in the paint medium when at least a thin varnish layer remains over the paint and when the optimum fluence is used. However, detection limitations of associated chromatographic techniques may not be sensitive to minor changes in bulk medium. On the other hand, when higher fluence is used, the oxidation by-products increase. In more detail, Figure 7.8 presents the C18:1/C16:0 and C9/C18:0 ratios of the GC-MS peaks for different values of ablated resin thickness. Based on the results described in the previous sections, the optimal fluence for the ablation of the particular aged dammar layer is 0.3 J/cm^2, while a fluence of 0.8 J/cm^2 is rather high for this

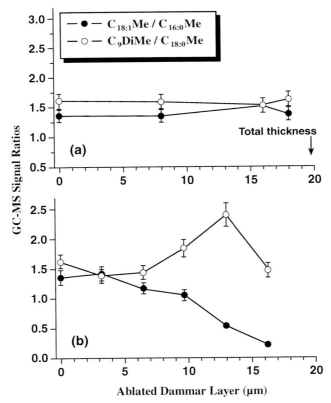

FIGURE 7.8 GC-MS signal ratios as a function of removed layer thickness using KrF excimer laser at fluence 0.3 J/cm^2 (a) and 0.8 J/cm^2 (b).

specific sample (see Figure 7.2b). Figure 7.8a and Figure 7.8b correspond to the results using these two values of fluence, respectively. Figure 7.8a shows there is no evidence of any oxidation change at a fluence of 0.3 J/cm^2, even when the remaining (inmost) varnish layer is only 2 μm thick. On the contrary, when using a fluence of 0.8 J/cm^2, the concentration of the oleic component sensitive to oxidation (C18:1) decreases as the thickness of removed varnish increases (see Figure 7.8b). At the same time, the concentration of C9 oxidation by-product increases: the exception to the last data point may be due to the further oxidation of C9 to lower molecular weight by-products. Therefore it is concluded that the chemical alteration of the binding medium can be avoided by using the optimal fluence, which is found to be different for each resin and its polymerization parameters. At the same time, the innermost resin must be left to reduce the fluence of the transmitted light to the pigmented layers. An additional reason for leaving the inmost varnish on the paint is the following: It has recently been proven [36] that owing to the existing gradient in resin degradation [11], the resin nearest the paint has a chemical structure almost identical to the unaged resin (see insert). This is due to the lack of oxygen and ambient UV photons, which may not penetrate the varnish nearest the paint layer.

The combination of several analytical techniques has been used to characterize and quantify the chemical and physical effects induced by KrF excimer laser cleaning of egg-yolk tempera paint systems [52]. An understanding of the effects of the UV laser radiation reaching the paint layer is necessary in order to define the consequences that could result in a worst-case scenario of laser cleaning, namely, complete removal of a protective varnish (or overpaint) layer. The direct interaction at high fluence can result in irreversible unwanted changes, although these effects are constrained to the sample surface and are strongly dependent on the nature and chemical characteristics of the paint. It was found that organic egg-yolk tempera paint systems are very stable even under intense laser treatment. In contrast, direct laser irradiation of inorganic paint systems induces various types and degrees of modifications in the properties of both pigment and binding medium. The observed discoloration could be due to a modification of the chemical composition of the pigment through a change in its oxidation state or a change in the pigment crystalline phase. On the other hand, the direct ablation of organic material from the substrate leads to formation of a very thin layer of charred pigmented material that covers the original paint surface. This layer is not produced in the absence of the inorganic pigment, indicating that pigment particles mediate the charring through an energy transfer mechanism. Clearly, the actual paint composition will determine which of these effects, change of pigment composition or charring of the outermost paint layer, is dominant. Finally, it was found that modifications induced by irradiation at fluence levels below the ablation threshold in both organic and inorganic paint systems were consistently less severe or undetectable. In conclusion, the results obtained by Castillejo et al. [52] further confirm the adequate strategy for laser cleaning consisting of a controlled partial removal of the outermost varnish layer, leaving a thinned protective resin layer that prevents discoloration or other chemically harmful effects to the underlying paint.

7.6 LASER ABLATION OF AGED RESIN LAYERS FROM PAINTINGS

The application of varnish plays a significant role in protecting painted surfaces. During aging, the varnish degrades through oxidation, auto-oxidation, and cross-linking processes, catalyzed by the absorption of light [40,54]. In particular, the absorption of ambient UV light leads to the formation of by-products playing a significant role in the polymer phase formation (cross-linking). For aged natural resins, it has been found [11,36,37,55] that certain optical properties such as absorption relating to polymerization are scalar with depth from the surface. This is owing to the exponential attenuation of ambient light as it propagates into the resin. Note that UV light catalyzes the formation of radicals and therefore the polymerization processes. The gradient of oxidation in the film is directly related to an equivalent gradient of the solubility throughout the thickness of varnish [37]. Consequently, the deeper one goes into the varnish layer, the less the concentration of the active solvent required to remove the remaining varnish. A number of test case studies have shown that the required concentration drops by a factor of about two when going from the original varnish surface to a depth of about 8 to 10 μm [37].

Sometimes it is necessary to remove the degraded varnish that can affect the underlying media [56]. Chemical cleaning is a complicated procedure that employs organic solvents potentially able to initiate a process of swelling and leaching of the film [57]. The outcome may be a significant alteration in paint layer constituency. It has been found that laser-assisted ablation of the topmost varnish (~8 to 10 μm) is not only helpful but also unique, since it cannot be matched in precision and control by any other cleaning method [13]. Moreover, it offers the choice of either leaving the innermost varnish intact (preserving the touch of time) [13,36] or further removing it by using less aggressive solvents than in an entirely traditional chemical treatment [13,36,37].

The laser-assisted removal of the outermost aged varnish must be performed at the optimal fluence, as described in the previous sections. When working on actual paintings, it is not trivial to determine the optimal fluence. Previous knowledge of the stratigraphy is necessary. The only established method for determining the optimal fluence [20] is to go through an ablation rate/efficiency study using a small area of the aged varnish (typically at the edge of the painting) to generate the ablation data. An area of 5×50 mm^2 is enough, where typically 7 to 10 ablation spots are created, using a different fluence level for each. When the painting is on canvas, the test area is chosen at the edges around the wooden panel. Even if not, the test spots disappear after the cleaning. Depending on the surface roughness and varnish thickness, each ablated spot created by several pulses should have a depth of about 5 to 10 μm in order to be measured with minimal error and, at the same time, not penetrate too deeply into the varnish where the resin is basically unaged (see insert in this section). Profilometry is applied across the gaussian direction of the rectangular spot, and its integrated area is divided by the spot size along the same direction. This method provides the mean ablation rate across the entire beam profile. The procedure is repeated for at least three parallel paths spanning the top-hat direction of the excimer beam, and a final mean ablation depth is derived, which in turn is divided by the number of laser pulses fired.

In this way the mean ablation rate is obtained. All ablation rates presented in this as well as in the following chapter are mean ablation rates measured following the method described above. Although this method is far more complicated than just measuring the maximum ablation rate, usually at the center of the spot (almost all literature data are obtained in such a way), it has a unique advantage: it provides an accurate ablation rate that can be used to calculate the true number of laser pulses needed to remove a certain layer of a given area and thickness also considering the spot overlap. Finally, the ablation efficiency is calculated from the mean ablation rate. A description of such measurements is given elsewhere [36,45]. A second method for obtaining the optimal fluence, which still needs further investigation, is based on the onset of plasma generation. A plasma begins to appear at a fluence level near or just above the optimal fluence.

The plasma observation has also been associated with the well-known monitoring technique LIBS (laser-induced breakdown spectroscopy), described in Chapter 3. Here, however, the usually observed plasma emission during laser ablation is a useful "by-product" that has been used for on-line control of the exact etch depth advancement [4–6,13,19–21]. During laser ablation, the emission produced is characteristic of the ablated material composition. In most cases, the composition changes in a predicted way as the material removal advances toward the final depth, and it is exactly this spectrum alteration that can be used to guide and control the laser ablation process.

For automatic control during the laser cleaning process, the light emission spectra are recorded on each pulse in a preselected spectral region. The data acquisition processor is programmed to calculate certain peak intensity ratios used as inputs to a computing algorithm, which produces an output value after each laser shot. When this value falls outside certain boundaries, the laser stops firing and the motorized X–Y stage moves the painting or the laser beam to a new position where the cleaning process starts again. At this new position, laser pulses are delivered until the output value again moves out of the chosen domain. The distance that the sample or beam is moved in each step is determined by an overlapping protocol for the sequential sample movement in both the X and Y directions. A schematic of the workstation [21] is shown in Figure 7.9.

Besides the plasma analysis method described above, it has been found recently that the detection of the plasma intensity can be used as well to monitor the end point of laser ablation [58]. The method is based on an automated closed-loop laser cleaning where comparison of a reference level of plasma signal to the actual signal is defined by the operator. This can be found based either on experience or by a teaching protocol on test samples. During this laser cleaning process, the plasma signal of each laser pulse is compared with the reference signal. Laser removal continues until the desired reference signal is reached. In a next step, the X–Y stage controller moves the sample to a new position.

At this point it is worthwhile presenting another piece of evidence for the existing scalar properties in the resin, which has been used for on-line control during the laser-assisted removal of aged varnish. Apart from polymer phase formation [36], natural oxidation [40] takes place, especially in the outermost varnish [36]. In the case of an original varnish layer that has not been retouched, the elemental composition throughout its cross section does not change with aging. This fact could be a major obstacle in using LIBS as a monitoring technique, since LIBS is, in general, an atomic analytical method (see Chapter 3). However, in LIBS experiments that were performed on naturally aged

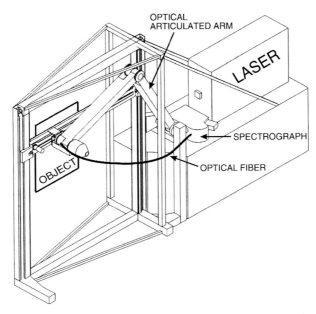

FIGURE 7.9 Sketch of the workstation for scanning the laser across the surface of a painting.

resin samples, we found that as the removal progresses toward the innermost varnish, there is a change in the plasma spectrum originating from excited dimers produced [36].

Figure 7.10 shows a typical series of LIBS spectra obtained when successive pulses were used to ablate aged varnish. The four spectra correspond to the 2nd, 3rd, 4th, and 5th pulse, respectively. A comparison of the four spectra reveals the gradual decrease of the CO $B^1\Sigma$-$A^1\Pi$ bands (Ångström bands) and the simultaneous increase and then stabilization of the two most intense C_2 $A^3\Pi_g$-$X'^3\Pi_u$ bands (Swan bands). The CO bands originate from the carbonyl groups as by-products of oxidation [38], whose density is expected to be higher in the outermost than the innermost varnish. This is verified by observing the micro-FTIR spectra (reflection mode) of the surface after each consecutive laser pulse. Figure 7.11 shows a representative FTIR spectrum [59]. The ratio of the peak absorbance attributed to carbonyl groups (1650 to 1730 cm^{-1}, marked as A) to the characteristic C–H bending of methyl and methylene groups (1400 to 1470 cm^{-1}, marked as B) was found to decrease after each laser shot (see small graph insert in Figure 7.11). This is in accordance with the LIBS observations shown in Figure 7.10. In this particular varnish system, the limiting value of A/B was found to be 1.4 ± 0.1.

Now we will present a few test case studies to demonstrate the correct application of laser-assisted outermost varnish removal on actual paintings. The research has been carried out in collaboration with Michael Doulgeridis, Director of the Conservation Department of the National Gallery of Athens, who provided the paintings and offered his guidance from the conservator's point of view. The first example presented here was an 18th century Flemish egg-yolk tempera painting on a 27 × 37 cm wooden panel. It suffered from an unsuccessful restoration some decades ago.

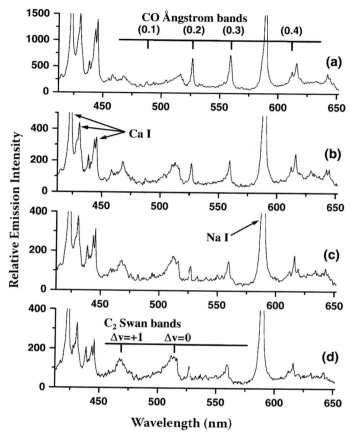

FIGURE 7.10 LIBS spectra of unknown naturally aged varnish using KrF excimer laser at a fluence value of 0.5 J/cm². The four spectra correspond to the 2nd (a), 3rd (b), 4th (c), and 5th (d) laser pulse.

The varnish that was added to protect the painting had undergone severe polymerization and was extremely hard. The optimal fluence was found to be 0.38 J/cm² with an ablation rate of 0.25 µm/pulse (λ_L = 248 nm, 25 nsec pulse duration). The final fluence used was 0.42 J/cm², while LIBS on-line control was used as described above with the mean thickness of removed varnish being 9 µm. Figure 7.12 a shows an initial stage of processing where a small area (left side, toward the bottom) has been laser cleaned, while in Figure 7.12b the top two thirds of the painting has also been processed. At the very bottom of Figure 7.12b and just below the small initially laser cleaned area, we can also see, for comparison, two small vertical zones that have been cleaned by the chief restorer using a mixture of solvents. The complete removal of the varnish with the solvent can be visually compared with the delicate laser-assisted outermost varnish removal. Figure 7.12c shows the color difference between the original oxidized resin (top) and the uncovered, less oxidized resin (bottom). Figure 7.12d shows the final cleaning result after completion of laser

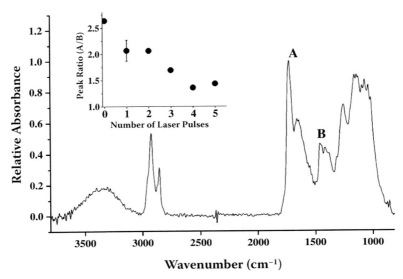

FIGURE 7.11 Typical FTIR spectrum from the surface of aged varnish (same as in Figure 7.10). The enclosed graph (top left) shows the intensity ratio of peaks A versus B after each laser pulse.

processing and subsequent removal of the exposed (softer) innermost varnish by the conservator using a mild solvent.

The second example is a 19th century oil painting of 49×77 cm size that has been severely damaged by an inhomogeneous layer of lime mortar applied over the surface of aged varnish (Figure 7.13a). The varnish was only a few microns thick (~7 μm) and spread evenly over the entire original painting. Any attempt to remove the lime particles or aged varnish using alternative techniques had been excluded. The strategy of KrF excimer laser-assisted cleaning in this case was to ablate the outermost aged resin together with the ejection of lime particles attached on the surface. In this particular case, a fluence level of 0.30 J/cm² (slightly below the optimal fluence of 0.33 J/cm²) was used in order to minimize the etching step depth. The thickness of removed varnish was ~2 μm with the simultaneous removal of superficial lime particles. The final result of the restoration is presented in Figure 7.13b.

In the case of evenly applied resins, we can detect evidence of the gradient of polymerization with depth in terms of crack formation. Figure 7.14 shows part of an egg-yolk tempera icon painting covered with an unknown varnish aged more than 100 years. A KrF excimer laser (fluence of 0.54 J/cm²) was used to remove outermost resin in ten rectangular zones in multiple passes. The zones are indicated by increasing numbers that represent an increased removal of varnish [37]. In zones 1 to 5, the incremental change in depth is 3.3 μm/step, i.e., zone 1 corresponds to an ablation depth of 3.3 μm into the original varnish, while zone 5 corresponds to a total depth of 16.5 μm. In zones 6 to 10, the depth-step intervals are larger, with a maximum depth of 82 μm in zone 10, where the remaining varnish thickness is ~20 μm. The observed cracks are due to the higher stiffness of the aged resin near the surface. The gradual decrease of cracks going from zone 1 to 10 is due to the polymer phase gradient as a function of

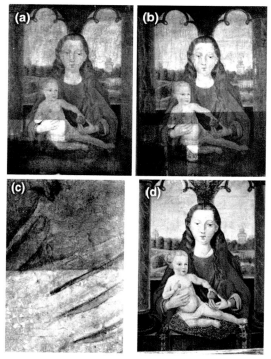

FIGURE 7.12 (See color insert following page 144.) (a) Original surface except a small laser-cleaned area at lower left. (b) Additionally, the top part has been laser cleaned. Two small zones (bottom) have been chemically cleaned for comparison. (c) A detail at the interface between original resin (top) and laser-cleaned surface (bottom). (d) After laser-assisted divestment followed by traditional chemical restoration.

increased depth into the varnish. The more varnish removed, the fresher the remaining varnish and therefore the fewer cracks are revealed. In this case, the end user can choose the end point according to the desired appearance or solubility of remaining varnish. More case studies on egg-yolk tempera icon paintings can be found in [60].

Multispectral imaging techniques can provide a powerful tool to optimize the cleaning process and achieve a homogeneous cleaning result. A specialized detection system appropriate for spectral imaging in the 0.35 to 1.6 μm region has been developed [61] and integrated with user-friendly software into a single unit. It has been used to monitor semiquantitatively the cleaning of paintings.

A characteristic test case example of the use of the multispectral imaging camera is shown in Figure 7.15. The 18th century oil painting on canvas (a French duke, M. Doulgeridis' collection) was subjected to KrF excimer laser cleaning. The painting had suffered paint loss at many points. A restoration was attempted some decades ago without first removing the old varnish and dirt. Even worse, a firmer varnish had been applied. In an attempt to remove this new varnish, the restorers met great difficulties since it had formed a rigid and resistant layer. For a proper removal of this resin using traditional chemical means, the application of chemicals had to be prolonged, while the results were of questionable quality. On the other hand, the

FIGURE 7.13 (See color insert.) (a) Original surface and (b) after laser intervention followed by traditional restoration.

laser-cleaned region presented several advantages: (1) the surface appearance was excellent; (2) in the undamaged areas where there was no previous restoration, a thin clean layer of the old varnish was left; (3) in the sections where the old restoration had been attempted, the new varnish was removed down to the original damaged layers, therefore with good retouching results; (4) the use of hazardous strong solvents, both for the restorer and for the artwork, was avoided; and (5) the whole process was totally controllable with the use of the multispectral camera. The laser parameters chosen allowed for a 0.5 μm depth-step removal for each scan over the surface. Figure 7.15a shows a visible spectra image (14 × 10 cm) in the fore-head–eyebrow region cleaned by laser, while Figure 7.15b shows a 20% expansion of this image area. Figure 7.15c shows the UV fluorescence image from the same area as in Figure 7.15b, but here the distinct marks of originally damaged sites are enhanced. For the laser cleaning, though, the main part of the process concentrated on the outermost resin. Therefore the UV reflectance image (Figure 7.15d) is helpful in displaying the homogeneity of the remaining resin after removal of the outermost portion. The last image, Figure 7.15e, presented as a pseudo-color display, helps to locate over-cleaned areas, which appear red in color.

Laser cleaning of varnish using a KrF excimer laser was also carried out [62] in the premises of Art Innovation b.v., Netherlands, where a prototype workstation was built in collaboration with FORTH. The Cooperative European Research project "Advanced Workstation for Controlled Laser Cleaning of Artworks" (ENV4-CT98-0787) yielded important information [52,62] on the application of UV laser cleaning

FIGURE 7.14 (See color insert.) Part of painting covered with naturally aged varnish. Ten laser-cleaned areas span a thickness of varnish removal from 3.3 μm (area 1) to 82 μm (area 10).

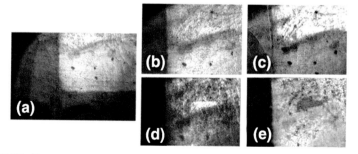

FIGURE 7.15 (See color insert.) Multispectral images of a laser-cleaned area.

to paint materials. In the project, in which conservators, researchers, and engineers participated, the viability of the laser technique as an additional tool in present conservation practice was investigated. The research was pointed at the definition of the boundary conditions in which laser cleaning can be safely applied. It included a systematic effect study of egg tempera paint systems. Physical and chemical changes, induced by exposure to UV (248 nm) excimer laser light under various conditions, were evaluated. The research focused on the integration of the results from various analytical techniques, yielding valuable information on the immediate and long-term effects of UV laser radiation on the paint materials. The analytical techniques include colorimetry, spectroscopic techniques, mass spectrometry and profilometry, as well as thermographic and UV transmission measurements.

From the first experiments of laser ablation on aged varnishes, it became clear that the removed material possessed a so-called gradient in physicochemical properties. As an example we mention its hard and glossy appearance. Even after only one scan of pulses, e.g., a depth of about 3 μm, the remaining resin is already much softer and less colored than the original aged surface. Such a gradient usually extends from the original surface down to a depth of about 8 to 15 μm depending on the resin and its aging. The gradient may continue to greater depth but with more gradual changes than in the first few microns. An example of visible (crack) alterations as we reduce the thickness of an aged varnish is presented in Figure 7.14. The existence of a bulk degradation gradient in the varnish was verified [10,11,13,20,36,37,55] mainly based on spectral evidence such as

- The decreasing CO band emission upon consecutive laser pulses (Figure 7.10).
- The decreasing FTIR carbonyl signal as a function of depth (Figure 7.11).
- The decreasing C_2 band emission upon consecutive laser pulses [36].
- The decreasing absorption coefficient as a function of depth and as well as a function of increasing film thickness [11].
- The change in interaction of aged resin with the UV radiation during cleaning [36].

The degradation gradient was finally established by Theodorakopoulos [36], where the chemical changes as a function of depth were uncovered. His work revealed that deterioration of aged dammar and mastic varnish films is critically dependent on their thicknesses. These films protect underlying photosensitive materials (e.g., pigments) from deteriorative UV light and from exposure to oxygen. The transition from the highly degraded outermost resin to the less affected or even unaffected underlying bulk resin is a matter of a depth-dependent gradient both in terms of oxidation and of photochemically induced high molecular weight (MW) condensed material. These chemical gradients have been observed via DTMS (Direct Temperature-Resolved Mass Spectroscopy) and HP-SEC (high performance–size exclusion chromatography).

Oxidation not only reduces gradually as a function of depth, but the transition from highly oxidative and polar states near the surface toward nonoxidative states in the bulk is qualitative. Based on MALDI-TOF-MS (matrix-assisted laser desorption/ionization time-of-flight mass spectroscopy) results, it has been found that UV-induced oxygenated triterpenoid compounds, such as oleanane/ursane molecules with oxidized A-rings, are formed only in the outermost 15 μm. Below 15 μm from the surface, only oleanane/ursane molecules with side chain oxidation are detected, which are typical compounds of oxidized varnishes. This finding is in line with the optical absorption length of short wavelength ambient light (λ < 350 nm) that is completely absorbed in the first 15 μm of resin. Although oxidation is considered as the main degradation mechanism of natural resin varnishes, the decreasing gradient in the high MW condensed fraction formed during photochemical degradation is also a reality. Provided that high MW cross-links, present only in the outermost region, are insoluble using traditional chemical treatments [40,63], laser ablation of the uppermost aged resin could be fully justified. Additionally, the fact that the 248 nm pulses make no detectable oxidative contribution to the remaining resin [36] is in favor of the excimer laser ablation of the outermost zone of aged varnishes.

Here we present some selected data taken from [36] that demonstrate the existence of the molecular weight (MW) gradient through the aged film thickness. The depth-dependent MW decrease of an aged mastic film is illustrated via the HP-SEC chromatograms at 400 nm in Figure 7.I. The SEC traces of the unaged mastic film and the aged film prior to ablation are plotted for comparison. There is an evident gradient in the absorption from the high MW fraction versus ablation depth. In particular, the abundances of the 900 to 1000 Da components and the condensed fraction in the high MW regime up to 80 kDa are gradually decreasing with depth. Sampling for HP-SEC included the remaining varnish (thickness ~55 μm) resulting in average molecular weight determination of the remaining films after the laser ablation of a certain thickness (indicated by each curve in Figure 7.1). Therefore the absolute molecular weight as a function of depth would most likely display an even sharper gradient than that demonstrated here.

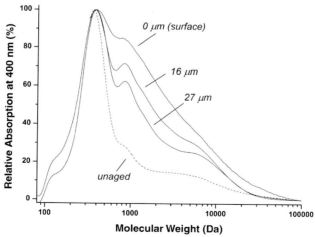

FIGURE 7.I HP-SEC traces of partially laser ablated aged mastic film, and of unaged mastic. Each trace corresponds to the remaining film after laser ablation of 0, 3, 16, and 27 μm, respectively. The apparent MW reduction across the depth profile is evident. (By kind permission of C. Theodorakopoulos.)

7.7 LASER ABLATION OF PAINT LAYERS FROM COMPOSITE MATERIALS

Short UV laser pulse ablation has been studied for a number of materials in terms of ablation rate/efficiency behavior and time-of-flight mass spectroscopy of the positively charged photofragments. One such case is the photoablation of polyurethane (PU) films used as paint systems on delicate composite materials [45]. The dependence of fluence on ablation rate was tested against existing photoablation models. The upper limit of the mean activation energy for desorption has been found to be considerably lower (~3 to 15 times) than the average energy required to break single covalent bonds, indicating the importance of photophysical types of processes [26,31–34]. At least in one case (for $\lambda_L = 193$ nm), it was found that photochemical processes predominate in the operative mechanism [46,48]. It was also found that coherent and incoherent multiphoton absorption must significantly contribute to the ablation in the case of femtosecond and picosecond laser pulses. Apart from these general conclusions, it is extremely difficult to know details of the operative ablation mechanisms when such complex materials are encountered.

A particular application of laser-assisted paint removal from composite materials is the so-called paint stripping from composite surfaces, e.g., aeronautical parts [64]. As in varnished paintings where the underlying paint layers are sensitive to light transmitted through the resin, the paint layers on composite structures can be removed only if the integrity of the composite resin is guaranteed. Therefore the use of on-line monitoring techniques is also essential for this type of application. From fatigue testing measurements it has been found [65] that there is always measurable fatigue of the composite when a laser ablates the entire paint system. Such an example is shown in the SEM pictures of Figure 7.16, where an epoxy/carbon fiber composite structure painted with an Si-based polymeric paint system has been divested using an XeCl excimer laser at 1.4 J/cm^2.

Figure 7.16a shows the reference painted surface, Figure 7.16b corresponds to a processing point where the paint has been partially removed, while in Figure 7.16c the breakage of carbon fibers is evident when the paint layer is entirely removed. Here, the epoxy primer — a 3 μm layer between the composite substrate and paint — had been applied as a continuation of the epoxy resin in which the carbon fibers were embedded. Therefore if the ablation is terminated only after the entire paint layer is removed, the laser disrupts the epoxy resin of the composite substrate. It is possible though to choose the end point so that almost complete removal of the paint layer takes place without fiber damage. Figure 7.17 shows the results of paint stripping a 26 × 10 cm composite material with the end point of the final processing corresponding to the state shown in Figure 7.16b.

In cases where the primer is a well-defined layer, removing the paint layer and leaving just part of the epoxy primer on the composite has been found to be safe for the substrate. Figure 7.18 shows the SEM image of a PU-based paint system applied on metallized Kevlar composite. In the laser-divested area (KrF excimer laser, 0.5 J/cm^2), the PU primer has been uncovered.

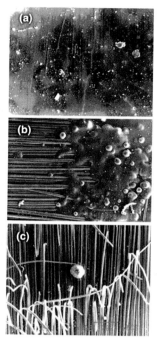

FIGURE 7.16 SEM pictures of painted epoxy/carbon fiber composite (×200): (a) reference surface, (b) after laser-assisted partial paint removal, and (c) after total paint removal.

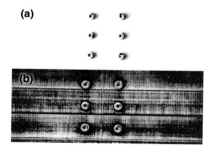

FIGURE 7.17 Epoxy/carbon fiber composite component: (a) reference surface and (b) after laser-assisted partial paint removal.

The optimal laser parameters for this particular system have been investigated by Zafiropulos et al. [45]. Figure 7.19 shows a 14 × 10 cm section of the composite where KrF excimer laser ablation has been done on eight different areas, each to a different depth. The area at the top right has been divested until the primer has been uncovered, while in the other areas the metallization of the composite has been uncovered to varying degrees. In the three sections of the middle row, the resin of the metallization layer has been partially ablated, while in the three areas of the lowest row, the ablation has stopped just after removal of the primer.

FIGURE 7.18 SEM picture of painted (PU) metallized Kevlar composite (×350, 35° angle of observation). In the center the material has been ablated down to the anticorrosive primer PU.

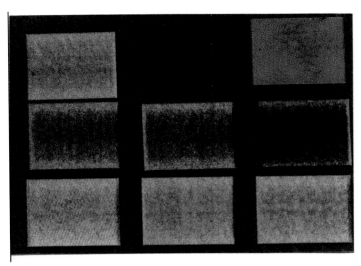

FIGURE 7.19 (See color insert.) Painted metallized Kevlar composite paint stripped down to eight different depths.

Finally, Figure 7.20 shows the result of paint stripping a 26 × 10 cm nonmetallized Kevlar composite — painted with the same PU paint system — down to the primer. In this particular example, the primer had a thickness of about 10 μm, showing the possibility of uncovering it without affecting the composite fibers.

Paint stripping investigations for industrial applications have provided the knowledge for the successful use of excimer lasers in the removal of superficial aged resins. Both types of applications require a delicate treatment owing to the sensitive

FIGURE 7.20 Painted nonmetallized Kevlar composite before (a) and after (b) paint stripping.

substrates. The main difference between these applications is the structure of the surface layers to be removed. In the case of painted composites, the surface layers are discrete and have a predetermined thickness, making the application of laser ablation somewhat straightforward. In the case of aged resins applied on paintings, on the other hand, the ablation process becomes much more complicated than in composites, since the thickness, the concentration, and the interface of the surface resins vary widely across the painting surface.

7.8 ER:YAG LASER IN THE REMOVAL OF SURFACE LAYERS

In the previous sections of this chapter we referred to laser ablation in terms of UV short (several nanoseconds or shorter) laser pulses that are able to photodissociate the covalent bonds of a polymeric matrix and remove molecular fragments from the surface. This dry process allows for a high depth resolution, i.e., in the order of submicrometers (see Figure 7.2, Figure 7.3, and Figure 7.4), owing to the strong absorption of UV radiation by polymeric matrixes. An alternative cleaning method that also relies on the strong absorption of laser radiation by the surface layers, primarily those containing OH bonds, is using the Er:YAG laser. The Er:YAG laser emits a train of pulses at 2.94 μm. The duration of each pulse is of the order of 1 to 2 μsec, while the whole train has a duration of about 250 μsec. The 2.94 μm wavelength is strongly absorbed by water, alcohols, and other OH-containing molecules that are present or sprayed onto the surface prior to laser cleaning. The ablation mechanisms operating at this wavelength are considerably different from those in nanosecond UV irradiation: direct, mostly electronic, excitation followed by dissociation [26–34] for the latter; explosive vaporization of the OH-containing molecules followed by parallel ejection of the contaminant species for the former [14,66]. Another important difference between the two processes is that UV irradiation is a clearly dry method, while Er:YAG laser usually requires the application of a liquid. Apart from the possible advantages and disadvantages of the one method over the other, wetting may prove to be a crucial consideration factor when choosing which method to use in various conservation problems.

Here let us consider more closely the energy deposition and thermal diffusion phases in Er:YAG laser cleaning. The pattern of energy deposition in the material depends on absorption and scattering. Both of these vary in each material type with wavelength, and together they define the limit for minimizing the volume of energy deposition and the following thermal diffusion. In the Er:YAG laser process at 2.94 µm, the important absorber as mentioned above is the OH bond in water or other molecules applied on the surface. Its absorption and scattering properties determine the pattern of energy deposition [67]. Pure water has an absorption peak at 2.94 µrn and an absorption coefficient of 10^4 cm^{-1}, corresponding to an absorption length of 1 µm [14]. At this wavelength, and for the absorption depths in question, scattering is insignificant. Thus the pattern of energy deposition is determined by the distribution of molecules containing OH bonds.

As explained in Chapter 6, the removal of superficial layers and at the same time the minimization of thermal damage to adjacent layers is limited by the material absorption characteristics, the thermal conductivity, and the time for the heat diffusion in the surface layers. The input energy needed for the phase change from water to steam without appreciable thermal diffusion into adjacent materials is reached only when the pulse length is shorter than the thermal relaxation time [14]. For the specific pulse duration of the Er:YAG laser the thermal relaxation criterion is essentially satisfied. Accordingly, an efficient ablation with the least surrounding thermal damage could be expected to be approached by a pulsed Er:YAG laser. With an exposure duration of 1 to 2 µsec (each pulse in the train), the absorbed energy available for heating will be essentially confined almost entirely in the volume where it is absorbed (in the top few micrometers) long enough to heat and vaporize just the desired portion of the material. This is in reality the case when the OH-containing liquid homogeneously penetrates into the surface layers. On the contrary, if this does not happen (e.g., owing to inhomogeneity across the surface, nonabsorbing or hydrophobic surface, vaporization by previous pulses, etc.), it is possible that the heated volume is much larger than a few microns.

Testing and evaluation of Er:YAG laser cleaning [68] revealed different results depending on the type of painted surface. Working on laboratory models, some drawbacks were noticed, probably due to the incomplete drying of some organic materials. Generally, the results were much better when working on old paintings. It was found that Er:YAG laser cleaning can be considered effective, selective, and safe when used at proper fluence levels, which are fixed for each category of surface material. Another important parameter was found to be the wetting of the surface with an OH-containing liquid, usually a dilution of alcohols or glycols. No significant chemical change in the surface materials was observed during testing. Finally, it was realized that a combined method (laser exposure followed by solvent or scalpel cleaning) facilitates the removal of hard and resistant overpaints and old varnishes owing to physical transformations induced by laser. The Er:YAG laser pulses cause the ablation of a portion of the top-layer components and disaggregate the remaining part, which can then be easily treated using mechanical tools and mild solvents. Er:YAG laser systems for the cleaning of painted surfaces, incorporating articulated arms for laser beam manipulation, are commercially available nowadays and easy to use.

7.9 CONCLUSIONS

It has been shown that, using a short-pulse UV laser or a long-pulse Er:YAG laser, it is possible to clean delicate substrates such as paintings or composite structures of polymeric layers (e.g., networks of polymerized resin or polymeric paints, respectively) when bearing in mind some fundamental criteria:

To achieve this,

1. A window of optimal laser parameters should be defined depending on the materials involved, for which any thermal photochemical or photomechanical effects are confined in the volume of the polymerized network and away from its interface with the delicate underlying layer.
2. In the case of UV cleaning, the surface layer nearest to the delicate original substrate — usually just a few micrometers thick — must not be removed, as it acts as a blocking filter (or attenuator) for the UV.
3. In certain demanding cases, on-line and *in situ* monitoring techniques such as LIBS or multispectral imaging or other diagnostic methods must be used for controlling the optimum cleaning depth.

REFERENCES

1. Asmus, J.F., More light for art conservation, *IEEE Circuits and Devices,* March: p. 6, 1986.
2. Watkins, K.G., Larson, J.H., Emmony, D.C., and Steen, W.M., Laser cleaning in art restoration: a review, in *Proceedings of the NATO Advanced Study Institute on Laser Processing: Surface Treatment and Film Deposition,* Kluwer Academic, 1995, p. 907.
3. Cooper, M., *Cleaning in Conservation: An Introduction,* Butterworth Heinemann, Oxford, 1998.
4. Maravelaki, P.V., Zafiropulos, V., Kylikoglou, V., Kalaitzaki, M., and Fotakis, C., Laser induced breakdown spectroscopy as a diagnostic technique for the laser cleaning of marble, *Spectrochim. Acta B,* **52**, 41, 1997.
5. Maravelaki-Kalaitzaki, P., Zafiropulos, V., and Fotakis, C., Excimer laser cleaning of encrustation on pentelic marble: procedure and evaluation of the effects, *Appl. Surf. Sci.,* **148**, 92, 1999.
6. Gobernado-Mitre, I., Prieto, A.C., Zafiropulos, V., Spetsidou, Y., and Fotakis, C., On-line monitoring of laser cleaning of limestone by laser induced breakdown spectroscopy, *Appl. Spectrosc.,* **51**, 1125, 1997.
7. Klein, S., Stratoudalsi, T., Marakis, Y., Zafiropulos, V., and Dickmann, K., Comparative study of different wavelengths from IR to UV applied to clean sandstone, *Appl. Surf. Sci.,* **157**, 1, 2000.
8. Fotakis, C., Zafiropulos, V., Anglos, D., Georgiou, S., Maravelaki, P.V., Fostiridou, A., and Doulgeridis, M., *Lasers in Art Conservation. The Interface between Science and Conservation,* S. Bradley, Ed., Trustees of the British Museum, 1997, p. 83.
9. Georgiou, S., Zafiropulos, V., Anglos, D., Balas, C., Tornari, V., and Fotakis, C., Excimer laser restoration of painted artworks: procedures, mechanisms and effects, *Appl. Surf. Sci.,* **127–129**, 738, 1998.
10. Zafiropulos, V., Galyfianali, A., Boyatzis, S., Fostiridal, A., and Ioakimoglou, E., UV-laser Ablation of Polymerized Resin Layers and Possible Oxidation Process in Oil-Based Painting Media. Optics and Lasers in Biomedicine and Culture — Series of the International Society on Optics Within Life Sciences, G.V. Bally, Ed., Vol. V, Springer-Verlag, Berlin, 2000, p. 115.

11. Zafiropulos, V., Manousaki, A., Kaminari, A., and Boyatzis, S., Laser ablation of aged resin layers: a means of uncovering the scalar degree of aging, in ROMOPTO 2000: Sixth Conference on Optics, SPIE, **4430**, 181, Washington, 2001.

12. Zergioti, I., Petrakis, A., Zafiropulos, V., Fotakis, C., Fostiridou, A., and Doulgeridis, M., Laser applications in painting conservation, in *Restauratorenblaetter, Sonderband — Lacona I*, Verlag Mayer, Vienna, 1997, p. 57.

13. Zafiropulos, V. and Fotakis, C., Lasers in the conservation of painted artworks, in *Laser Cleaning in Conservation: An Introduction,* Cooper, M., Ed., Butterworth Heineman, Oxford, 1998, chap. 6, p. 79.

14. De Cruz, A., Wolbarsht, M.L., and Hauger, S.A., Laser removal of contaminants from painted surfaces, *J. Cult. Heritage,* **1**, S173, 2000.

15. Koenig, E. and Kautek, W., Eds., Proceedings of the First International Conference LACONA I — Lasers in the Conservation of Artworks, October 4–6, 1995, Heraklion, Crete, Greece, Restauratorenblaetter, Sonderband, Verlag Mayer, 1997.

16. Salimbeni, R. and Bonsanti, G., Eds., Proceedings of the International Conference LACONA III — Lasers in the Conservation of Artworks, April 26–29, 1999, Florence, *J. Cult. Heritage,* 1-Suppl. 1, 2000.

17. Vergès-Belmin, V., Ed., Proceedings of the International Conference LACONA IV, Lasers in the Conservation of Artworks, September 11–14, 2001, Paris, *J. Cult. Heritage,* **4**-Supp. 1, 2003.

18. Dickmann, K., Fotakis, C., and Asmus, J.F., Eds., Proceedings of LACONA V, Lasers in the Conservation of Artworks, September 15–18, 2003, Osnabrueck, Germany, Springer Proceedings in Physics, **100**, Springer-Verlag, Berlin, Heidelberg, 2005.

19. Klein, S., Stratoudaki, T., Zafiropulos, V., Hildenhagen, J., Dickmann, K., and Lehmkuhl, T., Laser-induced breakdown spectroscopy for on-line control of laser cleaning of sandstone and stained glass, *Appl. Phys. A,* **69**, 441, 1999.

20. Zafiropulos, V., Laser ablation in cleaning of artworks, in *Laser Cleaning,* Luk'yanchuk, B., Ed., World Scientific, Singapore, 2002, Chapter 8, p. 343.

21. Scholten, J.H., Teule, J.M., Zafiropulos, V., and Heeren, R.M.A., Controlled Laser cleaning of painted artworks using accurate beam manipulation and on-line LIBS-detection, *J. Cult. Heritage,* **1**, S215, 2000.

22. McGlinchey, C., Stringari, C., Pratt, E., Abraham, M., Melessanaki, K., Zafiropulos, V., Anglos, D., Pouli, P., and Fotakis, C., Evaluating the effectiveness of lasers for the removal of overpaint from a 20th C minimalist painting, Lasers in the Conservation of Artworks, Dickman, K., Fotakis, C., and Asmus, J.F., Eds., Proceedings of LACONA V, September 15–18, 2003, Osnabrueck, Germany, Springer Proceedings in Physics, **100**, Springer-Verlag, Berlin, Heidelberg, 2005, p. 209.

23. Fogarassy, E. and Lazare, S., *Laser Ablation of Electronic Materials — Basic Mechanisms and Applications,* European Materials Research Society Monographs, North-Holland Elsevier, Amsterdam, 1992.

24. Dyer, P.E., *Photochemical Processing of Electronic Materials,* Academic Press, London, 1992.

25. Miller, J.C., *Laser Ablation — Principles and Applications,* Springer Series in Materials Science, **28**, Springer-Verlag, Berlin, 1994.

26. Bäuerle, D., *Laser Processing and Chemistry,* Springer-Verlag, Berlin, 2000.

27. Luk'yanchuk, B., *Laser Cleaning,* World Scientific, New Jersey, London, Singapore, Hong Kong, 2002

28. Srinivasan, R. and Braren, B., Ultraviolet laser ablation of organic polymers, *Chem. Rev.,* **89**, 1303, 1989.

29. Srinivasan, R., Ablation of polyimide (kapton) films by pulsed (ns) ultraviolet and infrared (9.17 μm) lasers; a comparative study, *Appl. Phys. A,* **56**, 417, 1993.
30. Srinivasan, R., Interaction of laser radiation with organic polymers, in Laser Ablation-Principles and Applications, Springer Series in Materials Science **28**, Miller, J.C., Ed., 1994, p.107.
31. Luk'yanchuk, B., Bityurin, N., Anisimov, S., and Bäuerle, D., The role of excited species in UV-laser materials ablation, Part I, Photophysical ablation of organic polymers, *Appl. Phys. A,* **57**, 367, 1993.
32. Luk'yanchuk, B., Bityurin, N., Anisimov, S., and Bäuerle, D., The role of excited species in UV-laser materials ablation, Part II, The stability of the ablation front, *Appl. Phys. A,* **57**, 449, 1993.
33. Luk'yanchuk, B., Bityurin, N., Anisimov, S., and Bäuerle, D., Photophysical ablation of organic polymers, Excimer Lasers, 1994 (L.D. Laude, NATO ASI Ser. E).
34. Luk'yanchuk, B., Bityurin, N., Anisimov, S., Arnold, N., and Bäuerle, D., The role of excited species in ultraviolet-laser materials ablation, III. Non-stationary ablation of organic polymers, *Appl. Phys. A,* **62**, 397, 1996.
35. Athanassiou, A., Lassithiotaki, M., Anglos, D., Georgiou, S., and Fotakis, C., A comparative study of the photochemical modifications effected in the UV laser ablation of doped polymer substrates, *Appl. Surf. Sci.,* **154–155**, 89, 2000.
36. Theodorakopoulos, C., Ph.D. thesis, Royal College of Arts, London, 2005.
37. Theodorakopoulos, C. and Zafiropulos, V., Uncovering of scalar oxidation within naturally aged varnish layers, *J. Cult. Heritage,* **4**, 216s, 2003.
38. Zumbóhl, S., Knochenmuss, R., Wólfert, S., Dubois, F., Dale, M.J., and Zenobi, R., A graphite-assisted laser desorption/ionisation study of light-induced ageing in triterpene dammar and mastic varnishes. *Anal. Chem.,* **70**, 707, 1998.
39. Van der Doelen, G.A., Van den Berg, K.J., and Boon, J.J., A comparrison of weatherometer aged dammar varnish and aged varnishes from paintings, in *Art Chimie: La Couleur: Actes du Congres,* Goupy, J. and Mohen, J.-P., Eds., CNRS Editions, Paris, 2000, p. 146.
40. De la Rie, E.R., Photochemical and thermal degradation of films of dammar resin, *Stud. Conserv.,* **33**, 53, 1988.
41. Feller, R.L., *Accelerated Aging: Photochemical and Thermal Aspects,* Berland, D., Ed., Getty Conservation Institute, Los Angeles, 1994.
42. Dietermann, P., Towards more stable natural resin varnishes for painintgs, Ph.D. thesis, Swiss Federal Institute of Technology, Zurich, 2003.
43. Van der Doelen, G.A., Molecular studies of fresh and aged triterpenoid varnishes, Ph.D. thesis, University of Amsterdam, 1999.
44. Scalarone, D., van der Horst, J., Boon, J.J., and Chiantore, O., Direct-temperature mass spectrometric detection of volatile terpenoids and natural terpenoid polymers in fresh and artificially aged resins, *J. Mass Spectrom.,* **38**, 607, 2003.
45. Zafiropulos, V., Petrakis, J., and Fotakis, C., Photoablation of polyurethane films using UV laser pulses, *Opt. Quantum Electron.,* **27**, 1359, 1995.
46. Pettit, G.H. and Sauerbrey, R., Pulsed ultraviolet laser ablation, *Appl. Phys. A,* **56**, 51, 1993.
47. Tornari, V., Fantidou, D., Zafiropulos, V., Vainos, N.A., and Fotakis, C., Photomechanical effects of laser cleaning: a long-term non-destructive holographic interferometric investigation on painted artworks, in Third International Conference on Vibration Measurements by Laser Techniques: Advances and Applications, SPIE, **3411**, International Society for Optical Engineering, Washington, 1998, p. 420.

48. Pettit, G.H., Ediger, M.N., Hahn, D.W., Brinson, B.E., and Sauerbrey, R., Transmission of polyimide during pulsed ultraviolet laser irradiation, *Appl. Phys. A*, **58**, 573, 1994.

49. De la Rie, E.R., The Influence of varnishes on the appearance of paintings, *Stud. Conserv.*, **32**, 1, 1987.

50. Zafiropulos, V., Stratoudaki, T., Manousaki, A., Melesanaki, K., and Orial, G., Discoloration of pigments induced by laser irradiation, *Surf. Eng.*, **17**, 249, 2001.

51. Luk'yanchuk, B.S. and Zafiropulos, V., The model for discoloration effect in pigments at cleaning of artworks by laser ablation, in Second International Symposium on Laser Precision Microfabrication, SPIE, International Society for Optical Engineering, **4426**, 326 Washington, 2002.

52. Castillejo, M., Martin, M., Oujja, M., Silva, D., Torres, R., Manousaki, A., Zafiropulos, V., van den Brink, O.F., Heeren, R.M.A., Teule, R., Silva, A., and Gouveia, H., Analytical study of the chemical and physical changes induced by KrF laser cleaning of tempera paints, *Anal. Chem.*, **74**, 4662, 2002.

53. Mills, J.S. and White, R., *The Organic Chemistry of Museum Objects*, Butterworths, London, 1987, and references therein.

54. De la Rie, E.R., Old master paintings: a study of the varnish problem, *Anal. Chem.*, **61**, 1228, 1989.

55. Theodorakopoulos, C., Zafiropulos, V., Fotakis, C., Boon, J.J., Horst, J.V.D., Dickmann, K., and Knapp, D., A study on the oxidative gradient of aged traditional triterpenoid resins using "optimum" photoablation parameters, Lasers in the Conservation of Artworks, Dickmann, K., Fotakis, C., and Asmus, J.F., Eds., in Proceedings of LACONA V, September 15–18, 2003, Osnabrueck, Germany, Springer Proceedings in Physics, **100**, Springer-Verlag, Berlin, Heidelberg, 2005, p. 255.

56. Stout, G.L., *The Care of Pictures*, Dover, New York, 1975.

57. Sutherland, K., Solvent extractable components of oil paint films, Ph.D. thesis, University of Amsterdam, 2001.

58. Lentjes, M., Klomp, D., and Dickmann, K., Sensor concept for controlled laser cleaning via photodiode, Lasers in the Conservation of Artworks, Dickmann, K., Fotakis, C., and Asmus, J.F., Eds., in Proceedings of LACONA V, September 15–18, 2003, Osnabrueck, Germany, Springer Proceedings in Physics, **100**, Springer-Verlag, Berlin, Heidelberg, 2005, p. 427.

59. Chryssoulakis, Y., Sotiropoulou, S., and Zafiropulos, V., Unpublished results, 1996.

60. Theodorakopoulos, C. and Zafiropulos, V., Laser Cleaning applications for religious objects, *Eur. J. Sci. Theol.*, **1**, 63, 2005.

61. Balas, C., An imaging colorimeter for the non contact color mapping, *IEEE Trans. Biomed. Eng.*, **44**, 468, 1997.

62. Teule, R., Scholten, H., van den Brink, O.F., Heeren, R.M.A., Zafiropulos, V., Hesterman, R., Castillejo, M., Martin, M., Ullenius, U., Larsson, I., Guerra-Librero, F., Silva, A., Gouveia, H., and Albuquerque, M.-B., Controlled UV laser cleaning of painted artworks: a systematic effect study on egg tempera paint samples, *J. Cult. Heritage*, **4**, 209s, 2003.

63. Stolow, N., Part II: Solvent action, in *On Picture Varnishes and Their Solvents*, Jones, E.H., Ed., Case Western Reserve University, Cleveland, Ohio, 1985.

64. Manz, C. and Zafiropulos, V., Laser ablation strips thin layers of paint, *Opto and Laser Eur. (OLE)*, **45**, 27, 1997.

65. Pantelakis, S., Despotopoulos, A., Lentzos, G., and Germanidis, T., Influence of the laser paint stripping process on the mechanical behaviour of fibre reinforced composites,

in Proceedings of the Fourth European Conference on Advanced Materials and Processes, 25–28 Sept., Padua/Venice, Italy, 1995, p. 345.

66. DeCruz, A., Hauger, S., and Wolbarsht, M.L., The role of lasers in fine arts conservation and restoration, *Opt. Photon News,* **10**, 36, 1999.

67. Wolbarsht, M.L., Laser surgery: CO_2 or HF, *IEEE J. Quant. Electron. QE,* **20**, 1427, 1984.

68. Bracco, P., Lanterna, G., Matteini, M., Nakahara, K., Sartiani, O., de Cruz, A., Wolbarsht, M.L., Adamkiewicz, E., and Colombini, M.P., Er:YAG laser: an innovative tool for controlled cleaning of old paintings: testing and evaluation, *J. Cult. Heritage,* **4**, 202s, 2003.

8 Laser Cleaning of Encrustations

8.1 INTRODUCTION

In the previous chapter, we described an application of laser-based ablation, namely, removal of mainly organic materials from a complex polymeric matrix, i.e., the surface of painted artifacts. Lasers have also been used for the removal of inorganic encrustations from marble sculptures and stonework [1–34]. Laser ablation mechanisms of inorganic surfaces differ from photoablation of polymerized material, with the main characteristic of the former being the self-limiting nature of the laser-assisted process. Current trends consider the need for laser-based approaches that may be complementary to or even replace some of the traditional methods used in art conservation. Extensive literature on laser-based conservation techniques exists in the LACONA conference proceedings [35–38].

The origins of the laser cleaning of stonework can be traced back to 1972, when the Italian Petroleum Institute funded John Asmus from the University of California, San Diego, to study laser holography for the recording of the decaying treasures of Venice. During this work, Asmus was asked to observe the effect of a focused ruby laser beam on an encrusted marble statue. It was found that the darker encrustations were selectively removed from the surface, resulting in no apparent damage to the underlying marble. Asmus returned to the United States and began researching the possibility of laser cleaning of artworks. His investigations laid the basis for several powerful techniques, particularly in using pulsed ruby and pulsed Nd:YAG lasers. Interestingly, the connection between the use of lasers in cleaning and in holographic recording of artworks remains, with the two technologies developing in parallel on diverse applications in the conservation field.

Laser cleaning is a physical process applied when there is a limited interaction with the substrate material. Thus the procedure has distinct advantages over traditional cleaning methods based on chemical and mechanical action. Such conventional techniques involve mainly particle/air abrasion techniques, chemical treatments, or direct mechanical contact using a knife or scalpel. In short, laser cleaning, properly done, has the following advantages over the traditional methods:

- It is a selective process that is entirely operator controlled for the precise removal of specific and diverse surface encrustations.
- It is a noncontact process.
- It is a photophysical and photoacoustic process that lasts much shorter than the time between adjacent laser pulses.

- It is a process that can preserve the underlying surface relief.
- It is quantitative, since a predetermined material thickness can be removed per laser pulse.
- It is versatile, since a wide variety of encrustation types can be removed by proper selection of the laser operating parameters.
- It is environmentally safe, since no fluid wastes are produced and no organic solvents are used.

Recently, lasers have proven to be very successful in the removal of submicron particles in the semiconductor industry [39–41]. The industrial sector has been motivated to evaluate laser techniques because of potential cost savings, yield enhancements, and environmental concerns. The successes of such techniques have brought about additional interest in using lasers in artworks conservation.

Laser cleaning of artworks and antiquities provides the opportunity for the selective removal of undesired surface materials. As in the case of removal of organic surface layers, it is possible, in principle, selectively to remove inorganic encrustations and uncover the original stone surface. It turns out though that the original inorganic surface has already been altered somewhat during the initial stage of encrustation formation, so the situation that the conservator faces is usually not simple. There are many different types of encrustation originating from a variety of mechanisms and environmental parameters. It is not the purpose of this book to discuss this issue. The interested reader may find relevant information in the literature (e.g., see [12,42,43]). Care is needed to avoid damage to the underlying substrate when removing contaminants. As in the case of organic and polymerized layers, the successful implementation of laser intervention relies on close collaboration of conservators with scientists to define and implement the most appropriate method and an acceptable process end point.

Pulsed Nd:YAG laser systems have two primary operating regimes, Q-switching and free-running. Q-switched Nd:YAG lasers have a pulse width in the 5 to 25 nsec range; free-running (or normal mode) Nd:YAG lasers have a pulse width (actually an envelope of many pulses) in the 100 to 200 μsec range. Salimbeni, Siano, and coworkers [28,44] have placed emphasis in laser cleaning of artifacts using a variable Q-switched pulse width ranging from about 175 to 1800 nanoseconds, thereby adjusting (and perhaps optimizing) the laser–matter interaction time on different materials. A Nd:YAG laser with variable pulse duration has been devised and constructed by means of interchangeable optical fibers in the resonator. Its output parameters demonstrate the possibility of optimizing the cleaning procedure for a wide variety of materials.

In general, it is important to continue the investigations on laser cleaning techniques that will not only improve the cleaning efficiency of artworks but also ensure the absence of any potential damages of the underlying original media. However, this by no means suggests that mechanical and/or chemical cleaning is safer than laser cleaning, since the acquired know-how on the latter is more than sufficient for the safe implementation of the proposed laser-based methodology.

Case studies have been presented from numerous countries on a wide range of substrates: marble, limestone, sandstone, terracotta, ivory, alabaster, plaster, gilding, and so forth, e.g. see [5, 35–38]. An important point that should not be overlooked is that the laser is potentially just one of many tools used by the conservator during the cleaning process. It is very often the case that the best cleaning results are obtained by combining laser cleaning with more traditional methods such as poultice, chemical, and micro-air abrasive cleaning. It is also sometimes the case that conventional laser cleaning at 1064 nm is ineffective or inappropriate for various reasons. In such cases, a different type of laser, an alternative method, or even a combination of methods may provide an acceptable solution. Investigations usually study the effect of wavelength, pulse length, fluence, and environment on the cleaning process. For instance, laser radiation at 355 nm appears well suited for removal of biological material or, when combined with 1064 nm, of particular inorganic encrustations. On the other hand, a pulse length of a few tens of microseconds appears to be advantageous for cleaning extremely fragile surfaces, and large-scale systems delivering >100 W average power may help make automated laser cleaning of buildings a reality. In any case, the close collaboration between scientists and conservators has improved our understanding of the advantages and limitations of laser cleaning and helped to provide viable solutions to previously intractable problems by using diverse forms of laser radiation in different ways.

One of the concerns of the conservation community is the high cost of a laser intervention. Cleaning is just one part of the conservation process, and in many cases the high cost of laser cleaning cannot be justified. The use of lasers in conservation has undoubtedly grown markedly during recent years, but it remains a major challenge for scientists and laser companies to make laser systems more accessible to the conservation community.

This chapter has been further divided into three parts, starting with the presentation of the scientific evidence for major operative mechanisms and associated optical phenomena. A selection of test case studies follows. Finally, we draw some general conclusions concerning the prospects and limitations of the laser-assisted removal of inorganic encrustation from marble and stone artifacts. By no means does this presentation cover the whole existing literature, which is extensive and diverse. On the contrary, the goal here is to present the main issues and entrigger the interested reader into further reading and searching.

8.2 MAJOR OPERATIVE MECHANISMS AND ASSOCIATED OPTICAL PHENOMENA

In pulsed UV laser ablation of complex layers of various polymeric phases, the dominant mechanism is the direct chemical bond breakage within the cross-linked macromolecules and the subsequent ejection of mostly small molecular fragments. As explained in Chapter 6, in the nanosecond laser removal of inorganic encrustations, the prevailing processes are (1) explosive vaporization, a phenomenon that is due to the rapid increase of temperature, and (2) spallation, a mechanism where layers of a few microns are ejected away from the surface by the generated shock wave. A mathematical description of the former has been given by Asmus et al. [1] and Watkins [45,46], and while the mechanics of shock wave action cannot be entirely quantified, the qualitative description is adequate [25,47,48]. This is because of the many uncontrolled variables due to the complex nature of the encrustation and the characteristics of the substrate interface that makes modeling difficult. Here we shall not attempt a detailed analysis of the ablation mechanisms prevailing in these complex surfaces. Rather, we shall elucidate the key features for laser cleaning by discussing successful remedies for undesirable effects that may appear.

Removal of dark encrustation from stone artifacts using a Q-switched Nd:YAG laser ($\lambda_L = 1064$ nm) has been done in conservation for many years [1–14,19,31]. Other pulse durations and wavelengths [1,15–18,20–30,33,47,49–50] have also been used. Actually, laser-assisted cleaning of stone sculptures and façades is by far the most widely implemented application in the field of lasers in the conservation of artworks. This is because of the so-called self-limiting removal process, i.e., the spallation threshold fluence for the encrustation is usually much lower than for the stone substrate. Thus by choosing proper fluence levels between the two thresholds, one can safely remove the encrustation without affecting the substrate. Shock wave formation, however, may induce some structural damage (e.g., microcracks) when excess fluence levels are used [1] or, for example, may lead to the ejection of underlying layers when using inappropriate exposure parameters [29].

For a better understanding of the self-limiting effect, let us consider some quantitative measurements. Figure 8.1 shows the ablation rate data for pentelic marble and for a thick dendritic encrustation [19]. In Figure 8.1a the two sets of data clearly show the spallation thresholds for marble and gypsum on the one hand and for a very hard type of encrustation such as the dendritic encrustation, formed by recrystallized stone material [19], on the other. In both cases a Q-switched Nd:YAG laser at 1064 nm was used. The two thresholds are designated T1 and T2. A fluence as low as 1.9 J/cm² (T2) will ablate the encrustation, while T1 (3.5 J/cm²) is the maximum safe fluence for cleaning this particular surface. So the difference between the fluence required to remove the crust and the fluence that may damage the substrate is quite large. Therefore, for this particular combination of encrustation and substrate, there is a self-limiting effect in using a Q-switched Nd:YAG laser. In other words, at optimum fluence levels, the cleaning proceeds in a controlled way until the encrustation is removed and then stops, i.e., it is self-limited, when the substrate is reached. In contrast, the corresponding fluence margin for a KrF excimer laser at 248 nm (based on T4 and T5 in Figure 8.1b) is unfavorable for cleaning. On the other hand, note that the third harmonic of a Q-switched Nd:YAG laser ($\lambda_L = 355$ nm) offers a substantial threshold margin *in this particular case* (T6 and T7 in Figure 8.1b). This issue will be discussed further in this chapter. It should be noted, however, that such a large threshold margin is not common in practice (e.g., in sandstone the corresponding margin is only a fraction of a J/cm² [15]) and depends strongly on the combination of materials involved, i.e., substrate/encrustation.

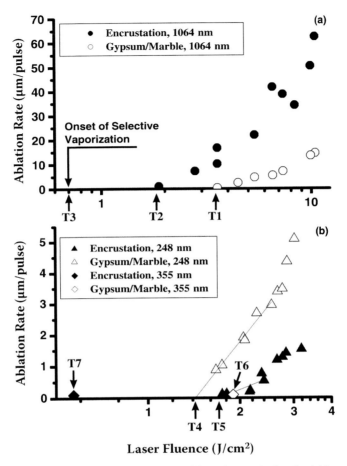

FIGURE 8.1 Ablation rate data for pentelic marble and a particular dendritic encrustation, a cross section of which is shown in Figure 8.2, obtained using a Q-switched Nd:YAG laser (a) and a KrF excimer laser and the 3rd harmonic of a Q-switched Nd: YAG laser (b). Arrows designate the various thresholds.

In general, it has been concluded [1] that normal mode pulses (~10^{-4} sec duration) clean by selective vaporization, whereas Q-switched pulses (~10^{-8} sec duration) clean mainly through spallation. In the latter case though, selective vaporization of highly absorbing (dark) particles can start at very low fluence values (see threshold T3 in Figure 8.1a), i.e., at fluences well below the spallation threshold.

At this point, a general consideration of the encrustation layer stratigraphy is appropriate. The crust thickness may be as thin as tens of microns to as thick as many millimeters. It is made up of various natural and man-made pollutants, atmospheric and biological constituents, coatings, and chemically reactive by-products bound together into a compact, amorphous black crust. From previous studies (e.g., [12,51,52]), it has been determined that there is almost always a thin section of the innermost crust located just on top of the marble/stone substrate, which should be

For comparison of the presented ablation rate data with those found in the literature, it should be considered that all fluence values presented here are obtained by dividing the laser pulse energy by the area of the spot on a PVC card (not photosensitive material) placed on top of the substrate. The choice of PVC over a common photosensitive material (e.g., a black photographic card) is documented by the size of the corresponding spot on the stone, which is the same as the size on the PVC card and certainly smaller than the corresponding size on a photosensitive material. Therefore someone can use ablation rate data to calculate processing rates in pulses per m^2 and thus in m^2 per hour (based on the working repetition rate). This type of calculation is problematic if the fluence measurement and the actual ablation refer to two different spot sizes, which is the case in many reports. Also the ablation rates refer to the mean ablation rates, averaged over the entire spot profile (see relative discussion in Chapter 7).

preserved. In other words, it is the outermost original surface that should be preserved, while all the other layers above this crust must be removed. In general, the resulting surface after cleaning at low fluence with a Q-switched Nd:YAG laser at 1064 nm has some unique characteristics. The first characteristic, which appears in certain cases, is a yellowish hue when compared to the uncleaned black crust. This impression might be misleading however, when comparing a laser cleaned with a reference surface. The puzzle that some conservators face, though, is that this yellowing is often not seen after cleaning with other wavelengths or pulse durations or after wetting the surface before laser cleaning, or even after using conventional methods such as microsandblasting. Another characteristic of the substrates after laser cleaning at 1064 nm is the remaining of the gypsum-rich outermost original surface layer while the dark particulates from the encrustation have been selectively vaporized. The preservation of this beneficial layer is the major advantage of Q-switched Nd:YAG laser cleaning of stone. Even in the cases of more complex encrustation, where the outermost black crust (rich in carbon particles and other pollutants) is completely removed, the remaining gypsum-rich underlying crust is left free of dark particles. Therefore we could consider such a layer as the usual outcome at the last stage of laser cleaning. A cross section of a 200 μm thick gypsum-rich encrustation on marble is shown in Figure 8.2a. At a macroscopic observation the impression of this encrustation is grey-black. Note the resemblance of such an encrustation with others appearing in the literature, for instance see Plate 5.4 in Reference [12]; the only difference is the non-appearance of a pure gypsum sublayer in Figure 8.2a, probably because the large scale does not allow a layer of a few microns (in this case) to be distinguished. Figure 8.2b shows a cross section at another point of the same sample after laser cleaning at 1064 nm at a fluence of 0.8 J/cm^2, which is just above the threshold of selective vaporization of the dark particles (shown in Figure 8.1a at 0.7 J/cm^2). In Figure 8.2b it is shown that the gypsum-rich layer is preserved, while the dark particles mainly consisting of carbon, iron, aluminum, and silicon [20,51] have been selectively vaporized.

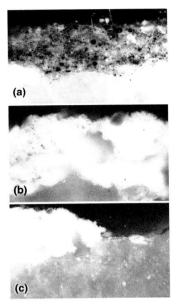

FIGURE 8.2 (See color insert following page 144.) Cross section (×200) of dendritic type of encrustation on pentelic marble: (a) reference encrustation, (b) after selective vaporization of particles using 50 pulses at 0.8 J/cm², and (c) after partial removal of the gypsum rich layer using 50 pulses at 1.2 J/cm² of a Q-switched Nd:YAG laser. The width of each picture corresponds to 500 μm.

In general, there are different mechanisms underlying the laser-assisted removal of encrustation, which may operate simultaneously [46,48]. For example, in the case of a multilayer encrustation, which is often found, the outermost black (usually not gypsum-rich) crust may be removed by simultaneous explosive vaporization and spallation (shock wave action). However, the above observations point instead to selective explosive vaporization of dark particles within the innermost gypsum-rich crust as the main mechanism responsible at the last stage of stone cleaning. This selectivity is due to the higher absorptivity of particles at 1064 nm, which is three to four times higher than for the gypsum or substrate [1]. The onset of this process occurs at a much lower fluence (T3) than the spallation threshold fluence of the composite crust of dark particles and gypsum-rich layer (T2), as shown in Figure 8.1a.

Asmus et al. [1] have applied a one-dimensional surface heating model to predict the fluence level required for selective vaporization. In this simplified model the temperature increase at the surface is found to depend on surface absorbance, incident laser fluence, and material properties. This model was applied mostly to a free-running laser (pulse duration of ~10^{-3} sec), and the results are in good agreement with the experimental fluences. In principle, it can also be applied to nanosecond pulses at fluences below the spallation threshold of the mixed crust, where selective vaporization of particles embedded in the gypsum-rich layer occurs. For the highly variable, inhomogeneous, and poorly characterized encrustations, the model is satisfactory in predicting laser effects and fluence requirements for particular laser cleaning projects.

In this simplified model, the temperature increase at the particle surface as a function of time, $\Delta T(0,t)$, is found to be

$$\Delta T(0,t) = \frac{2\beta I_0}{k}\left[\frac{\kappa t}{\pi}\right]^{1/2} \tag{8.1}$$

where β is the surface absorbance, I_0 is the incident laser flux, k is the thermal conductivity, and κ is the thermal diffusivity [$\kappa = k/(\rho_0 C_V)$]. If we replace t with the pulse duration t_L and $\Delta T(0,t)$ with the temperature increase from ambient temperature to the boiling point T_b, one can obtain the laser flux at which vaporization of different materials takes place. Equation (8.1) can be corrected for nonplanar laser/matter interaction, which is the case in vaporization of small particles. For very small ($<<\lambda$) and optically thin particles, the correction factor to Equation (8.1) is 0.5 [1].

Just to get an idea of the estimate of Equation (8.1) in the specific case presented above, we can assume that carbon particles (graphite) are present in the encrustation. Using properties of carbon ($T_b = 5100$ K, $C_p = 0.71$ J g^{-1} K^{-1}, density $\rho_0 = 2.25$ g cm^{-3}, $k = 1$ J sec^{-1} cm^{-1} K^{-1}, $\beta = 0.97$) and $t_L = 10^{-8}$ sec we find a vaporization threshold for carbon of 0.55 J/cm^2. The experimental fluence level for the onset of particle vaporization in the encrustation of Figure 8.2 is 0.7 J/cm^2 (T3 in Figure 8.1a), close to the calculated value. For fluence levels between 1.0 J/cm^2 and 1.8 J/cm^2, it was found [34] that the generated shock wave had an additive disruptive effect when an excess number of pulses was used. Figure 8.2c shows a cross section at another point of the same sample after irradiation of 50 pulses at 1.2 J/cm^2, where all particles are removed as well as part of the gypsum-rich layer (not an acceptable result). For fluence values greater than 1.8 J/cm^2, spallation of the mixed encrustation started at the first laser pulse.

Summarizing the results of the Q-switched pulses of 1064 nm in the particular case study, we can identify five fluence regimes: (i) below 0.7 J/cm^2, where no action occurs; (ii) between 0.7 and 1.0 J/cm^2, where only selective explosive vaporization of the dark particles occurs; (iii) between 1.0 and 1.8 J/cm^2, where the selective explosive vaporization is accompanied by accumulative stress in the material owing to the shock wave, finally resulting to uncontrolled spallation; (iv) between 1.8 and 3.5 J/cm^2, where the major mechanism is the spallation of the mixed encrustation with no possible direct damage to the underlying marble; and (v) above 3.5 J/cm^2, where spallation of both crust and marble takes place.

Figure 8.3 shows a cross section of the interface between the reference encrustation (left) and the laser-ablated area (right) using the third harmonic of a Q-switched Nd:YAG laser (355 nm) at a fluence of 0.7 J/cm^2 on a dry encrustation a few tens of microns thick. An excess of 50 pulses was chosen to demonstrate the self-limiting spallation process at this laser wavelength as well. Here, selective explosive vaporization is not observed; so the fluence gap between the two spallation thresholds (of mixed encrustation and marble) could be attributed to the loose adherence of the gypsum-rich encrustation on the substrate compared to the good coherence of the calcium calcite crystals. As a result, the gypsum-rich layer is not actually maintained on the marble if an excess number of pulses is used (see Figure 8.3). On the other hand, if the conservator retains control of the number of pulses fired per unit surface area, it is possible to maintain a thin gypsum

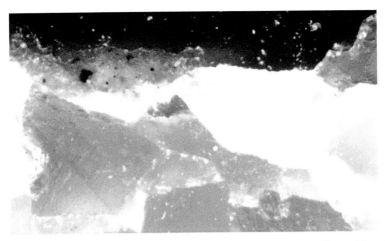

FIGURE 8.3 Cross section (× 200) of dendritic type of encrustation on pentelic marble: reference encrustation (left), after removal of encrustation (right) using 50 pulses of the third harmonic of a Q-switched Nd:YAG laser at 0.7 J/cm². The width of the picture corresponds to 500 μm.

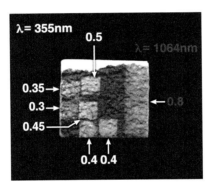

FIGURE 8.4 (See color insert.) Thin encrustation on pentelic marble divested using $\lambda_L = 355$ nm (left) and $\lambda_L = 1064$ nm just above the onset of selective vaporization (right). The numbers represent different fluence values used in units of J/cm².

layer, as can be seen in Figure 8.4. When working with 355 nm pulses, the best results are obtained on a loose soot-type encrustation or a thin compact crust [34,49]. In general, 355 nm alone is not a good choice when the encrustation varies in thickness and/or in coherence, in which case a continuous variation of fluence and number of pulses is needed, which is practically not feasible.

8.2.1 Laser-Induced Yellowing Effect

In certain cases, the application of a Q-switched Nd:YAG laser to remove pollution crusts from stonework has been reported to result in a final surface that gives the impression of discoloration toward yellow [50,53]. For quite some time this has been a drawback for the laser cleaning of stonework, sometimes leading to the use of alternative methods with inferior cleaning results. Although the phenomenon of yellowing is still under investigation (for instance, see References [54,55] presented in LACONA VI), it is now widely accepted that yellowing alone by no means should lead the conservators to condemn laser cleaning.

The yellowing of stone may be due to many different reasons, but it is attributed primarily to the preexisting yellow layers (patina) of the stonework, which is a characteristic of most of the naturally aged stone sculptures. Usually, simple macroscopic and colorimetric observation of stonework is enough for a proper assessment. If this is the case for a specific object, it is reasonable to deduce that 1064 nm pulses simply uncover the preexisting yellowish layers in the best possible way, compared to other wavelengths or even other techniques. This case should not be considered problematic by any means. Other reasons for yellowing may include the presence of residues left after the action of the laser pulse [12,53,56,57], especially ferric compounds; the staining of the original surface due to the migration of the yellowish fraction present in the pollution crust due to polar organic compounds [58]; the presence of water-soluble hydrocarbons [55]; etc. It has been also suggested [53] that yellowing may be an additive phenomenon originating from two or more of the above reasons. Although the above considerations are plausible, in the absence of preexisting yellow patinas or fractions present in the pollution crust (see Section 8.2.2 and Figure 8.10), they cannot explain the yellowing effect. Also based on the above considerations, it is not clear why yellowing is minimal or even absent when other laser wavelengths, pulse durations, or techniques are used. Another hypothesis based on light scattering theory has been suggested [32] that may explain the dependence of yellowing upon various parameters. This hypothesis together with its experimental evidence is described below, emphasizing the mechanistic aspects of laser ablation rather than the questionable importance of the phenomenon itself.

When 1064 nm Q-switched Nd:YAG is used on a surface to vaporize dark particles selectively, the crust color becomes yellowish [5,6,32,53], while using the third harmonic at 355 nm leaves the surface color unchanged (see Section 8.3) or sometimes gives a grey hue. This is shown in Figure 8.4, where both wavelengths have been used. In this example, the encrustation had a thickness of 25 ± 5 μm. Here the threshold of selective vaporization at 1064 nm was 0.7 J/cm^2, while the threshold of spallation at 355 nm was 0.3 J/cm^2 (compared to 0.57 J/cm^2 in the previous example). The lower fluence level is due to the thinner and looser cohesion of this encrustation, which can be easily removed by the generated shock wave. On the right side of Figure 8.4 we can see an area processed at 1064 nm, and on the left side there are small areas processed at 355 nm using different fluences. In the latter case, although there is a varying cleaning effect with fluence, the yellow hue does not appear in the 355 nm processed areas. Measuring the CIE–$L^*a^*b^*$ values of artificially aged marble samples processed with both wavelengths [57], it was found that 1064 nm pulses resulted in a b^* value increase of up to $\Delta b^* = 10.0$ compared to the gypsum reference surface, while almost no change ($\Delta b^* = -0.1$) resulted when using 355 nm pulses. Another series of measurements [24] on a compact black crust (from a corner addition of the Parthenon West Frieze) revealed that the corresponding b^* changes for the two wavelengths were $\Delta b^* = 8.6$ and $\Delta b^* = -0.2$, respectively (considering a reference b^* value of 11.4). The examples presented above as well as the results presented in Section 8.2.2 indicate that laser-induced yellowing is an optical phenomenon like light scattering [32,33].

As a result of the selective vaporization of dark particles, the preserved gypsum-rich portion of the crust becomes more porous than the same portion of the crust before irradiation because of the voids created. Furthermore, the partial dehydration of gypsum to anhydrites and hemihydrates [20,56] decreases the specific volume of remaining crust. When the removed particles are of the order of 10 to 20 microns, the voids in the gypsum-rich crust are visible in the cross section (Figure 8.2b). In such a situation we expect the elemental constituents of the dark particles to remain in the preserved clean crust even though its macroscopic appearance has changed from dark to light colored. This has been verified using LIBS and SEM-EDX analysis on the cleaned surface [19,51]. This is also in accordance with the results of a recent study [57], which found that the 1064 nm laser cleaned surface of an artificially soiled sample still contains iron as in the original layer.

All these observations lead to the concept of a simple light scattering model that consists of an optically thin layer (corresponding to the cleaned gypsum-rich crust) and a virtual top layer treated as a single scattering film. The light transport is diffusive, while the angular distribution of reflectance is influenced mostly by the last scattering event, a process that depends on the scattering anisotropy of these last scattering sites (voids). Although multiple scattering usually occurs, the final path of a ray is governed by the last scattering event. Figure 8.5 simplifies the possible path of two collinear rays of different color, where the higher λ ray finally escapes the surface while the lower λ ray scatters back into the bulk. This results in a longer path length within the encrustation and therefore higher absorption of the latter than the former. This figure schematically shows a representation of this simple model in relation to the two operative cleaning mechanisms of selective vaporization and spallation shown in Figure 8.6.

Following the approach by Popescu [59], the angular distribution of light scattered by scattering centers is related to the integrand of the product of the probability to scatter the light within the elementary solid angle and the angular transmission probability for the diffusive medium. Since the former term is a function of λ and refractive index, the spectral distribution of the light emerging from the sample can be calculated. Using this simplified scattering model, we can calculate the reflectance spectra in the backward direction for different sizes d of scattering voids ($d \gg \lambda$).

The resultant reflectance, monotonic in λ, can be further corrected for the spectral absorption of the diffusive medium by multiplying the scattering

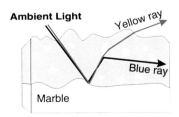

FIGURE 8.5 Possible path of a blue and a yellow ray that have collinearly entered the encrustation layer.

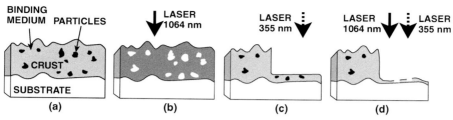

FIGURE 8.6 Schematic representation of encrustation cross section (a) and the two different ablative mechanisms: the selective vaporization using low fluence (0.7–1.0 J/cm²) at $\lambda_L = 1064$ nm (b), and the spallation at higher fluence (~ 2 J/cm²) or using $\lambda_L = 355$ nm (c). (d) represents the result when synchronously using both wavelengths at certain fluence values.

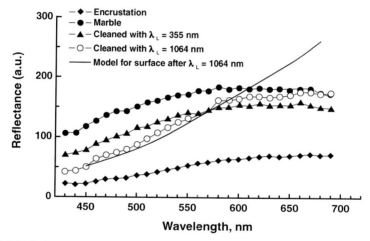

FIGURE 8.7 Reflectance spectra taken from the surface of the sample shown in Figure 8.4, recorded using a hyperspectral imaging apparatus. The error is <10%. The solid line represents the result of calculations for $d = 5$ μm multiplied by the squared transmission of the surface layer.

spectrum with the squared transmission of the thin absorbent layer [32]. The result for $d = 5$ μm is plotted in Figure 8.7 together with the measured reflectance spectra using a prototype hyperspectral imaging apparatus. Although much simplified, to a first approximation this model successfully predicts the measured relative reflectance spectra up to 600 nm. The higher gradient of the reflectance spectrum after laser cleaning at 1064 nm than the corresponding gradient after laser cleaning at 355 nm explains the yellowish hue of the former surface: less blue reflectance means more yellow appearance. In other words, a steep reflectance curve explains an increased yellow hue compared to a less steep reflectance curve.

At this point we should mention the absence of any yellowing when using wet surface or shorter wavelengths or longer pulse durations on a wet surface (see below). In these cases, different ablation mechanisms must be considered. UV (centered at 360 nm) reflection imaging [34] reveals that the black particles and the gypsum-rich crust absorb 355 nm equally. Figure 8.8 shows the UV

FIGURE 8.8 UV reflectance image of the sample presented in Figure 8.4. The window of observation is 320 to 400 nm centered at 360 nm.

reflectance image of the sample of Figure 8.4. It can be seen that the area corresponding to selective vaporization of particles (right side in Figure 8.4) is less apparent with UV imaging in Figure 8.8, owing to the similar UV absorption characteristics of particles and gypsum-rich crust. On the contrary, the areas corresponding to spallation (left in Figure 8.8) have a high reflectance owing to the almost complete removal of both gypsum-rich layer and particles. Therefore, at λ_L = 355 nm, the mechanism of selective vaporization of dark particles is not likely to occur as at λ_L = 1064 nm, owing to equal absorption of the laser photons from particles and gypsum. In such cases, cross sections have revealed the simultaneous and indiscriminate removal of black particles and gypsum-rich layer almost down to the substrate, as previously presented in Figure 8.3. Consequently, we expect the reflectance versus wavelength profile of the surface cleaned by λ_L = 355 nm qualitatively to resemble the profile of the encrusted surface, and even more, to resemble the reflectance profile of the marble, which is actually the case (see the upper two trends in Figure 8.7).

In the case of water-assisted, normal or Q-switched mode Nd:YAG laser cleaning, extensive studies have revealed that the superficial crust can be removed to a chosen depth, e.g., see [22,27,29,60]. In this case, stratigraphic examinations have also shown that controlled thinning of the encrustation occurs, leaving the remaining crust without any visible alteration, voids, or yellowing. A schematic representation of this phenomenon is shown in Figure 8.9, where we assume a gypsum-rich layer with dark particles just before the final pulse is fired. The areas to the left and right of the dashed line represent wet and dry regions, respectively (Figure 8.9, top center). The blue regions around the black particle represent diffused water. After the last pulse is fired (Figure 8.9, top left), the black particles are hot (shown with red color), having absorbed the radiation. In the next step (Figure 8.9, bottom left), the heat diffuses to the surroundings. In the wet area,

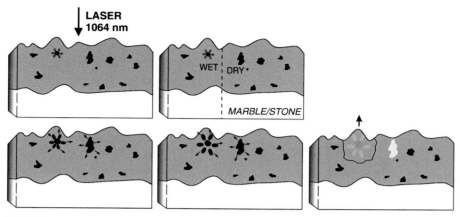

FIGURE 8.9 (See color insert.) Schematic representation of the explosive vaporization of a particle with and without the presence of water.

there is heat transfer from the hot black particle to the surrounding water molecules. When the temperature reaches 100°C, the water vaporizes, first sweeping along the surrounding material (Figure 8.9, bottom center). As a final result, the entire area that surrounds the particle is ejected away, leaving no void behind (Figure 8.9, bottom right). This explains the minimization of yellowing. On the other hand, the dry particle is explosively vaporized when its temperature increases and reaches the boiling point of the particle (limited energy transfer to the surrounding dry material), leaving a void behind.

8.2.2 SYNCHRONOUS USE OF FUNDAMENTAL AND THIRD HARMONIC OF A Q-SWITCHED ND:YAG LASER

In laser cleaning of stonework, the objective is the removal of superficial encrustation in an absolutely controlled way, without structural damage of the substrate and with optimal appearance. As mentioned in Section 8.1, the original inorganic surface has already been altered during the initial stage of encrustation formation. This means that there is not an absolute process end point, so the conservator must weigh up each individual situation, decide what the desired end point is, and finally try to achieve it throughout the intervention. This is not always trivial, especially when important conservation tasks are encountered. Recently, an innovative experimental procedure has been developed [24,33,61] to meet the optimal appearance objective when dealing with variable encrustation on the same relief. This was accomplished using simultaneously two laser beams at different wavelengths, whose pulses were temporally and spatially overlapped. For optimum results their fluences must be adjusted to a certain ratio, which may vary depending upon the particular encrustation type and substrate. A quantitative theoretical model is currently under development.

A Q-switched Nd:YAG laser may — by means of harmonic generation crystals — emit simultaneously in IR and UV wavelengths and thus offer a convenient basis [24,33,61] for exploiting the possibility of combining the two cleaning mechanisms described above. It turns out that the temporal and spatial overlapping of the two laser wavelengths is crucial to the synchronous operation of the different ablation mechanisms. In such an arrangement it is also possible to use either of the two wavelengths independently, should it be required for the specific application. The energy content of each beam can be adjusted independently by using variable attenuators in the individual beam paths. In this way the action of the two material removal mechanisms can be exploited either independently or in combination down to the same depth. Thus specific laser cleaning problems associated with laser-induced yellowing, required appearance, or inhomogeneous and incomplete cleaning can be tackled, e.g., where a biological crust was also present, The theoretical model for discoloration [32] may be extended to include resultant surface color when combining IR and UV laser pulses. The fluences of the two beams can be adjusted so that the two ablation mechanisms — explosive vaporization of dark particles and spallation of the encrustation at the final step of removal — operate to the same depth into the remaining crust. In this case, the observed cross section represented in Figure 8.6d corresponds to little or no color change.

To prove the scattering model, experiments were conducted on prepared encrustation samples [61] made of 90% pure analytical grade gypsum and 10% analytical grade carbon particles. The two components were well mixed before adding water. The samples were then left to dry, solidify, and stabilize. The absence of Fe^{2+}/Fe^{3+}, other minerals, and preexisting yellow layers excluded possible operation of all other suggested yellowing mechanisms except light scattering. When the Q-switched Nd:YAG laser at 1064 nm was used, the selective vaporization of carbon particles was observed, i.e., absence of particles on the surface, while the surface color became yellowish (Figure 8.10, left). Note that the spots at the left of Figure 8.10 are holes (not particles). When temporal–spatial overlap of 1064 and 355 nm pulses was

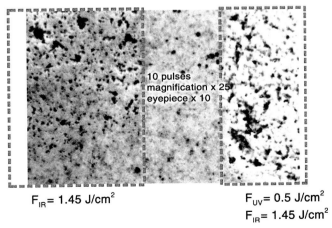

$F_{IR} = 1.45$ J/cm^2

$F_{UV} = 0.5$ J/cm^2
$F_{IR} = 1.45$ J/cm^2

FIGURE 8.10 (See color insert.) Prototype sample (90% gypsum–10% carbon) irradiated with 1064 nm pulses (left) and a temporal–spatial overlap of 1064 and 355 nm pulses (right). Reference material is in center.

used, the additional mechanism of spallation could be identified, while the resulting color was similar to the color of the reference surface, with no yellowish hue present (Figure 8.10, right). This test was made to provide evidence for the light-scattering model in relation to the applied ablation mechanisms, rather than assessing any final surface after laser cleaning (intentionally these samples had no reference substrate).

The prototype operating system shown in Figure 8.11 was developed at FORTH. It represents a functional solution for complete control of the combination of two beams of different wavelengths — here the fundamental and its third harmonic. The system allows selection of both the total energy and the energy ratio of the combined beams. The device has been successfully used in restoration of the Parthenon West Freeze [24,49].

8.2.3 DEPENDENCE OF ABLATION SIDE EFFECTS ON PULSE WIDTH

The photomechanical effect induced by Q-switched pulses produces peak pressure values from tens to hundreds of bars due to fast thermal expansion and shock wave

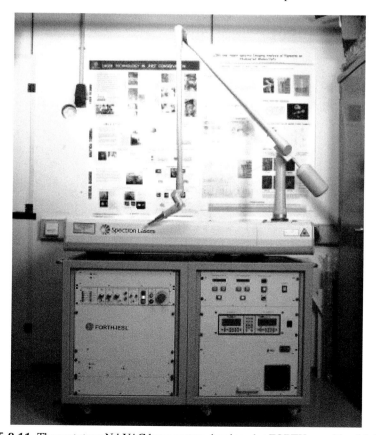

FIGURE 8.11 The prototype Nd:YAG laser system developed at FORTH, capable of delivering laser pulses at 1064 and 355 nm having variable relative intensities and temporal and spatial overlap.

dynamics [25]. With fragile substrates, e.g., highly sulphated marble or valuable layers weakly coupled to the substrate (Ca-oxalate films, gold leaf, etc.), such peak pressures may become excessive. Induced photomechanical transient effects may be strong enough to produce undesired detachment of small fragments even on a single pulse. In particular, this could explain the frequent whitening and spotting effects associated with the cleaning of marble by Q-switched pulses at relatively low fluence levels [44]. Shock pressure and all associated phenomena are proportional to laser intensity raised to a certain power [25,26,62]. The longer the pulse duration, the lower the peak intensity and therefore the effects induced by the pressure wave. Pulse durations in the range of tens of microseconds (short-free-running or SFR regime) minimize the pressure peak acting on the material, so that friable stone or fragile valuable layers can be safely cleaned, but at reduced efficiency in encrustation removal. On the other hand, the use of such pulses should be accompanied by wetting the surface prior to cleaning so as to lower the critical temperature for ablation, increase light coupling, improve efficiency, and reduce substrate heating. Thermal side effects are usually negligible when the laser pulse duration is less than the thermal conduction time, so that only a small fraction of deposited energy within the absorbing encrustation is lost outside the boundary of the material removed. The thermal diffusion length d_{therm}, representing the propagation distance of the thermal wave into the material during laser irradiation, is dependent on the laser pulse duration t_L and the thermal diffusivity of the material:

$$d_{\text{therm}} = 2(\kappa t_L)^{1/2} \qquad (8.2)$$

On the other hand, the optical penetration depth d_{opt} in saturated conditions can be assumed to be equal to the ablation depth. Hence, for insulating materials (depending on thermal diffusivity), use of pulse durations up to microseconds may obey the thermal confinement condition (i.e., $d_{\text{therm}} \ll d_{\text{opt}}$). Typical values of d_{therm} for different stone types may vary from 0.2 to 6 µm going from 10 nsec to 10 µsec pulse length [44].

The above considerations suggest that by controlling the pulse length and wetting the surface, it is possible to maintain thermally confined conditions and at the same time avoid high shock wave pressures. The lower cleaning efficiency of microsecond pulses with respect to a few nanosecond pulses usually allows a gradual removal of stratification encountered on stones. This can represent an important advantage in cleaning of high value surfaces, where an assessment of the optimum degree of cleaning represents a crucial phase and is usually a subject of thorough interdisciplinary discussions, whereas in these cases the efficiency is of minor importance [28,29]. A test case study is presented in section 8.3.2.

8.2.4 OTHER LASER CLEANING METHODS

Two new methods for the laser cleaning of dark encrustation using a Q-switched Nd:YAG laser have been developed and investigated by Watkins et al. [63]. These offer increased efficiency and reduced possible substrate damage for a wide range of substrate/encrustation combinations. In *angular laser cleaning*, it is shown that by controlling the angle of incidence of the cleaning beam, significant improvement in cleaning efficiency can be achieved over conventional methods using a normal angle of incidence. To explain this effect, a model has been proposed where the angular cleaning beam is basically trapped in the encrustation layer, causing multiple reflections, and is entirely absorbed by the material to be removed.

In *laser shock cleaning* [63], the second method, a completely different approach is presented. By aligning the incoming laser beam horizontally but very close to the surface and selecting operating parameters that lead to air breakdown just above the area to be cleaned, a laser-induced shock wave is produced that is much more effective than conventional normal incidence cleaning in removing surface pollutants. Since the beam is not directed onto the surface, this method significantly minimizes the potential for substrate damage if the original substrate does not present fracture problems. Aligning the beam parallel to the surface, though, is not technically easy, especially when dealing with sculpture, relief, etc. Again, a model for the cleaning process has been presented, where the shock wave pressure generated by rapid expansion of the plume and rapid evaporation of the material can remove even the small and strongly adhered particles from the surface. The results for the operation of both methods on polluted marble have been demonstrated [63].

Another laser cleaning method that has been recently presented [64] is using an Er:YAG laser on a wetted surface. The Er:YAG laser emits a train of pulses at 2.94 μm. The duration of each pulse is of the order of 1 to 2 μsec, while the whole train has a duration of about 250 μsec. This laser has been used safely and effectively to ablate contaminants from polychrome surfaces of marble and limestone sculptures [64]. The surface encrustations removed included mainly atmospheric deposits, organic films, lichen and other fungal growths, but also calcite, gypsum, whewellite, and soluble salts. A microscopic study of the polychrome surfaces before and after removal of the encrustations showed preservation of the polychrome pigments. Infrared absorption and x-ray fluorescence spectral analyses of the ablated materials and of the surfaces before and after laser ablation were used for evaluation of the mechanism of the laser action and for comparison of the results of Er:YAG laser treatment with traditional conservation methods.

Surface ablation at 2.94 μm using the Er:YAG laser relies on the strong absorption of OH bonds when present in the surface, which results in superficial heating [65]. Usually the removal is enhanced by dampening the surface with water containing hydroxyl (OH) bonds, since the OH stretching absorption band coincides with the laser wavelength. When surface moisture is rapidly heated by laser exposure, a phase transition of water to steam occurs. The expansion of the steam, in effective microexplosions, breaks up surface encrustations into small particles. The ablated material is near the phase transition temperature of water (100°C), which initiates little or no chemical activity, while the heat diffusion is in principle minimal. However, care is needed to avoid swelling of the surface owing to possible penetration of water into the substrate. This may lead to uncontrolled ablation of the original surface, which should be preserved. To avoid this side effect, the conservator must be skilled in applying the proper amount of water over the encrustation area and afterwards avoiding its overexposure as well as the exposure of the surrounding not encrusted area. Nevertheless, the Er:YAG laser has proven to be an excellent solution in certain conservation problems.

Elemental analysis by *Laser-Induced Breakdown Spectroscopy* (LIBS) has been applied, in some cases, as a real-time diagnostic technique for the cleaning process, with the aim of assessing the elemental abundance at different possible end points of intervention [14,16,19, 20,25,48,51,57,66–71]. In addition to the elemental loads, this technique can also provide information on the depth and morphology of the crust when combined with ablation rate studies [19]. In this way the choice of the optimum laser parameters for an efficient cleaning is facilitated. On some surfaces, the analytical and structural composition obtained by LIBS compares well with the results obtained by traditional analytical techniques such as scanning electron microscopy, energy dispersive x-ray analysis, Fourier transform infrared spectroscopy, and optical microscopy [19,51]. This indicates that, in some instances, LIBS can be used as an autonomous *in situ* off-line diagnostic technique, providing information necessary for achieving the proper end point in the laser cleaning of stonework and other materials. Here by off-line we mean the assessment of the end point by collecting elemental analysis data via LIBS at a separate stage adjacent to the cleaning stage. In practice this can be achieved by occasionally pausing the cleaning process, increasing the focus of the laser beam to a predetermined value, delivering one laser pulse on the area under assessment, and finally recording the LIBS spectrum.

In the example presented here, the elements Ca, Mg, Fe, Si, and Al have been detected by LIBS spectral analysis, as shown in Figure 8.I. The emission spectral have been obtained for consecutive pulses and therefore provide an in-depth profile of these elements to a total depth of 150 μm. For a better presentation of the spectra, the absolute intensity of each individual spectrum has been normalized, so that the strongest peak (Mg II) has the same peak intensity in every spectrum. In any case, the intensities of the strong lines are subject to optical thickness. The ablation rate study (Figure 8.1a) reveals that for the fluence used to collect the LIBS spectra, the first two pulses remove an encrustation thickness of 30 μm each, while additional pulses remove a gypsum layer of 5.5 μm each. Fe and Si almost disappear at a depth of approximately 80 μm. At this depth, Al starts to decrease and reaches barely detectable levels at a depth of approximately 100 μm, while Mg remains stable. Mg originates both from marble and from atmospheric pollution.

Ca concentration increases after the first pulses and reaches stable levels, while the loads of other elements such as Fe, Si, and Al (the presence of which is due primarily to contamination by aluminum-iron-silicates) decrease to undetectable levels.

Reference spectra were also obtained by varying the time delay between the laser pulse and the starting edge of the timing gate in order to verify the higher intensity of II lines against I lines in some cases (e.g., Mg and Fe). For instance, the Mg II doublet at 279.553 and 280.270 nm appears much stronger than the Mg I line at 285.213 nm, as is seen in Figure 8.I. The Mg II/Mg I intensity ratio, however, decreases with increasing delay and becomes insignificant for delays longer than 11 μs, where only the Mg I lines appear [19]. Other important information that the reference spectra reveal include the identification of spectral overlap under the specific experimental conditions, e.g., fluence, surface morphology, and plasma temperature. For example, there is an overlap between the second and strongest line of the Al I doublet at 309.3 nm and the Mg I line at the same wavelength. For this reason, the other Al I doublet line at 308.2 nm is used for the elemental identification of Al.

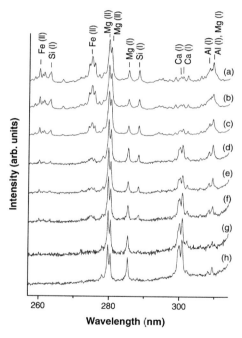

FIGURE 8.I Consecutive LIBS spectra from black encrustation and substrate (pentelic marble). Energy fluence was 6.3 J/cm², well above the fluence values used for cleaning. The time gate was 620 ns and the delay following the laser pulse was 122 ns. The different spectra correspond to (a) the 2nd pulse and a total depth of 60 μm; (b) the 4th pulse, 71 μm; (c) the 6th pulse, 82 μm; (d) the 8th pulse, 93 μm; (e) the 10th pulse, 104 μm; (f) the 12th pulse, 115 μm; (g) the 14th pulse, 126 μm; and (h) the 18th pulse, 148 μm.

8.2.5 DISCOLORATION OF POLYCHROMES

It is well known that some pigments discolor under certain conditions such as exposure to the atmosphere, mixing with certain other pigments or binding media, exposure to high temperature, and so forth. The question of discoloration of certain pigments after laser irradiation is a relatively new and challenging issue and applies to most cases of laser cleaning. In the conservation field, discoloration of some pigments was reported following attempts to clean polychromes using a Q-switched Nd-YAG laser [72], or tempera paints with no varnish layer using a KrF excimer laser [73]. When pigments might be directly exposed to laser radiation, laser cleaning cannot be applied.

In the case of varnished paintings, it was shown (see Chapter 7) that safe laser cleaning can be accomplished by maintaining the innermost part of the varnish, which serves as a light filter preserving the underlying pigment. In the case of polychromes, though, it is almost impossible to remove the encrustation without affecting the underlying pigment. This is why discoloration is addressed here rather than in Chapter 7, even though the problem is common to most laser applications.

Many pigments discolor when irradiated with a laser. This problem has hindered the use of lasers in many conservation applications including cleaning polychromes. Most pigments have very complex compositions, and an extensive analysis of the discolored pigments, especially interpretation of their XRD spectra, is a difficult task. In general, a detailed understanding of pigment discoloration has not yet been achieved. Experimental studies have shown that phase transformations take place upon laser irradiation to a varying extent, based on the type of pigment under consideration. The subject is too broad and escapes the aim of the current presentation. The interested reader may find extensive literature related to this subject [32,56,72–82].

The discoloration takes place in a very thin superficial (outermost) layer. Consequently, an XRD analysis could not detect significant changes at first. Therefore it is necessary to irradiate the pigments using hundreds of pulses for the alteration to be detectable. Although this excessive exposure is not realistic in practice, it gave the opportunity to detect the phase changes that took place in certain pigments [32,76]. It is well known that penetration of the XRD x-rays into the material is of the order of a few micrometers at the most, e.g., 0.7 μm in the case of cinnabar (HgS). Therefore after laser irradiation the XRD spectrum should have been dominated by the new crystalline phase. This, in fact, was not the case because of the low concentration of the new phase within the discolored crystal. It has been calculated [75] that, in the case of cinnabar irradiation, the resulting metacinnabar extends only a few hundreds of angstroms into the bulk. This finding agrees with the observed low intensities of the XRD peaks of the new phase formed. Finally, it was found that in some cases the discoloration is a reversible process. For instance, this happens in lead white pigment, which regains its original white color after long exposure to the air [74,76].

Possible ways of resolving this problem have been suggested [32,76]. Since the pigment alterations were found to be superficial, one way to solve the discoloration problem is by integrating the laser with traditional restoration techniques. This can be the case in polychromes covered by hard inorganic encrustation, which is easily removed using, for example, a Q-switched Nd:YAG laser on a dry or wet surface. Following the laser-assisted removal of the encrustation, the original pigments could be uncovered by removing the thin laser-affected superficial layer with a scalpel. This layer, composed of single crystals as well as amorphous material, is loosely bound to the original crystalline minerals. Although not ethically acceptable, this is an alternative that in certain cases could be considered by the conservator for uncovering the color of the original surface with minimal damage. Alternatively, someone may use a long pulse laser (e.g., Nd:YAG or Er:YAG) on a wetted encrustation surface [28,64] maintaining the interaction regime as far as possible from the pigment–encrustation interface. The best methodology remains to be proved on a case-by-case basis.

8.3 REMOVAL OF ENCRUSTATION: TEST CASE STUDIES

Cooper [12] has presented many test case studies on cleaning sculptures using Q-switched Nd:YAG lasers at 1064 nm. Here a few examples of laser cleaning are presented for demonstrating this particular application. In contrast to the fundamental, the third harmonic of a Q-switched Nd:YAG laser has been also used with very good results. The beam delivery is usually through a flexible optical articulated arm, although transmission through fibers has been also established for 1064 nm [83].

The marble sculpture shown in Figure 8.12 is from the Roman imperial period and is part of the Ince Blundell collection of classical Greek and Roman sculptures.

FIGURE 8.12 Sculpture after partial cleaning using the fundamental of a Q-switched Nd:YAG laser. (By kind permission of Martin Cooper and the National Museums and Galleries on Merseyside.)

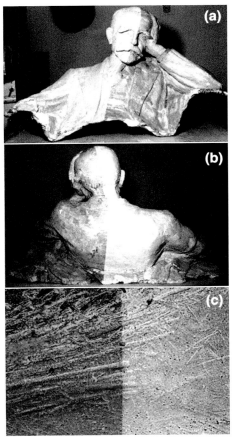

FIGURE 8.13 Front (a), backside (b), and a detail (c) of half-cleaned plaster sculpture using the 3rd harmonic of a Q-switched Nd:YAG laser at 0.55 J/cm². (Taken from Marakis, 2000.)

Years of exposure to a polluted environment led to the formation of an unsightly and potentially damaging black crust, which obscured the surface. In some areas, weathering led to material loss and severe degradation of the underlying structure. Figure 8.12 shows a sculpture after partial cleaning. No laser-induced yellowing effects were observed. A Q-switched Nd:YAG laser (1064 nm, 5 to 10 nsec) was used carefully to remove the hard black pollution encrustation [84] while preserving the underlying patina. Beam delivery was via an articulated arm, and a fluence in the range of 0.5 to 1.0 J/cm² was used. The repetition rate varied from 1 to 10 Hz, depending on the condition of the surface being cleaned. Alternative cleaning methods could not be used in this case owing to the hardness of the black crust and the fragile nature of the marble.

Another example, this time using the 3rd harmonic of a Q-switched Nd:YAG laser, is presented in Figure 8.13. In the National Gallery of Athens, the original surface of a plaster (made in 1937) was in good condition apart from the soot-type pollution covering it. There were also areas of organic nature, most probably fungi. Traditional cleaning methods would have been catastrophic, e.g., water-based poultices would have partially dissolved the gypsum, while microsandblasting would have destroyed the sculptural details owing to the material's softness. The spallation threshold for the plaster substrate using 355 nm was 0.62 J/cm^2, while the onset of dry soot removal was 0.3 J/cm^2 with the efficiency increasing with fluence [34]. Based on stereomicroscopic observations, the optimal fluence was found to be 0.55 J/cm^2. The repetition rate was 10 Hz. The whole operation was done on a dry surface. Here the front and back of the half-cleaned sculpture are shown in Figure 8.13a and Figure 8.13b, respectively, while a surface detail can be seen in Figure 8.13c showing preservation of its small features.

The photomechanical character of soot removal using the 3rd harmonic of a Q-switched Nd:YAG laser can be shown at very low fluence when particles are loosely attached on the surface. Figure 8.14 shows a modern plaster relief artificially covered with soot, which has been half-divested using a fluence of 0.2 J/cm^2. The same laser has also been successfully used to remove compact black encrustation from stone façades [34]. Figure 8.15 shows an example of a façade from the outer wall of a church in central Athens after decades of environmental pollution. In cases of uneven encrustation thickness and morphology, divestment is carried out in two or three steps with increasing fluence, but always being below the spallation threshold of the substrate. In the example of Figure 8.15, a first scan was made at 0.4 J/cm^2, while a second one followed in selected areas using 0.5 J/cm^2.

The 3rd harmonic of a Q-switched Nd:YAG laser can sometimes be used when the encrustation is of biogenic character in combination with inorganic encrustation. Owing to the high absorption of UV light by biological molecules, 355 nm is often very successful in removing fungi, lichen, and other biological encrustation. Figure 8.16 shows the results of dry laser cleaning at 1064 and 355 nm on biological encrustation using fluences of 1.85 and 0.84 J/cm^2, respectively. In this particular case, the marble spallation thresholds for the two laser wavelengths were 3.1 and 2.6 J/cm^2, respectively [34]. It is noticeable that in this particular case 355 nm gives superior results compared to 1064 nm.

FIGURE 8.14 Soot removal from plaster relief using the 3rd harmonic of a Q-switched Nd:YAG laser at 0.2 J/cm^2. (Taken from Marakis, 2000.)

FIGURE 8.15 Encrustation cleaning from stone façade using the 3rd harmonic of a Q-switched Nd:YAG laser at 0.4 and 0.5 J/cm². (Taken from Marakis, 2000.)

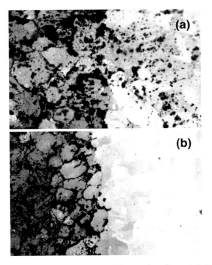

FIGURE 8.16 Divestment of biological encrustation using $\lambda_L = 1064$ nm (a) and $\lambda_L = 355$ nm (b), respectively. Magnification is × 50.

8.3.1 FUNDAMENTAL AND THIRD-HARMONIC TEST CASE STUDIES

A distinct case, in which laser cleaning is optimized by using two laser wavelengths simultaneously (wavelength blending method), is the laser cleaning of the Athens Parthenon West Frieze. The West Frieze blocks were removed from the monument and placed in the laboratory of the Acropolis Museum in 1993. The Acropolis Restoration Service (YSMA) of the Greek Ministry of Culture, which works under the scientific supervision of the Committee for the Conservation of the Acropolis Monuments (ESMA), conducted a detailed investigation on the optimum conservation strategies that lasted more than four years. Following these studies, the first phase of the conservation (2000–2002) focused on the consolidation of the surface, the removal

of bronze dowels and mortars from earlier treatments, and the reattachment of fragments. The second phase of the conservation (2002–2005) aimed at the aesthetic
retrieval, which included the sealing of cracks and gaps and the surface cleaning.

The preliminary studies that were carried out under a collaboration scheme between
YSMA and IESL-FORTH had shown that between many different — and finally four
prevailing — cleaning methods, laser cleaning was the most safe and efficient method
for the removal of all types of encrustation present on the Parthenon West Frieze [49].
Taking into account the advantages and limitations of the laser ablation mechanisms in
the IR and UV regime and in respect to the cleaning requirements and the nature of the
specific encrustations and substrates (pentelic marble covered in places with monochromatic surface layers of ancient origin), the FORTH research team suggested the combination of the two mechanisms in an attempt to reach an optimum cleaning result without
discoloration or surface alteration phenomena. A method for the synchronous (temporal
and spatial overlap) use of 1064 nm and 355 nm pulses was developed [33] and used
for the cleaning of the Parthenon West Frieze [24,49,61]. The simultaneous action of the
two wavelengths (1064 and 355 nm) was able to remove loose pollution deposits and
thin compact crusts in a controllable way. It was found [24,49,61] that for every combination of encrustation and substrate tested, parameters (UV-to-IR fluence ratio, total
fluence, and number of pulses) can be optimized to produce excellent results.

Figure 8.17 shows a series of laser cleaning preliminary tests on a recent corner
addition from the West Frieze that bear the same encrustation as the original surface
itself. The starting point was the measurement of the ablation thresholds at each
wavelength separately, on the thin compact crust [34]. The two areas at the top right
corner marked as IR and UV were cleaned using the corresponding wavelengths at

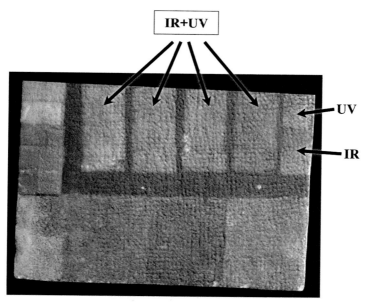

FIGURE 8.17 (See color insert.) Preliminary laser cleaning trials on a corner addition (dated
in the 1960s) from the Parthenon West Frieze block no. 11(ΔZ XI).

FIGURE 8.18 (See color insert.) Detail from the West Parthenon Frieze block no. 3 (ΔZ III) during the laser cleaning intervention with the combined beam methodology.

fluence levels just above the ablation threshold. Note the intense yellow color of the IR-treated area and, in contrast, the lighter area treated with UV pulses, which has a grey hue. The four areas marked IR+UV have been irradiated with different ultraviolet-to-infrared fluence ratios and fluence levels. The detailed information on the exposure parameters, the number of pulses applied, and the L*a*b* color measurement are given elsewhere [24,49,61]. The resulting color depends on three factors, namely the fluence at each wavelength, the UV-to-IR fluence ratio in the combined beam, and the total number of pulses (cleaning level). Figure 8.18 shows a detail of the West Frieze block III during the laser cleaning intervention using a combination of 1064 nm and 355 nm [24,49]. The results are quite satisfactory, as the original surface is fully revealed without any discoloration phenomena.

8.3.2 TEST CASE STUDY USING SHORT-FREE-RUNNING LASER PULSES

The *Santi Quattro Coronati* by Nanni di Banco, made between 1409 and 1417, is a marble sculptural group from the façade of the Orsanmichele church in Florence [85]. It is composed of three statues (two singles and a pair of saint figures), which represent four Christian martyrs.

Following a thorough characterization of the state of conservation, the masterpieces underwent conservation treatments at the beginning of 1999. The surface was covered by a typical gypsum–matrix black crust caused by urban pollution, which was superimposed onto a sulphated 18th century bronze-like patination and Ca-oxalate films resulting from the mineralization of ancient organic treatments. The cleaning was aimed at removing the black crust and the underlying brownish patination constituted by gypsum, earthy materials, pigments (ochres and black carbon), and organic binder. To reach this goal, an integrated cleaning technique using chemical poultices (7% tetrasodic EDTA solution) and mechanical abrasion by a small silicone wheel was designed. The procedure achieved satisfactory results on most of the marble statues, whereas it was not suitable to clean areas presenting traces of gilded decorations, which were indistinctly seen beneath the stratification of hair and robes of the statues. In particular, traces of a long gilded frieze were found at the robe borders of the three sculptures. SEM-EDX analyses identified the gilding as gold leaves (0.2 to 0.5 μm) applied using glue, which exhibited a very weak adhesion to the substrate. Since traditional techniques appeared unsuitable, an optimized laser cleaning approach was devised [29].

Cleaning tests were carried out using a Q-switched Nd:YAG laser (λ = 1064 nm, t_L = 11 nsec), a short-free-running (SFR) Nd:YAP laser (λ = 1340 nm, t_L = 50 μsec), and an SFR Nd:YAG laser (λ = 1064 nm, t_L = 20 μsec) under wetting conditions. The Q-switched laser provided a very efficient single-pulse removal of the entire stratification (200 to 300 μm) at a fluence value of 0.7 J/cm^2. At the same time, the cleaned areas had a somewhat yellow-orange color, according to spectral reflectance and color measurements. As displayed in Figure 8.19, no significant differences were observed in terms of ablation rate and surface hue when the fluence was increased to 2 J/cm^2.

This behavior is phenomenological evidence of a photomechanical ablation regime [48]. Conversely, slow vaporization regimes associated with SFR Nd:YAP laser irradiation at fluence values between 2.6 and 8.2 J/cm^2 provided in this case a very gradual removal of the stratification (area CA in Figure 8.20). A similar behavior was observed using the SFR Nd:YAG laser in the fluence range of 1.9 to 5 J/cm^2 (area CB in Figure 8.20). In the latter case, fluences are lower than when using the SFR Nd:YAP laser owing to the higher reflectance of ochres at 1064 nm.

At operating laser fluences, both SFR lasers provided a clear discrimination of the gilding. Using the SFR Nd:YAG and SFR Nd:YAP laser systems, a loss of gold fragments was observed at fluences above 5 J/cm^2 and 7 J/cm^2, respectively. This observation imposed using fluence lower than these limits, the results of which are shown in Figure 8.20. In contrast, the photomechanical regimes associated with the Q-switched laser did not allow safeguarding the gold leaf fragments, which were almost completely removed at fluences slightly above the cleaning threshold of 0.7 J/cm^2.

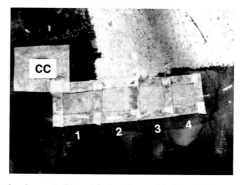

FIGURE 8.19 (See color insert.) Cleaning level provided by QS-Nd:YAG laser between 0.7 and 2 J/cm^2.

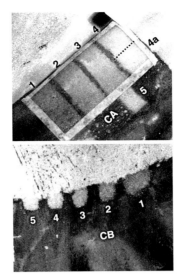

FIGURE 8.20 High removal control provided by SFR Nd:YAP (top) and SFR Nd:YAG (bottom). Sites CA_{1-4} cleaned at 2.6, 3.2, 4.2, and 8.2 J/cm^2, respectively. For comparison: site CA_5 by QS-Nd:YAG laser (0.7 J/cm^2) and CA_{4a} treated by SFR Nd:YAP at 8.2 J/cm^2 and then by QS-Nd:YAG at 2.5 J/cm^2.

The cleaning of the gilded decoration of the *Santi Quattro Coronati* by Nanni di Banco was finally performed by SFR Nd:YAG laser. To minimize any damage to the gilding fragments, a two-step cleaning treatment was applied: (a) a preliminary irradiation of the gilding traces at about 2 J/cm^2 and (b) a careful finishing at 3.5 to 4 J/cm^2, avoiding excess irradiation on gilding already exposed.

This case study shows an alternative way of proceeding in certain encrustations where the final color may be of major concern. SFR laser irradiation seems to be effective in removing stratified yellow pigments, which exhibit a high reflectance at 1064 nm, because of the relatively high operating fluences with respect to the Q-switched laser.

8.4 CONCLUSIONS

It has been shown that using lasers to clean marble and other stone surfaces by removing an inorganic encrustation attached to the already altered original surface can be a very delicate intervention. The mechanisms of encrustation removal vary depending on the wavelength, pulse duration, and fluence in particular, as well as other related parameters such as the condition of the surface (dry or wet) and its composition. Although a detailed understanding of all operating processes is not always possible, there are some general guidelines for using a particular type of laser for a certain form of intervention.

The two main cleaning mechanisms are (a) the explosive vaporization of the entire dark encrustation or selective explosive vaporization of the dark particles usually embedded in the gypsum-rich zone and (b) the shock wave following the pulse that may detach encrustation fragments or flakes. The first mechanism induces heat to the substrate, while the second one induces a pressure stress wave, both of which may cause undesired results to a friable substrate. An optimal contribution of these removal schemes must be found for each surface to be cleaned in the hands of a skilled conservator.

Discoloration phenomena are sometimes important when choosing the optimal laser system and exposure parameters. Different possibilities for a conservator include

- Careful adjustment of fluence and pulse rate for a certain range of pulse durations, e.g., nanosecond pulses
- Proper conditioning of the surface prior to laser treatment, e.g., wetting it with water or other solvent
- Increasing laser pulse duration to the microsecond regime and water conditioning the surface at the same time
- Using different wavelengths for different encrustations
- Using two wavelengths in a temporal–spatial overlapping mode for loose deposits and thin compact crusts
- Using a laser with constant fluence over the beam waist
- Attaining a comprehensive training program

Research in this important field continues in all of the above directions. Finally, it is now evident that the vast majority of conservation situations require an integration of laser-based and traditional conservation techniques for optimum results. Therefore a considerable part of the on-going research is to define the limitations and thus the possible combinations of different conservation methodologies.

REFERENCES

1. Asmus, J.F., Seracini, M., and Zetler, M.J., Surface morphology of laser cleaned stone, *Lithoclastia,* **1**, 23, 1976.
2. Asmus, J.F., Light cleaning: laser technology for surface preparation in the arts, *Tech. Conserv.,* Fall, 14, 1978.
3. Asmus, J.F., More light for art conservation, *IEEE Circuits Devices Mag.,* March, 1986, p. 6.
4. Asmus, J.F., Light for art conservation, *Interdisciplinary Sci. Rev.,* **12**, 171, 1987.
5. Bromblet, P. and Cooper, M., Cleaning of stone and ivory, *J. Cult Heritage,* **4**, 9s, 2003.
6. Bromblet, P., Laboure, M., and Orial, G., Diversity of the cleaning procedures including laser for the restoration of curved portals in France over the last 10 years, *J. Cult. Heritage,* **4**, 18s, 2003.
7. Calcagno, G., Koller, M., and Nimmrichter, H., Laserbased cleaning on stonework at St. Stephan's Cathedral, in *Restauratorenblaetter, Sonderband — Lacona I,* Verlag Mayer, Vienna, 1997, p. 39.
8. Calcagno, G., Pummer, E., and Koller, M., St. Stephen's Church in Vienna: criteria for Nd:YAG laser cleaning on an architectural scale, *J. Cult. Heritage,* **1**, S111, 2000.

9. Cooper, M.I., Emmony, D.C., and Larson, J.H., A comparative study of the laser cleaning of limestone, Proceedings of the 7th International Congress on Deterioration and Conservation of Stone, June 1992, Lisbon, p. 1307.

10. Cooper, M.I., Laser cleaning of stone sculpture, Ph.D. thesis, Loughborough University, 1994.

11. Cooper, M.I., Emmony, D.C., and Larson, J.H., Characterization of laser cleaning of limestone, *Opt. Laser Technol.,* **27**, 69, 1995.

12. Cooper, M., *Cleaning in Conservation: An Introduction,* Butterworth Heinemann, Oxford, 1998.

13. Gobernado-Mitre, I., Medina, J., Calvo, B., Prieto, A.C., Leal, L.A., Perez, B., Marcos, F., and de Frutos, A.M., *Appl. Surf. Sci.,* **96–98**, 474, 1996.

14. Gobernado-Mitre, I., Prieto, A.C., Zafiropulos, V., Spetsidou, Y., and Fotakis, C., On-line monitoring of laser cleaning of limestone by laser induced breakdown spectroscopy, *Appl. Spectrosc.,* **51**, 1125, 1997.

15. Klein, S., Stratoudaki, T., Marakis, Y., Zafiropulos, V., and Dickmann, K., Comparative study of different wavelengths from IR to UV applied to clean sandstone, *Appl. Surf. Sci.,* **157**, 1, 2000.

16. Klein, S., Dickmann, K., and Zafiropulos, V., Laser cleaning of natural stone with marble as an example, *LaserOpto,* **32**, 34, 2000.

17. Lanterna, G. and Matteini, M., Laser cleaning of stone artifacts: a substitute or alternative method? *J. Cult Heritage,* **1**, S29, 2000.

18. Marakis, G., Maravelaki, P.V., Zafiropulos, V., Klein, S. Hildenhagen, J., and Dickmann, K. Investigations on cleaning of black crusted sandstone using different UV-pulsed lasers, *J. Cult. Heritage,* **1**, S61, 2000.

19. Maravelaki, P., Zafiropulos, V., Kylikoglou, V., Kalaitzaki, M., and Fotakis, C., Laser induced breakdown spectroscopy as a diagnostic technique for the laser cleaning of marble, *Spectrochim. Acta B,* **52**, 41, 1997.

20. Maravelaki-Kalaitzaki, P., Zafiropulos, V., and Fotakis, C., Excimer laser cleaning of encrustation on pentelic marble: procedure and evaluation of the effects, *Appl. Surf. Sci.,* **148**, 92, 1999.

21. Maravelaki-Kalaitzaki, P., Zafiropulos, V., Pouli, P., Anglos, D., Balas, C., Salimbeni, R., Siano, S., and Pini, R., Short free running Nd:Yag laser to clean different encrustation on pentelic marble: procedure and evaluation of the effects, *J. Cult. Heritage,* **4**, 77s, 2003.

22. Pini, R., Siano, S., Salimbeni, R., Piazza, V., Giamello, M., Sabatini, G., and Bevilacqua, F., Application of a new laser cleaning procedure to the mausoleum of Theodoric, *J. Cult. Heritage,* **1**, S93, 2000.

23. Pouli, P., Zafiropulos, V., Balas, C., Doganis, Y., and Galanos, A., Laser cleaning of inorganic encrustation on excavated objects: evaluation of the cleaning result by means of multi-spectral imaging, *J. Cult. Heritage,* **4**, 338s, 2003.

24. Pouli, P., Frantzikinaki, K., Papakonstantinou, E., Zafiropulos, V., and Fotakis, C., Pollution encrustation removal by means of combined ultraviolet and infrared laser radiation: the application of this innovative methodology on the surface of the Parthenon West Frieze, Lasers in the conservation of artworks, Proceedings of LACONA V, Sept. 15–18, 2003, Osnabrueck, Germany, Springer Proceedings in Physics, **100**, Springer-Verlag, Berlin, Heidelberg, 2005, p. 333.

25. Siano, S., Margheri, F., Mazzinghi, P., Pini, R., and Salimbeni, R., Cleaning processes of encrusted marbles by Nd:YAG lasers operating in free running and Q-switching regimes, *Appl. Opt.,* **36**, 7073, 1997.

26. Siano, S., Pini, R., and Salimbeni, R., Variable energy blast modeling of the stress generation associated with laser ablation, *Appl. Phys. Lett.,* **74**, 1233, 1999.

27. Siano, S., Fabiani, F., Pini, R., Salimbeni, R., Giamello, M., and Sabatini, G., Determination of damage thresholds to prevent side effects in laser cleaning of Pliocene sandstone of Siena, *J. Cult. Heritage,* **1**, S47, 2000.

28. Siano, S., Salimbeni, R., Pini, R., Matteini, M., Porcinai, S., Giusti, A., and Casciani, A., The Santi Quattro Coronati by Nanni di Banco: cleaning of the gilded decorations, *J. Cult. Heritage,* **4**, 123s, 2003.

29. Siano, S., Giusti, A., Pinna, D., Porcinai, S., Giamello, M., Sabatini, G., and Salimbeni, R., The conservation intervention on the Porta della Mandorla, Lasers in the conservation of artworks, Proceedings of LACONA V, Sept. 15–18, 2003, Osnabrueck, Germany, Springer Proceedings in Physics, **100**, Springer-Verlag, Berlin, Heidelberg, 2005, p. 171.

30. Teppo, E. and Calcagno, G., Restoration with lasers halts decay of ancient artifacts, *Laser Focus World,* June, p. S5, 1995.

31. Vergès-Belmin, V., Comparison of three cleaning methods — micro-sandblasting, chemical pads and Q-switched Nd:YAG laser — on a portal of the cathedral Notre-Dame in Paris, France, in *Restauratorenblaetter, Sonderband* Lasers in the conservation of artworks, Proceedings of LACONA I, Oct. 4–6, 1995, Verlag-Mayer, 1997, Heraklion, Crete, Greece, p. 17.

32. Zafiropulos, V., Balas, C., Manousaki, A., Marakis, G., Maravelaki-Kalaitzaki, P., Melesanaki, K., Pouli, P., Stratoudaki, T., Klein, S., Hildenhagen, J., Dickmann, K., Luk'yanchuk, B.S., Mujat, C., and Dogariu, A., Yellowing effect and discoloration of pigments: experimental and theoretical studies, *J. Cult Heritage,* **4**, 249s, 2003.

33. Zafiropulos, V., Pouli, P., Kylikoglou, V., Maravelaki-Kalaitzaki, P., Luk'yanchuk, B.S., and Dogariu, A., Synchronous use of IR and UV laser pulses in the removal of encrustation: mechanistic aspects, discoloration phenomena and benefits, Lasers in the conservation of artworks, Proceedings of LACONA V, Sept. 15–18, 2003, Osnabrueck, Germany, Springer Proceedings in Physics, **100**, Springer-Verlag, Berlin, Heidelberg, 2005, p. 311.

34. Marakis, G., Pouli, P., Zafiropulos, V., and Maravelaki-Kalaitzaki, P., Comparative study on the application of the 1st and the 3rd harmonic of a Nd:YAG laser system to clean black encrustation on marble, *J. Cult. Heritage,* **4**, 83s, 2003; also see Marakis, G., Comparative study on the removal on encrustation from marble using the fundamental and the third harmonic of Q-switched Nd:YAG laser, B.S. thesis (in Greek), Technical Educational Institute of Athens & FORTH-IESL, Athens, Heraklion, 2000.

35. Konig, E. and Kautek, W., Eds., Proceedings of the First International Conference LACONA I — Lasers in the Conservation of Artworks, October 4–6, 1995, Heraklion, Crete, Greece, in *Restauratorenblaetter, Sonderband,* Verlag Mayer, Vienna, 1997.

36. Salimbeni, R. and Bonsanti, G., Eds., Proceedings of the International Conference LACONA III — Lasers in the Conservation of Artworks, April 26–29, 1999, Florence, *J. Cult. Heritage,* **1**(suppl. 1), 2000.

37. Vergès-Belmin, V., Ed., Proceedings of the International Conference LACONA IV, Lasers in the Conservation of Artworks, Sept. 11–14, Paris, *J. Cult. Heritage,* **4**(suppl. 1), 2003.

38. Dickmann, K., Fotakis, C., and Asmus, J.F., Eds., Proceedings of LACONA V, Lasers in the conservation of artworks, Sept. 15–18, 2003, Osnabrueck, Germany, Springer Proceedings in Physics, **100,** Springer-Verlag, Berlin, Heidelberg, 2005.

39. Luk'yanchuk, B., Ed. *Laser Cleaning,* World Scientific, New Jersey, London, Singapore, Hong kong, 2002.

40. Bäuerle, D., *Laser Processing and Chemistry,* Springer-Verlag, Berlin, 2000.

41. Miller, J.C., *Laser Ablation — Principles and Applications,* Springer Series in Materials Science **28**, Springer-Verlag, Berlin, 1994.

42. Honeyborne, D., Weathering and decay of masonry, in *Conservation of Building and Decorative Stone,* Ashurst, J. and Dimes, F., Eds., Butterworth-Heinemann, 1990, p. 153.

43. Maravelaki-Kalaitzaki, P. and Origin, B.G., Characteristics and morphology of weathering crusts on Istria stone in Venice, *Atmos. Environ.,* **33**, 1699, 1999.

44. Salimbeni, R., Pini, R., and Siano, S., A variable pulse width Nd:YAG laser for conservation, *J. Cult Heritage,* **4**, 72s, 2003.

45. Watkins, K.G., Larson, J.H., Emmony, D.C., and Steen, W.M., Laser cleaning in art restoration: a review, *Proceedings of the NATO Advanced Study Institute on Laser Processing: Surface Treatment and Film Deposition,* 1995, p. 907.

46. Watkins, K.G., A review of materials interaction during laser cleaning in art restoration, in *Restauratorenblaetter, Sonderband—LACONA I,* Lasers in the conservation of artworks, Verlag-Mayer, Vienna, 1997, p. 7.

47. Siano, S. and Pini, R., Analysis of blast waves induced by Q-S Nd:YAG laser photo-disruption of absorbing targets, *Optic. Commun.,* **135**, 279, 1997.

48. Salimbeni, R., Pini, R., and Siano, S., Achievement of optimum laser cleaning in the restoration of artworks: expected improvements by on-line optical diagnostics. *Spectrochim. Acta B,* **56**, 877, 2001.

49. Frantzikinaki, K., Panou, A., Vasiliadis, C., Papakonstantinou, E., Pouli, P., Ditsa, Th., Zafiropulos, V., and Fotakis, C., The cleaning of the Parthenon West Frieze: an innovative laser methodology, Proceedings of the 10th International Congress on Deterioration and Conservation of Stone (ICOMOS, Sweden, 2004), June 27–July 2, 2004, p. 801; also see Papaconstantinou, E., Frantzikinaki, K., Pouli, P., and Zafiropoulos, V., Study on the cleaning of the West Frieze. Study on the Restoration of the Parthenon (in Greek with English summaries), Vol. 7, Greek Ministry of Culture — Committee for the Preservation of the Acropolis Monuments, Athens, 2002.

50. Skoulikidis, T., Vassiliou, P., Papakonstantinou, P., Moraitou, A., Zafiropulos, V., Kalaitzaki, M., Spetsidou, I., Perdikatsis, V., and Maravelaki, P., Some remarks on Nd:YAG and excimer UV lasers for cleaning soiled sulfated monument surfaces, in book of abstracts *LACONA — Workshop on "Lasers in the Conservation of Artworks,"* Heraklion, Crete, 1995.

51. Maravelaki-Kalaitzaki, P., Anglos, D., Kilikoglou, V., and Zafiropulos, V., Compositional characterization of encrustation on marble with laser induced breakdown spectroscopy, *Spectrochim. Acta B,* **56**, 887, 2001.

52. Skoulikidis, T. and Papakonstantinou-Ziotis, P., The mechanism of sulfation by atmospheric SO_2 of limestones and marbles of the ancient monuments and statues, I, Observations *in situ* and measurements in the laboratory; activation energy, *Br. Corros. J.,* **16**, 63, 1981.

53. Vergès-Belmin, V. and Dignard, C., Laser yellowing: myth or reality? *J. Cult. Heritage,* **4**, 238s, 2003, and references therein.

54. Pouli P., Totou, G., Zafiropulos, V., Fotakis, C., Rebollar, E., Oujja, M., Castillejo, M., and Domingo, C., A comprehensive study of the coloration effect associated with laser cleaning of pollution encrustations from stonework, in *6th International Congress on Lasers in the Conservation of Artworks (LACONA VI),* Sept. 21–25, 2005, Vienna, Book of Abstracts, p. 100.

55. Laboure, M. and Vergès-Belmin, V., Study of different poultices to remove the laser cleaning yellowing effect, in *6th International Congress on Lasers in the Conservation of Artworks (LACONA VI)*, Sept. 21–25, 2005, Vienna, Book of Abstracts, p. 101.

56. Gracia, M., Gavino, M., Vergès-Belmin, V., Hermosin, B., Nowik, W., and Saiz-Jimenez, C., Mossbauer and XRD study of the effect of Nd:YAG-1064 nm laser irradiation on hematite present in model samples, Lasers in the conservation of artworks, Proceedings of LACONA V, Sept. 15–18, 2003, Osnabrueck, Germany, Springer Proceedings in Physics, **100**, Springer-Verlag, Berlin, Heidelberg, 2005, p. 341.

57. Klein, S., Ferksanati, F., Hildenhagen, J., Dickmann, K., Uphoff, H., Marakis, Y., and Zafiropulos, V., Discoloration of marble during laser cleaning by Nd:YAG laser wavelengths, *Appl. Surf. Sci.*, **171**, 242, 2001.

58. Gaviño, M., Castellejo, M., Vergès-Belman, V., Nowik, W., Oujja, M., Robollar, E., Hermosin, B., and Saiz-Jimenez, C., Black crusts removal: the effect of stone yellowing and cleaning strategies, in *Air Pollution and Cultural Heritage*, Saiz-Jimenez, C., Ed., A.A. Balkema Publishers, London, 2004, p. 239.

59. Popescu, G., Mujat, C., and Dogariu, A., Evidence of scattering anisotropy effects on boundary conditions of the diffusion equation, *Phys. Rev. E*, **61**, 4523, 2000.

60. Sabatini, G., Giamello, M., Pini, R., Siano, S., and Salimbeni, R., Laser cleaning methodologies for stone facades and monuments: laboratory analyses on lithotypes of Siena architecture, *J. Cult. Heritage*, **1**. S9, 2000.

61. Pouli, P., Zafiropulos, V., and Fotakis, C., The combination of ultraviolet and infrared laser radiation for the removal of unwanted encrustation from stonework: a novel laser cleaning methodology, Proceedings of the 10th International Congress on Deterioration and Conservation of Stone, ICOMOS, Stockholm, Sweden, June 27–July 2, 2004, p. 315.

62. Phipps, C.R., Turner, T.P., and Harrison, R.F., Impulse coupling to targets in vacuum by KrF, HF and CO_2 single-pulse lasers, *J. Appl. Phys.*, **64**, 1083, 1988.

63. Watkins, K.G., Curran, C., and Lee, J.-M., Two new mechanisms for laser cleaning using Nd:YAG sources, *J. Cult Heritage*, **4**, 59s, 2003.

64. deCruz, A., Wolbarsht, M.L., Palmer, R.A., Pierce, S.E., and Adamkiewicz, E., Er:YAG laser applications on marble and limestone sculptures with polychrome and patina surfaces, Laser in the conservation of artworks, Proceedings of LACONA V, Sept. 15–18, 2003, Osnabrueck, Germany, Springer Proceedings in Physics, **100**, Springer-Verlag, Berlin, Heidelberg, 2005, p. 113.

65. deCruz, A., Hauger, S., and Wolbarsht, M.L., The role of lasers in fine arts conservation and restoration, *Opt. Photon News*, **10**, 36, 1999.

66. Anglos, D., Melessanaki, K., Zafiropulos, V., Gresalfi, M.J., and Miller, J.C., Laser induced breakdown spectroscopy for the analysis of 150-year old daguerreotypes, *Appl. Spectrosc.*, **56**, 423, 2002.

67. Klein, S., Stratoudaki, T., Zafiropulos, V., Hildenhagen, J., Dickmann, K., and Lehmkuhl, T., Laser-induced breakdown spectroscopy for on-line control of laser cleaning of sandstone and stained glass, *Appl. Phys. A*, **69**, 441, 1999.

68. Georgiou, S., Zafiropulos, V., Anglos, D., Balas, C., Tornari, V., and Fotakis, C., Excimer laser restoration of painted artworks: procedures, mechanisms and effects, *Appl. Surf. Sci.*, **127–129**, 738, 1998.

69. Anglos, D., Laser-induced breakdown spectroscopy in art and archaeology, *Appl. Spectrosc.*, **55**, 186A, 2001.

70. Tornari, V., Zafiropulos, V., Bonarou, A., Vainos, N.A., and Fotakis, C., Modern technology in artwork conservation: a laser based approach for process control and evaluation, *Opt. Lasers Eng.*, **34**, 309, 2000.

71. Scholten, J.H., Teule, J.M., Zafiropulos, V., and Heeren, R.M.A., Controlled laser cleaning of painted artworks using accurate beam manipulation and on-line LIBS-detection, *J. Cult. Heritage,* **1**, S215, 2000.

72. Weeks, C., The conservation of the Portail de la Mere Dieu, Amiens Cathedral, France, in *Restauratorenblaetter, Sonderband — Lacona I,* Verlag Mayer, Vienna, 1997, p. 25.

73. Castillejo, M., Martin, M., Oujja, M., Silva, D., Torres, R., Manousaki, A., Zafiropulos, V., van den Brink, O.F., Heeren, R.M.A., Teule, R., Silva, A., and Gouveia, H., Analytical study of the chemical and physical changes induced by KrF Laser cleaning of tempera paints, *Anal. Chem.*, **74**, 4662, 2002.

74. Pouli, P., Emmony, D.C., Madden, C.E., and Sutherland, I., Studies towards a thorough understanding of the laser induced discoloration mechanisms of medieval pigments, *J. Cult. Heritage,* **4**, 271s, 2003.

75. Luk'yanchuk, B.S., Zafiropulos, V., On the theory of discoloration effect in pigments at laser cleaning, in *Laser Cleaning,* Luk'yanchuk, B, Ed., World Scientific, New Jersey, London, Singapore, Hong Kong, 2002, chap. 9, p. 393.

76. Zafiropulos, V., Stratoudaki, T., Manousaki, A., Melesanaki, K., and Orial, G., Discoloration of pigments induced by laser irradiation, *Surf. Eng.*, **17**, 249, 2001.

77. Pouli, P., Emmony, D.C., Madden, C.E., and Sutherland, I., Analysis of the laser-induced reduction mechanisms of medieval pigments, *Appl. Surf. Sci.,* **173**, 252, 2001.

78. Chappe, M., Hildenhagen, J., Dickmann, K., and Bredol, M., Laser irradiation of medieval pigments at IR-, VIS- and UV-wavelengths, *J. Cult. Heritage,* **4**, 264s, 2003.

79. Gordon Sobott, R.J., Heinze, T., Neumeister, K., and Hildenhagen, J., Laser interaction with polychromy: laboratory investigations and on-site observations, *J. Cult. Heritage,* **4**, 276s, 2003.

80. Sansonetti, A. and Realini, M., Nd:YAG laser effects on inorganic pigments, *J. Cult. Heritage,* **1**, S189, 2000.

81. Gaetani, M.C. and Santamaria, U., The laser cleaning of wall paintings, *J. Cult. Heritage,* **1**, S199, 2000.

82. Athanassiou, A., Hill, A.E., Fourrier, T., Burgio, L., and Clark, R.J.H., The effects of UV laser light radiation on artists' pigments, *J. Cult. Heritage,* **1**, S209, 2000.

83. Boquillon, J.P., Laser light transmission through optical fibers for laser cleaning applications, in *Restauratorenblaetter, Sonderband — LACONA I*, Lasers in the conservation of artworks, Verlag-Mayer, 1997, p. 103.

84. Cooper, M., and the National Museums and Galleries on Merseyside, 2001.

85. Giusti, A., Lalli, C., Lanterna, G., Matteini, M., and Rizzi, M., Documentary and analytical analogies in the study of patinas of the "Quattro Santi Coronati" by Nanni di Banco, Proceedings of 9th International Congress on Deterioration and Conservation of Stone, Elsevier Science, Amsterdam, 2000.

9 Laser Cleaning of Other Materials

9.1 INTRODUCTION

Laser cleaning of other materials, such as paper, parchment, metal, stained glass, textiles, pottery, concrete, and so forth, was tested as early as 1978 by John Asmus [1,2]. Ever since, there is a large number of case studies as well as considerable scientific work being performed in this field. The reader may find numerous relevant references in the LACONA conferences proceedings [3–6]. The scope of this chapter, though, is not to present all laser cleaning applications for different surfaces and materials. Instead, an attempt will be made to summarize the findings and discuss the prospects for two major categories of materials, namely parchment/paper and metals.

9.2 PARCHMENT AND PAPER

Laser cleaning of biogenetic surfaces such as parchment and paper has been addressed only in recent years. As in the case of aged varnish on tempera or oil paintings, paper and parchment belong to the most fragile substrates from both the chemical and mechanical point of view. The main constituents of paper and parchment are cellulose and collagen fibers, respectively. The overall structure is of aggregated fibrils with extensive pores capable of holding relatively large amounts of water by capillarity.

In contrast to the bond breakage mechanism in laser ablation of aged varnish, paper and parchment are cleaned according to a different strategy. The existing contrast in visible absorptivity between contaminant and substrate is exploited by using visible laser light (usually 532 nm, Q-switched Nd:YAG), where the substrate exhibits a minimum interaction. This guarantees minimal cellulose and collagen deterioration. The contaminant to be removed is usually dust and dirt particles, fungi, adhesives, or ink. The first prototype of a dedicated workstation was developed by Kautek and his co-workers, where the laser cleaning process may be programmed for automatic beam scanning operation [7]. The pulse duration is nanoseconds and the repetition rate could be chosen up to 1000 Hz. The workstation may include on-line diagnostic tools such as visible, ultraviolet, and fluorescence imaging for the identification and documentation of visible and chemical changes of the irradiated area. One of the major technical challenges of precision cleaning is to avoid beam contact with areas of original ink, printed letters, or pigments. Later on, Scholten and coworkers developed a commercially available laser cleaning system by solving most of the technical problems encountered [8]. One such difficulty is the accurate spatial positioning of the laser beam, which was achieved by combining imaging techniques with a high accuracy scanning mirror system.

9.2.1 CLEANING OF PARCHMENT

Laser cleaning of parchment [9–12,13–17] has similarities to the widespread laser applications in dermatology [18,19], where selective photothermolysis of pigmented subsurface structures takes place. The skin's natural physiological mechanisms break down and remove the laser-altered remnants. Parchment, on the other hand, can be seen as dead skin [20], where natural resorption is absent. Parchment consists of collagen fibers from the dermis layer of animal skin obtained after strong alkaline removal of the epidermis and the subcutaneous tissue layers. It consists of poly-peptide chains with various amino acid side chains. Two-dimensional reorientation of the collagen fibers takes place in a stretching and drying process controlled by $CaSO_4$, CaO, and $CaCO_3$ additives. Often, a plasticizing rehydration end treatment is applied using egg white and linseed oil.

The specific removal of contamination layers or phases is governed by the control of the operating fluence F in respect to the ablation threshold fluence F_{th} of the substances to be removed or maintained [14]. An essential requirement though is the higher optical density of contaminants compared to the original substrate [9,10]. Therefore fluence levels must lie below the ablation threshold of the parchment fibers, F_{p-th}. Moreover, F must be lower than the morphological modification thresh-old, F_{p-m}, and the photochemical or chemical conversion threshold, F_{p-chem}. Finally, F must exceed the threshold for the removal of the contamination matter, F_{c-th} [12,14,16]. This usually leads to a narrow working range of fluence F (Figure 9.1) that may vary considerably in every individual pair of contaminant and substrate. An additional complexity may arise when trying to determine F_{p-chem} in each particular case study, which is only possible by chemical analytical methods, whereas F_{c-th}, F_{p-m}, and F_{p-th} can be verified by optical and/or electron microscopy.

The oxidative breakdown processes of parchment are caused by heat and light [14,21]. Initially, hydroxyl radicals are generated by the laser irradiation. Parchment

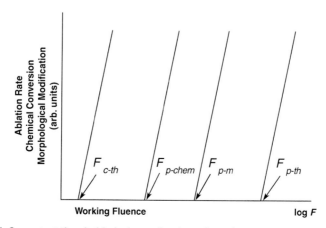

FIGURE 9.1 Important thresholds in laser cleaning of parchment.

starts degrading at tripeptides in clusters of charged amino acids, i.e., acidic breakdown causes hydrolysis of the peptide bonds in the peptide chains, and amino end groups are generated. These may attack carbon atoms in peptide side chains. Polar groups, particularly carboxylic acid functions (besides some conjugated double bonds), are finally formed.

Diffuse reflectance infrared Fourier transform (DRIFT) spectroscopy studies have shown [20] that laser-induced autoxidation reactions of parchment do not occur more or less up to the destruction threshold fluences, F_{p-m}, observed as melting and evaporation. This means that F_{p-chem} practically coincides with F_{p-m} in most cases. Therefore there exists a safe fluence range for laser cleaning at both 308 and 1064 nm. Nevertheless, spectroscopic diagnostics are necessary to guarantee the safe cleaning in all practical cases. At the same time it was also shown that the aging condition of the parchment artifact strongly affects the destruction thresholds. Ancient substrates older than a couple of centuries exhibited higher chemical stability than new model systems [14] but showed reduced physicochemical stability in respect to phase changes. Laser treatment of aged collagen led to conformational changes on the surface at lower fluences than the ablation threshold. Second, the substantial gelatin content of the aged parchment evaporated at lower fluences than that of the intact collagen matrix. Thus aging status of parchment artifacts plays a major role in assessing the laser cleaning limits. This means that the laser processing behavior of model systems can be compared with that of original fibrous artworks to only a very limited extent, and that original artifacts have to be treated rather as individual specimens.

Photometric determination of the water-soluble degradation by-products of collagen is a sensitive method of detecting laser-induced alterations of parchment [22], which are indicative of changes on a molecular level. UV laser light at 308 nm causes photochemical degradation, i.e., molecular degradation at fluencies as low as 0.1 J/cm² far below the ablation threshold ($F_{p-chem} \ll F_{p-th}$). At higher fluences ($F > F_{p-chem}$), the photothermal and morphological alteration starts. At the visible (532 nm) and the infrared (1064 nm) wavelengths, the photochemical phenomena are practically absent. Only thermally induced cross-linking of the collagen fibers is observed as the fluence increases.

The hydrothermal stability and shrinking temperature, T_s, measurements by the micro-hot-table technique [22], rendered themselves sensitive only to laser irradiation that caused photochemical reactions (i.e., UV, 308 nm). Laser wavelengths that are efficient in inducing thermal reactions (e.g., cross-linking and melting) do not affect T_s even at fluences above the ablation threshold. The photochemical deterioration at 308 nm observed by the increase of soluble by-products and the decrease of the shrinking temperature may be related to oxidative breakdown processes, which occur only in the presence of UV light.

9.2.2 Cleaning of Soiled Paper

The first experiments on laser cleaning of paper [23–26] followed the debate on wet and dry traditional cleaning techniques [27–29]. Later, a project on laser cleaning of paper and parchment was launched, aiming at development of a prototype laser cleaning system for historical paper and parchment. Three different lasers were evaluated determining the negative effects that may result during laser-assisted cleaning of cellulosic surfaces [9,10,30–34]. Irradiation of cellulose by using an XeCl excimer laser at 308 nm resulted in photooxidative degradation of the substrate, accompanied by an increase in carbonyl or carboxyl concentration and a rigorous decrease in the degree of polymerization (DP). On the contrary, irradiation of cellulose at 1064 nm resulted in the formation of inter- and intramolecular cross-links of ether origin, which was manifested by an increase of the DP of the irradiated cellulose. Finally, no chemical changes were observed when using the second harmonic of an Nd:YAG laser (532 nm), although use of high fluence levels resulted in degradation of the treated samples during accelerated aging experiments. Nevertheless, laser cleaning at 532 nm has proven to be far more efficient in dirt removal than using a mechanical eraser; however, a distinct discoloration of the sample accompanies the laser treatment [34,35].

Yellowing can be attributed to several factors, including redeposition of transition metals such as iron, discoloration of noncellulosic organic constituents of dirt, cellulose degradation, or even light scattering effects induced by laser–substrate interaction (see also Chapter 8). In studying the origin of discoloration, laser cleaning was applied on purified cellulose containing carbon powder [36]. Addition of carbon lowered the L* value, while it had no marked effect on b* values of the L*a*b* color space. However, laser cleaning of paper containing carbon resulted in a strong increase in b* values, denoting formation of "yellow chromophores". Due to the absence of transition metals and noncellulosic organic material in purified cellulose and analytical grade carbon, the observed discoloration was ascribed to degradation of cellulose, strengthened by heat transfer from carbon particles to cellulose or possible plasma formation. This hypothesis was supported by the observed decrease in DP, which takes place during the irradiation of soiled paper with laser light.

In order to gain deeper insight into the changes in the chemical composition of the treated surface, diffuse reflectance FTIR (DRIFT) spectroscopy has also been performed [36,37]. Laser treatments resulted in degradation of cellulose, demonstrated by the DRIFT difference spectra. The increase of C-O stretching and C-O-C asymmetric stretching bands indicates the formation of intra- or intermolecular ether bonds, while the decrease of the band at 1647 cm^{-1} designates the decrease of water content after the treatment. Considering though the low intensity of the DRIFT difference spectra, either the overall degradation is limited or volatile degradation products are formed. Finally, there was little DRIFT difference between low (near threshold) and higher fluence levels, as well as between dense and thin deposits of carbon.

The above findings suggest that although pulsed laser light at 532 nm is superior in removing soiled contamination from paper, some degradation of cellulose is unavoidable. Nevertheless, conservators should assess this degradation in comparison to the effects caused by mechanical distortion or breaking of fibers in alternative traditional dry cleaning or in comparison to the chemical and structural modification of fibers in wet cleaning.

9.2.3 REMOVAL OF FUNGI

Biodeterioration of organic cultural heritage materials is a common problem. Cellulose is probably the most abundant material of biological origin on the earth and is a prime nutrient for many fungi. While the conventional sanitizing techniques for removal of fungal material from paper using chemical or physical means have proven sufficient in many cases, the removal of discoloration caused by fungal pigments is still a problem in paper conservation.

The deterioration is often initiated by soil saprophytes, such as aspergillus or penicillium. Further degradation by cellulolytic fungi such as the Trichoderma or Chaetomium can cause marked destruction of the paper substrate [38,39]. Fungi include both molds and yeasts. Molds form often-segmented hyphae, which in turn can become long and intertwined, creating a hairy mass called mycelium (Figure 9.2). The mycelium can be vegetative, lying beneath the surface, or it can differentiate into reproductive forms above the surface. Yeasts are more primitive without a mycelial structure and are much less problematic. The adhesion between the hyphae and the paper is often strong, because the molds can form anchorlike holdfast structures and secrete gluelike mucopolysacharides. Reproductive fruiting bodies such as picnidia, perithecia, and arcervulia may also become embedded in the paper substrates [38]. These may be darkly colored, further contaminating the artifact as well as promoting the growth of new fungal colonies. Stains may be produced by some of the fungi because trace metallic elements contained within the paper substrate are concentrated and altered during fungal growth. These stains may be secondary to pigments produced in the fruiting bodies, pigments secreted into the paper substrate, or pigments located within the mycelia. Colorless products of fungal metabolism may also react with the paper substrates to produce stains. Alternatively, stains may also develop as part of oxidation of the paper itself. The color of many of the above stains is often dictated by the type and concentration of the metallic elements within the paper [38]. Fungal growth is promoted by high humidity. Storage of paper substrates in controlled humidity environments is probably the most important step that can be taken to avoid mold overgrowth [40]. Sterilization can be achieved with fumigants or with ethylene oxide. These chemical methods, however, pose a biohazard to the conservator [41]. Furthermore, some investigators have found that

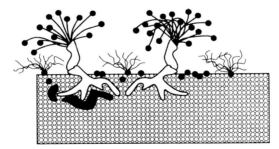

FIGURE 9.2 Molds may grow into the paper substrate, produce stains by metabolizing trace metals in the paper, produce fruiting bodies that can imbed themselves in the paper and develop into new mold colonies, or initiate cellulolytic processes that destroy the paper matrix.

after initial sterilization, ethylene oxide treated substrates may be more susceptible to regrowth of mold than if no sterilization had been done [40].

Therefore there are several types of fungal damage on paper that can be categorized as follows:

- Discoloration may be caused by pigments produced in fruiting bodies, located in mycelium, or secreted into the paper substrate.
- Surface damage caused by obstruction of any image by the growth of colonies, and embedded fruiting bodies.
- Structural deterioration of the paper may be due to the enzymatic digestion of cellulose.

Removal of the mold is desirable if the growth degrades the aesthetic balance of the work or if the paper or overlying art is being damaged. The decision to clean the piece should ultimately be left to a conservator in collaboration with the owner. If removal is desired, mechanical methods utilizing chemicals, suction pipettes, or brushes are often employed. Fruiting bodies can sometimes be dug out of the paper with a fine scalpel. However, the hyphae and vegetative bodies of the fungi may stubbornly remain, and such methods may damage the work and pose a relative biohazard to the conservator by dispersing spores into the air.

Considering these shortcomings, the role that lasers may play in the removal of fungi and stains from paper substrates has been examined extensively. The majority of studies employed Nd:YAG laser [1,2,23,42] and excimer laser [9,24]. A pulsed dye laser may also be considered as a potential light source if tuned to a wavelength (color) that amplifies the cleaning effect of the laser on the mold while minimizing the effect on the underlying substrate [24,32]. In principle, though, the excimer laser could be more useful in fungi removal because of its ability to remove microscopic layers of thickness of one micron or less (see Chapters 6 and 7). In addition, pulsed UV lasers can disintegrate fungi via the bond-breakage mechanism. Besides, it is well known that the excimer laser is used in some medical applications with apparent safety, including the sculpting of the human cornea to correct a patient's refractive error, eliminating the need for spectacle correction. However, as mentioned above, irradiation of cellulose with an excimer laser resulted in photooxidative degradation of the paper substrate, accompanied by an increase of carbonyls and carboxyls and a decrease in degree of polymerization [32,34]. Nevertheless, in certain occasions where there is a lack of alternatives, the conservator may weigh the possibilities and find that an excimer laser is the less harmful tool. An example of such is presented below.

A case presented in Reference [24] is that of an old poster from the National Gallery of Athens, with heavily overgrown fungus, both on its front surface and on its cardboard backing. The contamination was so heavy that in some areas the mold surface was several millimeters thick (Figure 9.3, top). Both the poster front and the support material were treated with the excimer laser for removing the biogenic contamination. To characterize the molds, the fungi were grown on Sabaroud's agar cultures, isolating *Aspergillus niger* and penicillium. The object to be cleaned was translated across a KrF excimer laser beam stepwise with an 80% overlap between adjacent laser spots (see Chapter 7 for details). Several energy fluences were used

FIGURE 9.3 (See color insert following page 144.) (Top) Detail of the poster before laser treatment. (Bottom) After excimer laser treatment the underlying image is revealed.

ranging from 0.4 J/cm^2 to 1.5 J/cm^2 with repetition rates from 2 to 8 Hz. Monitoring of the cleaning level was performed via reflectance spectrophotometry. Fungi were incompletely removed from the poster art without destruction of the underlying image if low laser fluence levels (~0.5 J/cm^2) were used at a repetition rate of 2 to 3 Hz. Without careful monitoring, removal of the surface art together with the fungi could occur. Some streaking of the poster inks illustrates this hazard. A detail of the poster near the face of the stewardess is presented in Figure 9.3 before (top) and after (bottom) the laser cleaning. In a parallel study, the biohazards during laser cleaning were investigated [43]. It was found that during laser–fungi interaction, the generated shock wave released viable spores and mold fragments into the atmosphere, suggesting proper safety measures are necessary.

In a recent study [42], a combination of ethanol and laser treatment was compared with chemical bleaching by $KMnO_4$. Here the laser used was the second harmonic of an Nd:YAG laser (532 nm) that had previously given the most prominent results in terms of minimal photooxidative degradation of cellulose. Remnants after the removal of hyphae and mycelia and the deactivation of the remaining conidia by ethanol could be removed using the laser. Spectral imaging served to register the visible reflection, infrared reflection, visible fluorescence, and ultraviolet reflection in a spectral range from 320 up to 1550 nm. Semiquantitative color change measurements according to the L*a*b* color space allowed the identification of irreversible changes caused by the chemical treatment, alcohol vs. $KMnO_4$. Both paper types showed an increase of fluorescence in both liquid treatments, suggesting irreversible chemical alterations. However, the comparison of ΔL of the infected areas after and before the laser interaction and after and before the $KMnO_4$ bleaching was a measure for the efficiency of the cleaning processes. The laser turned out more efficient than the $KMnO_4$ treatment, particularly on penicillium, alternaria, and cladosporium. A further criterion was the comparison of the brightness ΔL of the infected areas after laser irradiation with pure uninfected samples. Again laser cleaning of alternaria and the penicillium types showed the best success.

9.3 METAL

Laser cleaning of metal was first reported by Asmus [1,2] in a number of case studies including tarnish on silver threads in textiles and minerals, corrosion on bronze, and calcareous deposits on lead. Since the first LACONA conference a series of laser cleaning studies on metal have been reported [3–6].

In the conservation of metal artifacts, two types of surface coatings often require removal: inorganic corrosion by-products, such as rust and inorganic deposits from various metallic surfaces; and organic coatings such as paints and lacquers.

Inorganic corrosion products are the crystalline salts formed during the exposure of metals to natural or hostile environments, usually over a long period of time. Corrosion products are commonly inhomogeneous and polycrystalline in nature whose chemical and physical properties can vary greatly over a small area on a given object. Not all corrosion layers are detrimental to surface quality; often an aesthetically pleasing patina is formed, and some types of corrosion create a protective, passivating layer that inhibits further corrosion of the metal substrate.

Whereas inorganic layers spread and thicken over time, paints and lacquers degrade through physical [44] and chemical [45] aging. The action of sunlight on polymer coatings induces cross-linking between molecules, lowering their solubility in organic solvents, making them resistant to removal by chemical means (see relevant aging of resins in Chapter 7). Cross-linking also induces brittleness in polymers that, through the effect of a combination of temperature cycling and differing thermal expansion between the polymer and the metal, causes cracking and peeling. Coatings damaged in such a way are not only unpleasant to the eye, but also provide sites for water retention and electrochemical attack that can accelerate the rate of corrosion of the metal substrate. The aging of organic coatings can also influence their absorptivity of radiation at a given wavelength.

Cottam et al. presented one of the first studies on laser cleaning of metallic artifacts using a TEA CO_2 laser [46]. TEA stands for transverse excitation of atmospheric mixture [47]. It was used first in conjunction with the CO_2 system but has been successfully applied to other lasers too. This laser is very efficient for metal cleaning as it emits pulses of several tens of nanoseconds in the far infrared (10.6 µm) with a sufficient pulse energy to vaporize surface contaminants. TEA CO_2 lasers have already found some industrial metal cleaning applications to remove organic coatings from a variety of surfaces [48]. All cleaning was performed at fluences insufficient to cause plasma formation. The plasma plume formed may shield the surface from most of the energy of the laser pulse, thus lowering the efficiency of the process. Additionally, the high temperature obtained in the plasma is sufficient to produce ultraviolet radiation that can heat the metal substrate and cause surface damage [49]. No visible damage was inflicted upon any of the metal samples subjected to the laser at subplasma formation fluences. This is to be expected, because of the high reflectivity of metals in the far-infrared, and it would be reasonable to extrapolate this result to include most metals and alloys.

Summarizing the experimental results of Cottam et al. [46], TEA CO_2 laser cleaning of metals is possible without damage or significant heating of the metal substrate. All organic coatings were removed successfully with no visible residue remaining, while cleaning rates of the order of 1 m^2/h are feasible. The effect of laser cleaning on inorganic corrosion products varies greatly; cleaning down to the base metal was avoided owing to the presence of valuable oxide layers. Often thin underlying corrosion layers, as opposed to the metal substrate, defined the limit of the self-regulating cleaning process. This has the advantage of avoiding exposure of a bare metal surface to further attack from a hostile environment. Inorganic coatings show a great variation in their absorptivity and thus their susceptibility to ablation. Water is strongly absorbing of 10.6 μm radiation, and it could be the O–H group, whenever present, that dominates the absorption process at this wavelength. On the other hand, thin layers often existing between the metal substrate and other layers of corrosion, are resilient to the photoacoustic shock action of the laser. TEA CO_2 lasers have the potential to tackle a combination of aged organic coatings and inorganic corrosion products. A later study by Koh and Sárady [50] confirmed the results by Cottam et al. concerning the removal of organic coatings (wax from metal surfaces).

Another laser that can be used to remove organic paint from bronze sculpture is the KrF excimer laser [51]. Liverpool's civic sculpture was painted with an epoxy-based black paint over 30 years ago. This has formed a very hard, unattractive surface layer, beneath which corrosion of the bronze continues. X-ray diffraction analysis had revealed the corrosion layer (20 to 50 μm thick) to be brochantite, $Cu_4(OH)_6SO_4$, which is a typical by-product resulting from exposure to a polluted atmosphere. The black paint layer must be removed without damaging the underlying corrosion layers. Once this has been achieved, the conservator can decide on the appropriate treatment for stopping the corrosion process and preserving the sculpture. Using a Q-switched Nd:YAG laser, incorporating a handheld articulated arm delivery system, it is possible to remove the paint layer. However, the fluence required for paint removal exceeds that required to alter the underlying corrosion layer, which is sensitive to laser radiation at 1.06 μm, 532 nm, and UV wavelengths. Irradiation of the corrosion layer leads to a color change from light blue/green to dark green. The surface is also uneven and the thickness of the paint and corrosion varies, so that some areas are overcleaned while other areas are undercleaned.

To tackle this problem, initial tests using a KrF excimer laser and on-line LIBS control were carried out [51] on a sheet of corroded copper roof which had been painted with a black epoxy-based paint. A summary of the results obtained is presented here for helping the reader in tackling similar complex cases. The experiment was performed in five stages:

1. The sensitivity of the substrate was established. As expected, the corrosion layer was found to be extremely sensitive to radiation at 248 nm. Discoloration (light blue/green to dark green) was still evident at a fluence of 0.15 J/cm^2 (too low to remove the paint layer).
2. The fluence was increased to obtain a clear LIBS spectrum from both the paint layer and the corrosion layer (Figure 9.4).
3. The approximate number of pulses required to remove the paint layer was established. It was then possible to remove the majority of the paint layer relatively quickly by scanning the laser over the sample (see Chapter 7 on the similar case of scanning over aged varnish) and initiating the LIBS on-line control for the last few layers, thereby speeding up the process.
4. The remaining paint layers were removed by scanning the laser over the surface and on-line recording the spectra feeding back the LIBS data into the scanning loop code. Peaks chosen for the on-line control were paint layer; C_2 peak (fragments of the carbon chain) at 470.2 nm (P_1); corrosion layer; and Cu I peak at 324.5 nm (P_2). The ratio of P_1:P_2 was selected and used to control the cleaning process, i.e., once that ratio dropped below a certain value, the laser stopped firing until the translation stage had moved the sample by a predefined increment. An area approximately 19×5 mm was cleaned in this way. A second area, approximately 19×6 mm, was cleaned more thoroughly by adjusting the controlling ratio.
5. At the end of the scan, the cleaned area had a brown coloration. This was removed relatively easily using a cotton wool swab and water, revealing a

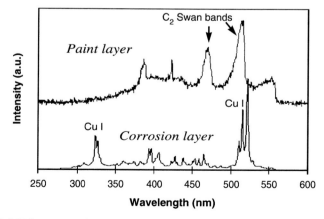

FIGURE 9.4 LIBS spectra obtained from paint and corrosion layers on copper roofing. Fluence, 1.2 J/cm^2.

dark green surface. The dark green color was similar to the color obtained by irradiating the light blue/green corrosion layer directly. Using a scalpel, it was possible to remove gently the surface layer and reveal a light blue/green layer (see Figure 9.5) similar to that found by scraping away an equal amount of the original corrosion layer. In addition, approximately 1/5 of the surface was still covered by paint (in one corner, where the original paint had been thicker), and in a few small areas (barely visible to the naked eye) the corrosion layer had been removed completely to reveal the metal surface. Similar results were obtained for the second, more thoroughly cleaned area, although less paint remained and more metal was exposed.

Removal of the paint layer appeared to be relatively controlled. Although the surface layer was discolored, the underlying material appeared unaffected (infrared spectroscopy failed to detect any differences between the brochantite layer and the layer revealed by laser cleaning). Provided the corrosion layer is continuous, this would suggest that the underlying metal is also unaffected by the cleaning. The maximum thickness of the affected outermost region of the corrosion layer has been estimated as 5 μm. There was no noticeable change to the LIBS spectrum with increasing depth into the paint layer, which means that it is not possible to avoid irradiating the corrosion layer. Using LIBS as an on-line control technique, however, should minimize the number of pulses that reach the corrosion layer. The initial brown coloration of the cleaned area is believed to be caused by the lower power tail of the laser pulse, which causes heating rather than ablation, i.e., different mechanisms operate at the center and tail of the pulse. Even cleaning could be improved by reducing the beam size on the sample. The LIBS spectrum is an average over the whole of the irradiated region, since all parts contribute to the plume. Reducing the beam size improves the resolution of the technique so that changes in the paint thickness can be more successfully compensated for. On the other hand, as expected, the latter minimizes the process speed.

In conclusion, the combination of a UV wavelength, on-line control, an accurate positioning system, and an alternative cleaning method has led [51] to more controlled removal of the paint layer than was possible using a handheld Nd:YAG

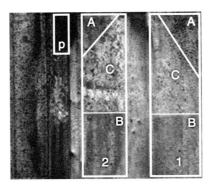

FIGURE 9.5 Removal of paint from corroded copper roofing using LIBS for on-line control. p: original paint; 1: test area 1; 2: test area 2; A: after laser; B: after laser and wet swab; C: after laser, wet swab, and scalpel.

system. Although cleaning at the longer wavelength led to more effective removal of the paint layer, it also led to a greater degree of overcleaning and discoloration of the corrosion layer to increased depth. Alternative all-mechanical removal methods were tested without any success.

Although a Q-Switched Nd:YAG laser is not the best choice for paint removal, it proved to be very efficient in removing pollution layers from aluminum artifacts. More specifically, laser cleaning of an aluminum alloy statue was carried out by Cooper with excellent results [17]. In this case, a uniform dark grey layer was accumulated on the surface formed through exposure of the metallic surface to polluted atmosphere and complex white and brown deposits. By using low fluence levels, it was possible to remove the deposits completely and preserve the original tool markings [17].

Another important application of laser cleaning is the removal of corrosion from iron artifacts. Often archaeological ironwork is covered by a thick corrosion layer. In many cases the corrosion crust exceeds the volume of the original ironwork by several times. Removal of the corrosion layers is necessary both for conservation and for exposing the original topography containing information on the former manufacturing process. Investigations were carried out [52] using a Q-switched Nd:YAG laser at 1064, 532, 355, and 266 nm. The results showed that a combination of conventional cleaning methods with laser cleaning opens new possibilities in restoration of strongly corroded archaeological ironwork. More specifically, the results of this work [52] can be summarized as follows.

First, there are different fluence thresholds for corrosion crust removal, for original iron removal, and for modification (e.g., blackening) of the original surface. It was found that the threshold for removal of corrosion crust was lower than the other two thresholds in most case studies. All fluence thresholds also showed a distinct decreasing tendency with shorter wavelength. Besides the blackening effect, for some ironworks being excavated from silica sand ground, remaining silica grains were also observed on the surface after laser cleaning. Obviously this is because of the transparency of silica at the wavelengths employed. The maximum removal rate using a constant spot size area of 0.25 cm^2 was >10 μg/pulse at 1064 nm, <10 μg/pulse at 532 nm, and <<10 μg/pulse at 355 and 266 nm. Thus with regard to practical applications for further studies, only 1064 and 532 nm were used. Also, wetting with an ethanol film prior to laser processing increased the removal rate up to 60% and at the same time removed the adhering silica grains from the surface. This can be explained by the laser-induced vaporization of ethanol sweeping along the surrounding single grains.

Analytical investigations using a pyrometer and SEM analysis revealed a minor thermal load on the exposed iron surface [52]. High-resolution imaging by SEM, however, gave obvious indications of tiny molten areas in the range of some microns. Using EDX and XPS, an increasing content of iron on the surface was detected, going from 65 to 96% after laser cleaning. Finally, in cooperation with the Museum for Archaeology in Muenster, Germany, laser cleaning was applied to several originals of corroded ironwork with a good deal of success [52].

Comparisons of laser cleaning with microsandblasting by different groups have shown inconsistent results. The previously presented study [52] showed that the laser-cleaned surface maintains more or less the original roughness; the microsandblasted surface shows a more smoothened face owing to the homogeneously abrasive effect of corundum microparticles. In addition, the processing time for laser cleaning was found to be about three times shorter than that for sandblasting. In contrast, another study [50] showed increased roughness and decreased processing time for microsandblasting over laser cleaning. Both studies, however, agree that superior results are obtained by laser cleaning, as conservators' evaluations have indicated. As for the inconsistency, we should bear in mind that there are many parameters that could alter the resultant surface roughness or determine the processing time. Such parameters may include the corrosion layer roughness, the original metal roughness, the laser beam quality, the exposure conditions (fluence, wetting, number of pulses, etc.), the operator's skill, and so on.

A parallel detailed study [53] was based on model copper and iron substrates. The investigation was carried out using an Nd:YAG laser at 1064, 532, and 355 nm. It was found that the removal of iron corroded layers and the reduction of dirt and black crust on bronze and brass seem possible without any damage to the underlying substrates.

Another comprehensive investigation on laser cleaning of metal by Batishche et al. [54] was focused on the comparison between short-pulse (SP) Q-switched Nd:YAG laser at 532 nm (15 nsec), 1064 nm (15 nsec), and 1320 nm (40 nsec) and long-pulse (LP) Nd:YAG laser (40 to 60 μsec) at 1064 and 1320 nm. The results showed that the removal of contamination occurs in both LP and SP regimes at 1064 and 1320 nm. For thick objects when surface heating is insignificant, the results for LP are practically identical to those for SP, a substantial benefit of the former being its higher efficiency in fiber transmission for long distances (tens of meters). On the surface, a layer of patina always remained in LP mode. The layer of patina could completely disappear in SP mode. It should be stressed that removal of patina occurs in general with melting bulk bronze material. In the case of thin metal objects, better results are obtained in the SP regime, which agrees with the results of Pini et al. [55], especially when coatings of silver or other metals are dealt with. In all cases, the conservators found laser cleaning of metal artworks at 1320 nm less damaging. This was attributed to the low ablation threshold of the superficial layer and therefore the more delicate cleaning of the surface at this wavelength. Besides, at 1320 nm in LP mode, there exists a wider working fluence range. The outcome of this study [54] was that both 1064 and 1320 nm in LP and SP modes are needed to obtain the best results for metal cleaning, depending on the various types of encrustation encountered.

A recent pilot study by Burmester et al. [56] has taken advantage of the use of a titanium–sapphire femtosecond laser for metal cleaning. The removal of corrosion products or pigment coatings from original objects made of copper, bronze, and silver was investigated. In addition, the influence of laser fluence and repetition rate on the removal efficiency of the various corrosion products was analyzed. Specific fluence thresholds have been found for the removal of different types of corrosion products and pigment coatings. As a result, a sequential removal by nonthermal ablation of discrete corrosion layers has been achieved by varying the laser fluence, with excellent results. However, owing to beam handling limitations of the laser system (motorized focusing and 2D scanning), the corrosions inside cavernous areas or on a highly porous surface were not entirely removed, limiting the quality of the final conservation result. Other current limitations on using this femtosecond laser system include poor availability, immobility, pulse duration instability, nonrobustness, long processing time, and low pulse energy. Especially the latter limitation leads to the need of using a very small spot size (strong focusing), which makes manual beam handling impossible by the conservator: a motorized scanning system is required. Upon future improvement of femtosecond lasers, however, they will certainly play a significant role in fine cleaning of certain metal artifacts.

Removing the encrustation from gilded bronze artworks is another challenging conservation task that requires cross-disciplinary knowledge and effort. A successful case study was carried out [57] on a panel of the *Gate of Paradise,* a masterpiece by Lorenzo Ghiberti at the Baptistery of Florence. The stratification consists of the bronze substrate, a copper oxide layer, the gilding film, and the encrustation. On top of the encrustation a thin water film was applied to keep the temperature low. At the beginning of the laser–surface interaction — an Nd:YAG laser with a pulse duration of 28 nsec was used — most of the energy was converted to heat. Depending on the fluence and on the distribution of optical absorbers within the encrustation, different material removal regimes could be engaged. Just above the cleaning threshold, the most important contributions to the material removal are from localized vaporization of water sweeping away surrounding material, as well as from photoacoustic generation. At this first phase, laser light is entirely absorbed by the encrustation, with the gilding not experiencing any irradiation. After the removal of the most of the encrustation and during the finishing phase, there is a nonnegligible direct heating of the gold film. A detailed modeling of the process [57] has shown that the temperature rise at the surface of the gold film is not more than 213°C at the maximum operating laser fluence of 0.5 J/cm^2. Finally, considering the thermal wave propagation inside the gold film, the model predicts at the most several tens of degrees Celsius at the inner gold–copper oxide interface. The experimental results were satisfactory without any damage to the gilding, working though at a repetition rate below 5 Hz.

Cleaning of textiles with tarnished metal threads made of silver, gilt silver, or copper is a difficult task, as treatments commonly applied to textiles and metals are incompatible. Mechanical cleaning removes the plating. If the threads are made of silver or gilt silver, chemical or electrolytic techniques can be used, but the immersion process may damage the fibers and dissolve any dye [58]. These problems have led conservators to look for other cleaning techniques including dry methods. Laser cleaning using an Nd:YAG laser (Figure 9.6) seems to be promising, having been applied with good results [59,60], compared to traditional cleaning techniques. In these studies, the UV wavelengths showed better results than IR-VIS wavelengths in terms of both efficiency and the safety of the composite materials. An example is shown in Figure 9.7.

Heating of the metal threads is the main concern when a composite made of textile and metal is considered. Side effects may modify the appearance of the materials. The question is to what extent textile conservators in charge of textile collections will accept this change. Whitening of silver threads causes a loss of brightness, but the original color of the metal is recovered through the cleaning process [59]. If a lower level of cleaning is applied, less damage occurs, but the surface may take on a gilt appearance that appreciably modifies the appearance of the material. Before going further it is then essential to obtain advice from textile professionals, since they may give precise guidelines for optimizing the cleaning parameters. Another important

FIGURE 9.6 (See color insert.) Laser cleaning of fringes with silk core and tarnished gilt silver threads.

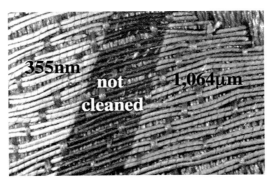

FIGURE 9.7 (See color insert.) Silver threads on a red satin substrate treated with different wavelengths of an Nd:YAG laser: 355 nm (0.08 J/cm^2) and 1064 nm (0.35 J/cm^2), in comparison to an area not cleaned.

question concerns the effect of the laser impact on the long-term conservation of the materials. No tests have been conducted to determine how reactive the metal is after cleaning. If it does become more reactive, a cleaning process is perhaps not really needed unless the artifact is afterward placed in a very pure environment. In addition, the effect of laser cleaning on the textile has to be clarified, but again, in comparison with the effects induced by other alternative techniques.

9.4 CONCLUSIONS

For cleaning parchment/paper and metal artifacts, conservators often use techniques such as chemicals, erasers, scalpels, dental tools, and microblasting to assist with the removal of corrosion. When using these methods, very small fragments of the original artifact are often removed from the surface, or the chemicals penetrate into the body of the artifacts with unforeseeable effects. Also, in several cases these methods are difficult to control, which easily results in over-cleaning of the surface, as well as possible reduction of the visual appeal of a surface and in extreme cases even acceleration further decay.

Laser cleaning is a modern conservation technique that has been taken from industrial and medical uses of lasers. Laser cleaning uses no chemicals and leaves almost no residues, making it ideal for use in the conservation and restoration of archaeological artifacts and artwork. Attempts at cleaning of artworks using lasers began in 1970s. The laser technique has become an important tool for the precise removal of surface pollutants without damaging the underlying material.

The most common type of laser used in conservation of parchment/paper is the Q-switched Nd:YAG laser at both 532 nm and 1064 nm. The most important parameter when using the Nd:YAG laser for cleaning is the fluence. The fluence should be high enough to remove the dirt layers but low enough to ensure that the substrate surface is not damaged or altered. This is ensured by the high contrast between high absorption of the surface contaminants and low absorption of the substrate. The excimer laser may also prove a helpful tool for fungi removal.

In metal cleaning, the Nd:YAG laser can be used either in the Q-switched mode or in the long-pulse mode, depending mainly on the thickness of the corrosion layer. Another important laser for metal cleaning is the TEA CO_2 laser, providing short pulses of infrared radiation at a wavelength of 10.6 μm, which has proven to be very efficient mostly in the removal of organic coatings as well as corrosion layers. Excimer lasers in combination with LIBS on-line control may also become helpful in special cases, where fine removal of organic coatings is necessary. Finally, femtosecond lasers provide promising results but for the moment are not a practical choice for conservators.

Laser cleaning appears to be an effective technique for assisting the conservation of parchment/paper and metal artifacts, since it provides a high degree of control during cleaning, allowing fragile objects or items with much surface detail to be effectively cleaned. This degree of control is essential when preserving items with ink illustrations or surface relief, original tool markings, and surface patina.

The technique must be used with care, however, and process parameters must be set carefully for preventing over-cleaning. At present, process parameters must be set on a case-by-case basis. It is therefore important to use proper monitoring, controlling, and analysis techniques and skilled conservators with experience in the process to achieve the most satisfactory results. This by no means hinders the use of lasers compared to other conservation methods, which many times must be used in parallel for optimum results.

The most important barrier in introducing the laser as a common tool in the conservation of other materials is the perception of the conservators in what they consider a clean substrate or an acceptable final stage of conservation. This is often the case in metal cleaning, where various different oxide layers are encountered and it is difficult to decide what is the optimum cleaning. In conclusion, the laser should be viewed as a powerful addition to the conservator's toolkit whose application must be studied, carefully depending on the surface/substrate materials involved. In many cases, it can be exploited in conjunction with other available conservation methods.

REFERENCES

1. Asmus, J.F., Light cleaning: laser technology for surface preparation in the arts, *Tech. Conserv.,* Fall, 1978, p. 14.
2. Asmus, J.F., More light for art conservation, *IEEE Circ. Devices Mag.,* Mar. 1986, p. 6.
3. König, E. and Kautek, W., Eds., Proceedings of the First International Conference LACONA I — Lasers in the Conservation of Artworks, October 4–6, 1995, Heraklion, Crete, Greece, in *Restauratorenblaetter, Sonderband,* Verlag-Mayer, Vienna, 1997.
4. Salimbeni, R. and Bonsanti, G., Eds., Proceedings of the International Conference LACONA III — Lasers in the Conservation of Artworks, April 26–29, 1999, Florence, Italy, *J. Cult. Heritage,* vol. **1**-suppl. 1, 2000.
5. Verges-Belmin, V., Ed., Proceedings of the International Conference LACONA IV — Lasers in the Conservation of Artworks, Sept. 11–14, Paris, France *J. Cult. Heritage,* vol. **4**-suppl. 1, 2003.
6. Dickmann, K., Fotakis, C., and Asmus, J.F., Eds., Lasers in the Conservation of Artworks, in Proceedings of the International Conference LACONA V, Sept. 15–18, 2003, Osnabrueck, Germany, *Springer Proceedings in Physics,* vol. **100**, Springer-Verlag, Berlin, Heidelberg, 2005.
7. Kautek, W. and Pentzien, S., Laser cleaning system for automated paper and parchment cleaning, in Lasers in the Conservation of Artworks,—LACONA V, *Springer Proceedings in Physics,* vol. **100**, Springer-Verlag, 2005, p. 403.
8. Scholten, H., et al., Laser cleaning investigations of paper models and original objects with Nd:YAG and KrF laser systems, in Lasers in the Conservation of Artworks, *Springer Proceedings in Physics,* vol. **100**, Springer-Verlag, Berlin, Heidelberg, 2005, p. 11.
9. Kautek, W., Pentzien, S., Kruger, J., and König,E., Laser cleaning of ancient parchments, in Lasers in the Conservation of Artworks I, Kautek, W. and König, E., Eds., *Restauratorenblaetter,* Verlag Mayer, Vienna, 1997, p. 69.
10. Kautek, W., Pentzien, S., Rudolph, P., Kruger, J., and König, E., Laser interaction with coated collagen and cellulose fibre composites: fundamentals of laser cleaning of ancient parchment manuscripts and paper, *Appl. Surf. Sci.,* **127–129**, 746, 1998.
11. Rudolph, P., Pentzien, S., Kruger, J., Kautek, W., and König, E., Laserreinigung von Pergament und Papier: Experimente an Modellsystemen und historischen Originalen, Lasertechnik in der Restaurierung, *Restauro,* **104**(6), 396, 1998.
12. Kautek, W., Pentzien, S., Rudolph, P., Kruger, J., Maywald-Pitellos, C., Bansa, H., Grosswang, H., and König, E., Near-ultraviolet pulsed laser interaction with contaminants and pigments on parchment: spectroscopic diagnostics for laser cleaning safety, in *Optics and Lasers in Biomedicine and Culture, Optics within Life Science (OWLS V),* Optics within Life Science Series, Fotakis, C., Papazoglou, T., and Kalpouzos, C., Eds., Springer-Verlag, Heidelberg, 2000, p. 100.

13. Sportun, S., Cooper, M., Steward, A., Vest, M., Larsen, R., and Poulsen, D.V., An investigation into the effect of wavelength in the laser cleaning of parchment, *J. Cult. Heritage,* **1**, S225, 2000.

14. Kautek, W., Pentzien, S., Rudolph, P., Kruger, J., Maywald-Pitellos, C., Bansa, H., Grosswang, H., and König, E., Near-ultraviolet laser interaction with contaminants and pigments on parchment: laser cleaning and spectroscopic diagnostic, *J. Cult. Heritage,* **1**, S233, 2000.

15. Müller-Hess, D., Troschke, K.K., Kolar, J., Strlic, M., Penzien, S., and Kautek, W., Reinigung ist so gut wie ihre Kontrollierbarkeit: ein interdisziplinares Forschungsprojekt zum Thema Laserreinigung von Papier und Pergament-LACLEPA, *Konservieren Restaurieren (Mitteilungen des Osterreichischen Restauratorenverbandes),* **7**, 64, 2000.

16. W. Kautek, Pentzien, S., Müller-Hess, D., Troschke, K., and Teule, R., Probing the limits of paper and parchment laser cleaning by multispectral imaging, *Laser Tech. Syst. Art Conserv.,* SPIE vol. **4402**, 2001, p. 130.

17. Cooper, M., *Cleaning in Conservation: An Introduction,* Butterworth Heinemann, Oxford, 1998.

18. Manni, J.G., Lasers in dermatology: they are on the way to a billion-dollar market, *Biophotonics Int.,* May/June 1998, p. 40.

19. Hogan, H., Technology is being fine-tuned for erasing body art, *Biophotonics Int.,* Nov., 2000, p. 62.

20. Kautek, W., Pentzien, S., Conradi, A., Leichtfried, D., and Puchinger, L., Diagnostics of parchment laser cleaning in the near-ultraviolet and near-infrared wavelength range: a systematic scanning electron microscopy study, *J. Cult. Heritage,* **4**, 179s, 2003.

21. O'Flaherty, F., Roddy, W.T., and Lollar, R., *The Chemistry and Technology of Leather,* vol. **4**, Reinhold, New York, 1965.

22. Puchinger, L., Pentzien, S., Koter, R., and Kautek, W., Chemistry of parchment-laser interaction, in Lasers in the Conservation of Artworks, *Springer Proceedings in Physics,* vol. **100**, Springer-Verlag, Berlin, Heidelberg, 2005, p. 51.

23. Szczepanowska, H.M. and Moomaw, W.R., Laser stain removal of fungus-induced stains from paper, *JAIC,* **33**, 25, 1994.

24. Friberg, T.R., Zafiropulos, V., Petrakis, Y., and Fotakis, C., Removal of fungi and stains from paper substrates using laser cleaning strategies, in *Lasers in the Conservation of Artworks, LACONA I,* Kautek, W. and König, E., Eds., Restauratorenblaetter, Sonderband, Verlag Mayer, Wien, 1997, p. 79.

25. Caverhill, J., Latimer, I., and Singer, B., An investigation into the use of a laser for the removal of modern ink marks from paper, *Paper Conservator,* **20**, 65, 1996.

26. Caverhill, J., Stanley, J., Singer, B., and Latimer, I., The effect of ageing on paper irradiated by laser as a conservation technique, *Restaurator,* **20**, 57, 1999.

27. Moroz, R. and Roever, E.A., A new approach to systematic dry cleaning with technical devices, *Restaurator,* **14**, 172, 1993.

28. Pearlstein, E.J., Cabelli, D., King, A., and Indictor, N., Effects of eraser treatment on paper, *JAIC,* **22**, 1, 1993.

29. Sterlini, P., Surface cleaning products and their effects on paper, *Paper Conserv. News,* **76**, 3, 1995.

30. Kautek, W., Laser-Reinigung von Pergament-Handschriften und Papier, Bibliotheksdienst, *Organ Der Bundesvereinigung Deutscher Bibliotheksverbande (BDB),* **31**, 1942, 1997.

31. Kautek, W., Pentzien, S., Rudolph, P., Kruger, J., Kolar, J., Strlic, M., and Troschke, K., Laser cleaning and diagnostics of paper and parchment, Abstracts of Papers Presented at the 28th AIC Annual Meeting, Presession 1–2, 2000.

32. Kolar, J., Strlic, M., Müller-Hess, D., Gruber, A., Troschke, K., Pentzien, S., and Kautek, W., Near-UV pulsed laser interaction with paper, *J. Cult. Heritage*, **1**, S221, 2000.

33. Kolar, J., Strlic, M., and Marincek, M., Effect of Nd:YAG laser radiation at 1064 nm on paper, *Restaurator*, **21**, 9, 2000.

34. Kolar, J., Strlic, M., Pentzien, S., and Kautek, W., Near-UV, visible and IR pulsed laser light interaction with cellulose *Appl. Phys.*, **A71**, 87, 2000.

35. Kaminska, A., Sawczak, M., Cieplnski, M., and Sliwinski, G., The post-processing effects due to pulsed laser ablation of paper, in Lasers in the Conservation of Artworks, *Springer Proceedings in Physics*, vol. **100**, Springer-Verlag, Berlin, Heidelberg, 2005, p. 35.

36. Kolar, J., Strlic, M., Müller-Hess, D., Gruber, A., Troscke, K., Pentzien, S., and Kautek, W., Laser cleaning of paper using Nd:YAG laser running at 532 nm, *J. Cult. Heritage*, **4**, 185s, 2003.

37. Ochocinska-Komar, K., Kaminska, A., Martin, M., and Sliwinski, G., Observation of the post-processing effects due to laser cleaning of paper, in Lasers in the Conservation of Artworks, *Springer Proceedings in Physics*, vol. **100**, Springer-Verlag, Berlin, Heidelberg, 2005, p. 29.

38. Szczepanowska, H.M., Biodeterioration of art objects on paper, *Paper Conservator*, **10**, 31, 1986.

39. Szczepanowska, H.M. and Lovett, C.J., A study of the removal and prevention of fungal stains on paper, *JAIC*, **31**, 147, 1992.

40. Craig, R., Alternative approaches to the treatment of mold biodeterioration — an international problem, *Paper Conservator*, **10**, 27, 1986.

41. Valentin, N., Biodeterioration of library materials disinfection methods and new alternatives, *IBID*, 40, 1986.

42. Pilch, E., Pentzien, S., Madebach, H., and Kautek, W., Anti-fungal laser treatment of paper: a model study with a laser wavelength of 532 nm, in Lasers in the Conservation of Artworks, *Springer Proceedings in Physics*, vol. **100**, Springer-Verlag, Berlin, Heidelberg, 2005, p. 19.

43. Friberg, T.R., Zafiropulos, V., Kalaitzaki, M., Kowalski, R., Petrakis, J., and Fotakis, C., Excimer laser removal of mold contaminated paper: sterilization and air quality considerations, *Lasers Med. Sci.*, **12**, 55, 1997.

44. Struik, I.C.E. *Physical Aging in Amorphous Polymers and Other Materials*, Elsevier Scientific, 1978.

45. Stevens, M.P., *Polymer Chemistry, An Introduction*, 2nd ed., Oxford University Press, 1990.

46. Cottam, C.A., Emmony, D.C., Larson, J., and Newman, S., Laser cleaning of metals at infra-red wavelengths, in Lasers in the Conservation of Artworks I, Kautek, W. and König, E., Eds., *Restauratorenblaetter (special issue)*, Verlag–Mayer, Vienna, 1997, p. 95.

47. Verdeyen, J.T., *Laser Electronics*, Prentice-Hall, Englewood Cliffs, NJ, 1981.

48. Lovoi, P., Laser paint stripping offers control and flexibility, *Laser Focus World*, Nov., 75, 1994.

49. Ready, J.F., *Effects of High-Power Laser Radiation*, Academic Press, 1971.

50. Koh, Y.S. and Sárady, I., Surface cleaning of iron artefacts by lasers, in Lasers in the Conservation of Artworks, *Springer Proceedings in Physics*, vol. **100**, Springer-Verlag, Berlin, Heidelberg, 2005, p. 95.

51. Cooper, M.I., Fowles P.S., and Zafiropulos, V., Applications of laser-induced break-down spectroscopy in conservation, Ultraviolet Laser Facility, Internal Report, FO.R.T.H. — I.E.S.L., Heraklion, Greece, 1998.

52. Dickmann, K., Hildenhagen, J., Studer, J., and Musch, E., Archaeological ironwork: removal of corrosion layers by Nd:YAG-laser, in Lasers in the Conservation of Artworks, *Springer Proceedings in Physics,* vol. **100**, Springer-Verlag, Berlin, Heidelberg, 2005, p. 71.

53. Mottner, P., Wiedemann, G., Haber, G., Conrad, W., and Gervais, A., Laser cleaning of metal surface — laboratory investigations, in Lasers in the Conservation of Artworks, *Springer Proceedings in Physics,* vol. **100**, Springer-Verlag, Berlin, Heidelberg, 2005, p. 79.

54. Batishche, S., Kouzmouk, A., Tatur, H., Gorovets, T., Pilipenka, U., and Ukhau, V., 1320 nm range Nd:YAG laser in restoration of artworks made of bronze and other metals, in Lasers in the Conservation of Artworks, *Springer Proceedings in Physics,* vol. **100**, Springer-Verlag, Berlin, Heidelberg, 2005, p. 87.

55. Pini, R., Siano, S., Salimbeni, R., Pasquinucci, M., and Miccio, M., Tests of laser cleaning on archeological metal artefacts, *J. Cult. Heritage,* **1**, S129, 2000.

56. Burmester, T., Meier, M., Haferkamp, H., Barcikowski, S., Bunte, J., and Ostendorf, A., Femtosecond laser cleaning of metallic cultural heritage and antique artworks, in Lasers in the Conservation of Artworks, *Springer Proceedings in Physics,* vol. **100**, Springer-Verlag, Berlin, Heidelberg, 2005, p. 61.

57. Siano, S. and Salimbeni, R., The gate of paradise: physical optimization of the laser cleaning approach, *Stud. Conserv.,* **46**, 269, 2001.

58. Degrigny, C. and Majda, J., Mise au point d'un traitement de conservation-restauration de composites plomb/textile, influence des traitements electrolytiques sur les lacs teints reliés aux bulles des documents écrits, Internal Report, Arc Antique, Nantes, France, 1999.

59. Degrigny, C., Tanguy, E., Le Gall, R., Zafiropulos, V., and Marakis, G., Laser cleaning of tarnished silver and copper threads in museum textiles, *J. Cult. Heritage,* **4**, 152s–156s, 2003.

60. Lee, J.-M., Yu, J.-E., Koh, Y.-S., Experimental study on the effect of wavelength in the laser cleaning of silver threads, *J. Cult. Heritage,* **4**, 157s, 2003.

Index

A

FIGURE 3.7 Paintings analyzed by LIBS. (a) 19th century miniature on ivory, (b) 19th century colored daguerreotype, (c) 18th century oil painting, (d) 19th century Byzantine icon.

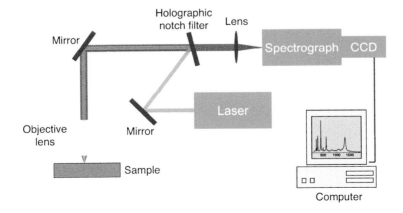

FIGURE 4.5 Schematic diagram of a Raman microspectrometer.

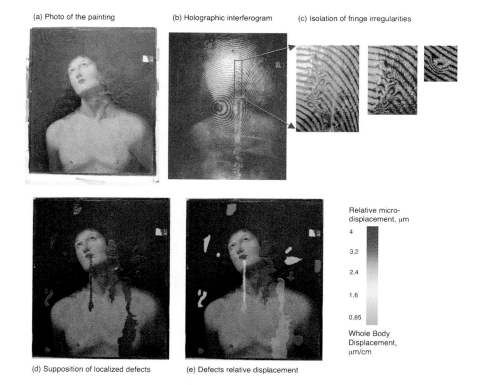

(a) Photo of the painting   (b) Holographic interferogram   (c) Isolation of fringe irregularities

Relative micro-displacement, μm

4

3,2

2,4

1,6

0,85

Whole Body Displacement, μm/cm

(d) Supposition of localized defects   (e) Defects relative displacement

FIGURE 5.20 (a-e) Example of application on real painting on conservation direct structural diagnosis: Saint Sebastian (painting attributed to Rafael, National Gallery of Athens). (b) Holographic interferograms provided a variety of small local discontinuities that were isolated one by one and classified as a type to defect according to the classification table. Calibration of whole-body fringes allows setting of priority risk indicator for structural restoration. Superposition of defects on painting with fringes shown on left, and on right the priority for restoration.

FIGURE 5.22 The interferogram revealed highly irregular fringe distribution with extended superficial detachments and defect propagation affecting the whole front surface to react as loosely attached to frame membrane.

FIGURE 5.33 Laseract Multitask system for onfield investigation of wall painting.

FIGURE 7.12 (a) Original surface except a small laser-cleaned area at lower left. (b) Additionally, the top part has been laser cleaned. Two small zones (bottom) have been chemically cleaned for comparison. (c) A detail at the interface between original resin (top) and laser-cleaned surface (bottom). (d) After laser-assisted divestment followed by traditional chemical restoration.

FIGURE 7.13 (a) Original surface and (b) after laser intervention followed by traditional restoration.

FIGURE 7.14 Part of painting covered with naturally aged varnish. Ten laser-cleaned areas span a thickness of varnish removal from 3.3 μm (area 1) to 82 μm (area 10).

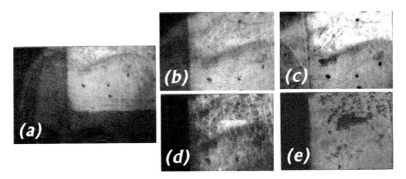

FIGURE 7.15 Multispectral images of a laser-cleaned area.

FIGURE 7.19 Painted metallized Kevlar composite paint stripped down to eight different depths.

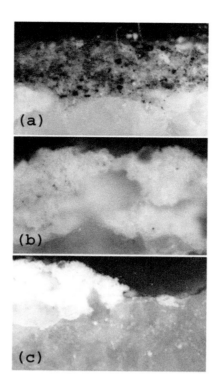

FIGURE 8.2 Cross section (×200) of dendritic type of encrustation on pentelic marble: (a) reference encrustation, (b) after selective vaporization of particles using 50 pulses at 0.8 J/cm², and (c) after partial removal of the gypsum rich layer using 50 pulses at 1.2 J/cm² of a Q-switched Nd:YAG laser. The width of each picture corresponds to 500 μm.

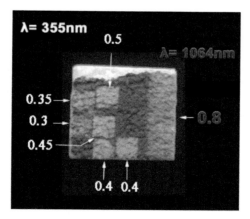

FIGURE 8.4 Thin encrustation on pentelic marble divested using λ_L = 355 nm (left) and λ_L = 1064 nm just above the onset of selective vaporization (right). The numbers represent different fluence values used in units of J/cm².

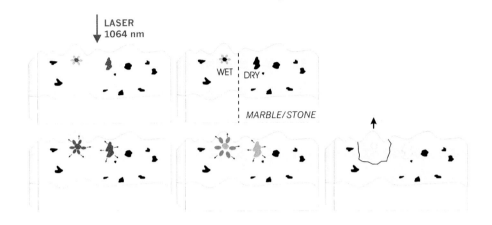

FIGURE 8.9 Schematic representation of the explosive vaporization of a particle with and without the presence of water.

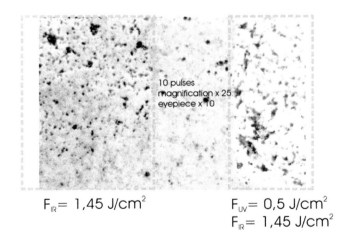

FIGURE 8.10 Prototype sample (90% gypsum–10% carbon) irradiated with 1064 nm pulses (left) and a temporal–spatial overlap of 1064 and 355 nm pulses (right). Reference material is in center.

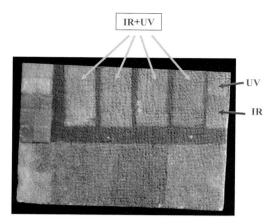

FIGURE 8.17 Preliminary laser cleaning trials on a corner addition (dated in the 1960s) from the Parthenon West Frieze block no 11(ΔZ XI).

FIGURE 8.18 Detail from the West Parthenon Frieze block no. 3 ((ΔZ III) during the laser cleaning intervention with the combined beam methodology.

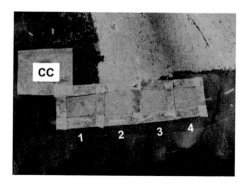

FIGURE 8.19 Cleaning level provided by QS-Nd:YAG laser between 0.7 and 2 J/cm².

FIGURE 9.6 Laser cleaning of fringes with silk core and tarnished gilt silver threads.

FIGURE 9.7 Silver threads on a red satin substrate treated with different wavelengths of an Nd:YAG laser: 355 nm (0.08 J/cm^2) and 1064 nm (0.35J/cm^2), in comparison to an area not cleaned.